Ubiquitin

Ubiquitin

Edited by
Martin Rechsteiner
University of Utah School of Medicine
Salt Lake City, Utah

Plenum Press • New York and London

Library of Congress Cataloging in Publication Data

Ubiquitin.

Includes bibliographies and index.
1. Ubiquitin. I. Rechsteiner, Martin
QP552.U24U25   1988   574.19′245   88-9809
ISBN 0-306-42850-4

Contributors

ANDREAS BACHMAIR • Department of Biology, Massachusetts Institute of Technology, Cambridge, Massachusetts 02139

BONNIE BARTEL • Department of Biology, Massachusetts Institute of Technology, Cambridge, Massachusetts 02139

WILLIAM M. BONNER • Laboratory of Molecular Pharmacology, Division of Cancer Treatment, National Cancer Institute, National Institutes of Health, Bethesda, Maryland 20892

AARON CIECHANOVER • Unit of Biochemistry, Faculty of Medicine, Technion–Israel Institute of Technology, Haifa 31096, Israel

DANIEL FINLEY • Department of Biology, Massachusetts Institute of Technology, Cambridge, Massachusetts 02139

STEPHEN A. GOFF • Department of Physiology, Tufts Medical School, Boston, Massachusetts 02111. *Present address:* Plant Gene Expression Center, U.S. Department of Agriculture, Albany, California 94710

ALFRED L. GOLDBERG • Department of Physiology and Biophysics, Harvard Medical School, Boston, Massachusetts 02115

DAVID GONDA • Department of Biology, Massachusetts Institute of Technology, Cambridge, Massachusetts 02139

ARTHUR L. HAAS • Department of Biochemistry, The Medical College of Wisconsin, Milwaukee, Wisconsin 53226

CHRISTOPHER L. HATCH • Laboratory of Molecular Pharmacology, Division of Cancer Treatment, National Cancer Institute, National Institutes of Health, Bethesda, Maryland 20892

AVRAM HERSHKO • Unit of Biochemistry, Faculty of Medicine, Technion–Israel Institute of Technology, Haifa 31096, Israel

RONALD F. HOUGH • Department of Biochemistry, School of Medicine, University of Utah, Salt Lake City, Utah 84132

STEFAN JENTSCH • Department of Biology, Massachusetts Institute of Technology, Cambridge, Massachusetts 02139. *Present address:* Friedrich Miescher Institute, Max Planck Society, 7400 Tübingen, Federal Republic of Germany

JOHN P. MCGRATH • Department of Biology, Massachusetts Institute of Technology, Cambridge, Massachusetts 02139

ENGIN ÖZKAYNAK • Department of Biology, Massachusetts Institute of Technology, Cambridge, Massachusetts 02139. *Present address:* Creative Biomolecules, Hopkinton, Massachusetts 01748

MICHAEL PAZIN • Department of Biology, Massachusetts Institute of Technology, Cambridge, Massachusetts 02139. *Present address:* Department of Biochemistry and Biophysics, School of Medicine, University of California, San Francisco, California 94143

CECILE M. PICKART • Department of Biochemistry, SUNY/Buffalo, Buffalo, New York 14214

GREGORY W. PRATT • Department of Biochemistry, School of Medicine, University of Utah, Salt Lake City, Utah 84132

MARTIN RECHSTEINER • Department of Biochemistry, School of Medicine, University of Utah, Salt Lake City, Utah 84132

IRWIN A. ROSE • Institute for Cancer Research, Fox Chase Cancer Center, Philadelphia, Pennsylvania 19111

MARK SIEGELMAN • Laboratory of Experimental Oncology, Department of Pathology, Stanford University School of Medicine, Stanford, California 94305

ROBERT M. SNAPKA • Department of Biology, Massachusetts Institute of Technology, Cambridge, Massachusetts 02139. *Present address:* The Ohio State University, Division of Radiobiology, Department of Radiology, Columbus, Ohio 43210-1214

ALEXANDER VARSHAVSKY • Department of Biology, Massachusetts Institute of Technology, Cambridge, Massachusetts 02139

RICHARD VOELLMY • Department of Biochemistry and Molecular Biology, University of Miami School of Medicine, Miami, Florida 33136

IRVING L. WEISSMAN • Laboratory of Experimental Oncology, Department of Pathology, Stanford University School of Medicine, Stanford, California 94305

KEITH D. WILKINSON • Department of Biochemistry, Emory University School of Medicine, Atlanta, Georgia 30322

ROY S. WU • Laboratory of Molecular Pharmacology, Division of Cancer Treatment, National Cancer Institute, National Institutes of Health, Bethesda, Maryland 20892

INGRID WÜNNING • Department of Biology, Massachusetts Institute of Technology, Cambridge, Massachusetts 02139

Preface

Higher eukaryotes express at least 2000 proteins and contain enough genetic information to encode many times that number. It is therefore a bit frightening to imagine a book devoted to each and every one of them. Fortunately, not all polypeptides are created equal. Some, like actin, myosin, and calmodulin, play such crucial roles in metabolism or the structural organization of cells that they do in fact deserve extensive discussion. Believing that ubiquitin falls within this special category, I accepted the invitation from Plenum to edit a book on this remarkable eukaryotic protein. Had Avram Hershko's laboratory been located in the United States, the task might well have fallen to him. At times, particularly when trying to speed delivery of a late chapter, I envied Avram's distance from the project. Still, these frustrations were minor, and the experience was an enjoyable one, for both me and my wife, Florence, who made major contributions to the editing. The final product, *Ubiquitin*, should serve as a useful guide to biochemists or cell biologists who wish to become familiar with this important cellular protein. I take this opportunity to thank the contributors, the editing staff at Plenum, especially Mary Born, and all those colleagues who sent preprints to the various authors.

Martin Rechsteiner

Salt Lake City, Utah

Contents

Chapter 2

Molecular Genetics of the Ubiquitin System

Daniel Finley, Engin Özkaynak, Stefan Jentsch, John P. McGrath,
Bonnie Bartel, Michael Pazin, Robert M. Snapka, and
Alexander Varshavsky

Chapter 3

Ubiquitin Activation and Ligation

Cecile M. Pickart

Chapter 4

Ubiquitin/ATP-Dependent Protease

Ronald F. Hough, Gregory W. Pratt, and Martin Rechsteiner

Chapter 5

Ubiquitin Carboxyl-Terminal Hydrolases

Irwin A. Rose

Chapter 6

Ubiquitinated Histones and Chromatin

William M. Bonner, Christopher L. Hatch, and Roy S. Wu

Chapter 7

Immunochemical Probes of Ubiquitin Pool Dynamics

Arthur L. Haas

Chapter 8

Protein Breakdown and the Heat-Shock Response

Stephen A. Goff, Richard Voellmy, and Alfred L. Goldberg

Chapter 9

Lymphocyte Homing Receptors, Ubiquitin, and Cell Surface Proteins

Mark Siegelman and Irving L. Weissman

Chapter 10

Role of Transfer RNA in the Degradation of Selective Substrates of the Ubiquitin- and ATP-Dependent Proteolytic System

Aaron Ciechanover

Chapter 11

The N-End Rule of Selective Protein Turnover: Mechanistic Aspects and Functional Implications

Alexander Varshavsky, Andreas Bachmair, Daniel Finley, David Gonda, and Ingrid Wünning

Chapter 12

Selectivity of Ubiquitin-Mediated Protein Breakdown: Current Status
of Our Understanding

Avram Hershko

Introduction

Martin Rechsteiner

Ubiquitin (Ub) approaches actin, tubulin, and the histones in intracellular abundance. Thus, it is surprising that the protein was discovered just 13 years ago when Gideon Goldstein and his colleagues[1] isolated it as a protein capable of inducing B-lymphocyte differentiation. Because the antibodies prepared against Ub reacted with similar proteins in organisms as diverse as mammals, yeast, and celery, the protein was named ubiquitous immunopoietic peptide. Later in 1975, Schlesinger *et al.*[2] published the sequence for ubiquitin, as it was now called. Ubiquitin reappeared in the literature in 1977. That year, two groups[3,4] reported that A24, a rat liver chromosomal protein identified by two-dimensional electrophoresis, contained histone H2A covalently linked to Ub. The two proteins were joined by an isopeptide bond between the ε-amino group of lysine 119 on histone H2A and the carboxyl terminus of Ub. In 1980, Ub surfaced for a third time in yet another guise. Hershko and his colleagues showed that the ATP-dependent proteolysis factor in reticulocyte lysates, APFI, was covalently linked to various proteins, including exogenous proteolytic substrates.[5,6] It was soon demonstrated that APFI was indeed ubiquitin.[7]

During the past eight years there have been a number of exciting developments regarding Ub. It now seems clear that the protein plays a central role in intracellular proteolysis. The enzymes for activating and

MARTIN RECHSTEINER • Department of Biochemistry, School of Medicine, University of Utah, Salt Lake City, Utah 84132.

1

conjugating Ub to a variety of proteins have been identified[8,9] and, in some cases, purified.[10] A large ATP-dependent protease that degrades ubiquitinated substrates has also been purified.[11] These results confirm central features of the marking hypothesis of Hershko and his co-workers.[6] The x-ray structure of Ub has been resolved to 1.8 Å,[12] and its NMR spectrum has been assigned.[13] Site-directed mutagenesis by Ecker et al.[14] has revealed the importance of specific residues in ubiquitin's C-terminal tail and core, but interestingly enough, changes at several residues on the surface produce molecules with considerable residual activity. Ubiquitin has been shown to be a heat-shock protein,[15] and it has been found on the cell surface.[16] Analysis of ubiquitin genes produced several surprises. One locus encodes tandem arrays of Ub without intervening stop codons[17]; presumably Ub monomers are generated by cleavage from a polyprotein. Remarkably, processing can occur in E. coli[18] even though Ub conjugation apparently does not. Other loci encode Ub with C-terminal extensions up to 80 residues in length.[19,20] The extensions have a number of interesting features, including zinc finger motifs and histone-like regions.

Fried et al.[21] have recently reported that Ub has proteolytic activity. Because an enzymatic cleft is not readily apparent in the x-ray structure, this is a most unexpected result. Moreover, the sequence of the Ca^{2+}-activated protease inhibitor, calpastatin, has just been published,[22] and we noticed that Ub and calpastatin share similar sequences:

$$\text{Calpastatin—EEEREKLGEKEETIPPD–YRLEEA}$$
$$\bullet \quad \circ \quad \bullet\circ \quad \circ\bullet \quad \bullet \quad \bullet\bullet\bullet\bullet \quad \bullet\bullet \quad \bullet$$
$$\text{Ubiquitin—ENVKAKIQDK–EGIPPDQQRLIFA}$$

Although the "homology" only borders on statistical significance, the fact that both molecules are members of cytosolic proteolytic pathways adds physiological significance. This result, which blurs the distinction between Ca^{2+}-mediated and Ub-mediated degradation pathways, would suggest that Ub interacts with proteases rather than being one. Still, the observation by Fried et al. may prove correct since it is conceivable that Ub adopts several physiologically important conformations. The existence of several conformations is one explanation for the extreme conservation of Ub's sequence.

Besides being a biological object of considerable interest, Ub also serves a sociological function. The molecule has become a common topic for discussion among students of the heat-shock response, intracellular proteolysis, post-translational modifications, and transcriptional regulation. The exchange of information between investigators in these fields can only speed our understanding of protein interactions within living cells. Finally, I believe that Ub is a "lucky" molecule. Almost everyone who has studied it has made fascinating observations. Many of these are presented in the chapters that follow.

REFERENCES

1. Goldstein, G., Scheid, M., Hammerling, U., Boyse, E. A., Schlesinger, D. H., and Niall, H. D., 1975, Isolation of a polypeptide that has lymphocyte-differentiating properties and is probably represented universally in living cells, *Proc. Natl. Acad. Sci. U.S.A.* **72:** 11–15.
2. Schlesinger, D. H., Goldstein, G., and Niall, H. D., 1975, The complete amino acid sequence of ubiquitin. An adenylate cyclase stimulating polypeptide probably universal in living cells, *Biochemistry* **14:** 2214–2218.
3. Goldknopf, I. L., and Busch, H., 1977, Isopeptide linkage between nonhistone and histone 2A polypeptides of chromosomal conjugate–protein A24, *Proc. Natl. Acad. Sci. U.S.A.* **74:** 864–868.
4. Hunt, L. T., and Dayhoff, M. O., 1977, Amino-terminal sequence identity of ubiquitin and the nonhistone component of nuclear protein A24. *Biochem. Biophys. Res. Commun.* **74:** 650–655.
5. Ciechanover, A., Heller, H., Elias, S., Haas, A. L., and Hershko, A., 1980, ATP-dependent conjugation of reticulocyte proteins with the polypeptide required for protein degradation, *Proc. Natl. Acad. Sci. U.S.A.* **77:** 1365–1368.
6. Hershko, A., Ciechanover, A., Heller, H., Haas, A. L., and Rose, I. A., 1980, Proposed role of ATP in protein breakdown: Conjugation of proteins with multiple chains of the polypeptide of ATP-dependent proteolysis, *Proc. Natl. Acad. Sci. U.S.A.* **77:** 1783–1786.
7. Wilkinson, K. D., Urban, M. K., and Haas, A. L., 1980, Ubiquitin is the ATP-dependent proteolysis factor I of rabbit reticulocytes, *J. Biol. Chem.* **225:** 7529–7532.
8. Hershko, A., Heller, H., Elias, S., and Ciechanover, A., 1983, Components of ubiquitin–protein ligase system: Resolution, affinity purification, and role in protein breakdown, *J. Biol. Chem.* **258:** 8206–8214.
9. Pickart, C. M., and Rose, I. A., 1985, Functional heterogeneity of ubiquitin carrier proteins, *J. Biol. Chem.* **260:** 1573–1581.
10. Ciechanover, A., Elias, S., Heller, H., and Hershko, A., 1982, "Covalent affinity" purification of ubiquitin-activating enzyme, *J. Biol. Chem.* **257:** 2537–2542.
11. Hough, R., Pratt, G., and Rechsteiner, M., 1987, Purification of two high molecular weight proteases from rabbit reticulocyte lysate, *J. Biol. Chem.* **262:** 8303–8311.
12. Vijay-Kumar, S., Bugg, C. E., Wilkinson, K. D., Vierstra, R. D., Hatfield, P. M., and Cook, W. J., 1987, Comparison of the three-dimensional structures of human, yeast, and oat ubiquitin, *J. Biol. Chem.* **262:** 6396–6399.
13. Weber, P., Brown, S., and Mueller, L., 1987, Sequential ^{1}H-NMR resonance assignments and secondary structure identification of human ubiquitin, *Biochemistry* **26:** 7282–7290.
14. Ecker, D. J., Butt, T. R., Marsh, J., Sternberg, E. J., Margolis, N., Monia, B. P., Jonnalagadda, S., Khan, M. I., Weber, P. L., Mueller, L., and Crooke, S. T., 1987, Gene synthesis, expression, structures, and functional activities of site-specific mutants of ubiquitin, *J. Biol. Chem.* **262:** 14213–14221.

15. Bond, U., and Schlesinger, M. J., 1985, Ubiquitin is a heat shock protein in chicken embryo fibroblasts, *Mol. Cell. Biol.* **5**: 949–956.
16. St. John, T., Gallatin, W. M., Siegelman, M., Smith, H. T., Fried, V. A., and Weissman, I. L., 1986, Expression cloning of lymphocyte homing receptor cDNA: Ubiquitin is the reactive species, *Science* **231**: 845–850.
17. Finley, D., and Varshavsky, A., 1985, The ubiquitin system: Functions and mechanisms, *TIBS* **10**: 343–347.
18. Jonnalagadda, S., Butt, T., Marsh, J., Sternberg, E. J., Mirabelli, C. K., Ecker, D. J., and Crooke, S. T., 1987, Expression and accurate processing of yeast pentaubiquitin in *E. coli, J. Biol. Chem.* **262**: 17750–17756.
19. Lund, P. K., Moats-Staats, B. M., Simmons, J. G., Hoyt, E., D'Ercole, A. J., Martin, F., and Van Wyk, J. J., 1985, Nucleotide sequence analysis of a cDNA encoding human ubiquitin reveals that ubiquitin is synthesized as a precursor, *J. Biol. Chem.* **260**: 7609–7613.
20. Ozkaynak, E., Finley, D., Solomon, M. J., and Varshavsky, A., 1987, The yeast ubiquitin genes: A family of natural gene fusions. *EMBO J.* **6**: 1429–1439.
21. Fried, V. A., Smith, H. T., Hildebrandt, E., and Weiner, K., 1987, Ubiquitin has intrinsic proteolytic activity: Implications for cellular regulation, *Proc. Natl. Acad. Sci. U.S.A.* **84**: 3685–3698.
22. Emori, Y., Kawasaki, H., Imajok, S., Imahori, K., and Suzuki, K., 1987, Endogenous inhibitor for calcium-dependent cysteine protease contains four internal repeats that could be responsible for its multiple reactive site, *Proc. Natl. Acad. Sci. U.S.A.* **84**: 3590–3594.

Chapter 1

Purification and Structural Properties of Ubiquitin

Keith D. Wilkinson

1. INTRODUCTION

This chapter discusses the purification and characteristics of ubiquitin (Ub),[1] the highly conserved and widely distributed peptide that participates in a variety of cellular functions. As this book makes clear, Ub has a role in protein degradation,[2-4] chromatin structure,[5-7] the heat-shock response,[8,9] cell surface receptors,[10-12] and perhaps even immunological response.[13] The one unifying theme in all these functions is the formation of an amide bond between the carboxyl terminus of Ub and amino groups of a variety of proteins. Thus, Ub can be thought of as a marker molecule that targets proteins for any of a variety of metabolic fates. One of the critical unanswered questions about its mode of action involves the definition of how Ub contributes to the partition of various Ub–protein conjugates between these metabolic fates.

I believe that the structural chemistry of Ub plays some role in this partitioning and that by discerning structure–function relation we will learn more about the functions of this molecule and its vital role in cellular metabolism. To date, ATP-dependent protein degradation has been the

KEITH D. WILKINSON • Department of Biochemistry, Emory University School of Medicine, Atlanta, Georgia 30322.

only pathway where we have the types of detailed information and purified enzymes that allow the establishment of structure–function relation. Thus, most of the information presented in this chapter derives from the elucidation of its role in that system.

The early portions of this chapter deal with the purification and characterization of Ub, while the latter portions deal with experiments that probe the roles of Ub by manipulating its structure in various ways. I conclude with some suggestions on possible contributions from future studies on the structural chemistry of Ub.

2. PURIFICATION OF UBIQUITIN

2.1. Assays for Ubiquitin

A number of suitable assays are available to detect and to quantitate the amounts of Ub present at various stages of purification or in cell extracts. The choice of assay method depends primarily on the capabilities of the individual laboratory. These will be discussed in approximate order of sensitivity for detection of the 76 residue active protein.

2.1.1. Radioimmunoassay for Ub

A solution phase radioimmunoassay has been described by Haas et al.,[14] which is very specific for the intact 76 residue protein. Rabbit antibodies directed against the carboxyl terminus of Ub have been reproducibly obtained in a number of laboratories by immunization with Ub cross-linked to horse gamma globulin. These antibodies are nonprecipitating and the immune complex must be precipitated with polyethylene glycol. Using [^{125}I]ubiquitin, which is not precipitated by polyethylene glycol, it is possible to detect 1 pmole of free Ub. Cleavage of the carboxyl-terminal glycylglycine, conjugation of Ub to other proteins, or inactivation of Ub in crude extracts all abolish the cross-reactivity with these antibodies. The limiting factors in this assay are the affinity of the antibodies for Ub (10^{-9} M) and the specific radioactivity of Ub.

Other antibody preparations have been described that are useful in detecting Ub conjugated to a variety of proteins.[15,16] Antibodies to SDS-denatured cross-linked Ub can be used to quantitate endogenous conjugates of Ub by Western blot techniques.[16] Monoclonal antibodies to Ub have proved useful in detecting ubiquitinated cell surface molecules and receptors.[10,11]

2.1.2. ATP:Pyrophosphate Exchange Catalyzed by the Ub Activating Enzyme

This assay depends on the isotope exchange reaction catalyzed by the purified Ub activating enzyme (E1),[17] the enzyme responsible for adenylating the carboxyl terminus of Ub. This enzyme has been purified to homogeneity and its mechanism determined by Haas and Rose[18] to be ordered, with ATP adding first. Ub is activated by adenylation of the carboxyl-terminal glycine and subsequent release of pyrophosphate. The tightly bound Ub adenylate can undergo nucleophilic attack by an enzyme thiolate to form bound AMP and the thiol ester of Ub. In the presence of a high concentration of AMP, no further reactions take place and Ub is not released from the enzyme. Incubation of this enzyme with ATP, AMP, [^{32}P]pyrophosphate, and Ub results in exchange of the radiolabel into ATP, which can be quantitated by adsorption of the nucleotide on charcoal and either extracting the ATP or directly counting the charcoal. This assay is rapid, quantitative, and quite specific for the 76 residue peptide. The apparent K_m for stimulation of isotope exchange is 6×10^{-7} M,[18] and the assay can reproducibly detect 3 pmole of Ub. Sensitivity can be enhanced several-fold by running the assay under V/K conditions.

Two cautionary notes should be mentioned. First, in crude samples, there may be significant amounts of nonspecific ATP–pyrophosphate exchange activity, resulting in a high background for this assay. Thus, it is important to run controls in the absence of E1. Second, the ordered mechanism of this enzyme results in inhibition of exchange at higher concentrations of Ub. This means that samples must be assayed at several dilutions in order to obtain quantitative determination of the amount of Ub present. Finally, this assay requires purification of the adenylating enzyme.

2.1.3. Precipitation of Ubiquitin Adenylate

Some of the problems associated with the isotype exchange assay can be surmounted while still retaining the specificity and sensitivity of this assay. It has been observed that iodoacetamide-inactivated E1 can still form a very tightly bound Ub adenylate.[18] Recently, Rose and Warms have described an assay that makes use of this fact.[19] Ub and iodoacetamide-treated enzyme are incubated with [^{3}H]ATP and the resultant Ub adenylate is precipitated with TCA. In the presence of an excess of the E1, Ub is quantitatively adenylated and the amount in an unknown sample can be determined from the radioactivity in the pellet and the known specific radioactivity of the ATP. It appears that this assay can accurately detect about 1 pmole of Ub. A method has been described by these workers for preparation of crude Ub and E1, neither of which requires pure Ub for standards or affinity columns.

A few precautions need be observed with this assay. First, samples must be freed of any ATP (which would dilute the specific activity of the $[^3H]ATP$) and phosphate (which inhibits stoichiometric adenylation). This can be done by repeated precipitation of the proteins in the unknown sample by TCA or by dialysis. Second, it is important to inactivate any Ub carboxyl-terminal hydrolytic activity by heat treating the TCA pellet in order to prevent the degradation of the Ub adenylate.[19]

2.1.4. Stimulation of Protein Degradation

Historically, the stimulation of ATP-dependent proteolysis catalyzed by a crude peparation (fraction II) from reticulocytes has been important.[20] Fraction II which is prepared from fresh reticulocytes depleted of ATP, contains all the enzymes necessary to reconstitute ubiquitin-dependent proteolysis. The best substrate for these purposes is reduced, carboxymethylated $[^{125}I]BSA$.[21] Substrate, fraction II, and Ub are incubated and the reaction is terminated by the addition of trichloroacetic acid. Ub stimulates the rate of release of TCA-soluble radioactivity by ten- to 20-fold. Since the apparent K_m for Ub stimulation[21] is higher than that of isotope exchange, this assay method is somewhat less sensitive but still capable of detecting 30 pmole of Ub. The difficulties with this assay include the extensive preparation of the assay reagents and the fact that crude samples of Ub contain many proteins, some of which can act as competitive inhibitors of the degradation of the radiolabeled substrate.

2.1.5. HPLC Detection of the Protein

A less sensitive but more rapid and convenient assay for Ub in column fractions or cell lysates is afforded by HPLC.[22] This assay takes advantage of the fact that Ub is not precipitated by 5% perchloric acid,* whereas many other proteins are. The sample to be examined is subjected to precipitation by 5% perchloric acid and centrifugation to clarify the sample, and then it is injected directly onto a C-8 reverse phase HPLC column equilibrated with 25 mM sodium perchlorate and 12.5 mM perchloric acid in 42% acetonitrile. By monitoring the effluent at 205 nm, as little as 50 pmole of Ub can easily be detected. There are a number of advantages to this procedure. First, it takes only 10 min, and the result of each de-

---

* I would like to thank Koko Murakami for first pointing this out.

termination is immediately available. This facilitates the selection of appropriate fractions for assay. The sensitivity is such that the Ub from approximately 10^6 cells can be quantitated. Second, the HPLC separation is very sensitive to damage to or alterations of the Ub structure. Oxidation, proteolysis, deamidation, and chemical modifications of Ub all result in different chromatographic behavior in this system. The obvious shortcoming of this assay is the lack of specificity for Ub, as opposed to other compounds with the same retention time. Whereas this complication has not been a problem, it may be necessary to characterize the putative Ub by one or more of the assays described above.

2.2. Tissue Source

Ub has been isolated from a wide variety of mammalian tissues, including reticulocytes,[2] erythrocytes,[4,23,24] thymus,[25] liver,[14] testis,[26] pituitary,[27] parathyroid,[28] and hypothalamus.[29] An identical protein has been isolated from trout testis,[30] Mediterranean fruit fly eggs,[31] and the blowfly.[32] The cDNA of *Xenopus laevis*[33] and chicken[9] indicate that an identical protein is made in these organisms. *Tetrahymena* Ub has been isolated[34] and shows a slightly different amino acid analysis and appears to vary somewhat in the region between residues 12 to 27.[35] Finally, Ub has been isolated from yeast[22] and from oat[36] and the sequences of these proteins[37,38] differ from each other and from animal Ub (Figure 1).

The amount of Ub in tissues has been only briefly examined using radioimmunoassay[16,25,32] and varies from 2 to 100 μg/g of tissue. For the routine preparation of large amounts of animal Ub, human or bovine erythrocytes are the preferred source. These cells contain between 0.1 and 0.2 mg Ub/ml packed cells and low levels of conjugated Ub. They can be obtained economically in large amounts from outdated human blood or a slaughterhouse. Cell lysates are easily obtained and the relatively simple composition of proteins present in this cell type simplifies purification. Finally, erythrocytes have a very low level of proteolytic activity which could inactivate the protein. Since it will sometimes be desirable or necessary to purify Ub from other sources, the following sections deal with a few recommended precautions.

2.2.1. Use of Protease Inhibitors

Ub is quite stable to a variety of proteases but is rapidly inactivated by trypsin[4] and tryptic-like activity.[4,14,36] Cleavage of the carboxyl-terminal Gly-Gly has been observed in most preparations of Ub, indicating the presence of these activities in tissue homogenates. In two cases this

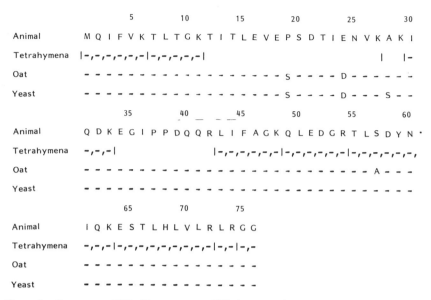

Figure 1. Sequences of Ub. The sequence of Ub from a variety of animal sources is given in the top line. For the other entries, a dash indicates identity with the animal sequence. The *Tetrahymena* sequence is based on compositional analysis of several tryptic peptides, and constituent amino acids are separated by commas. A blank entry indicates that no corresponding peptide has been analyzed.

was a cathepsin-like activity that was completely inhibited by TLCK or pCMB.[14,36] Partial protection was afforded by other thiol reagents, chelating agents, pepstatin, leupeptin, or soybean trypsin inhibitor. Thus, the inclusion of a trypsin inhibitor and a thiol reagent is recommended when Ub is to be isolated from a complex tissue. In trout testis, carboxypeptidase activity was apparently a problem since Des-Gly Ub was also isolated.[30] We have been unable to demonstrate cleavage of Ub by carboxypeptidases a, b, or y, but it would seem prudent to include a zinc chelator to minimize this complication.

2.2.2. Depletion of Endogenous Conjugates

Recent immunochemical quantitation of Ub in cells by Haas and Bright[16] has demonstrated that between 50 and 80% of the cellular Ub is conjugated to other proteins in the steady state. In reticulocytes, these conjugates can be depleted by incubation of the intact cells with inhibitors of energy metabolism. In the absence of ATP, the Ub is released by the action of proteases of the system, as well as enzymes such as A24 lyase and the isopeptidase (see Section 5.2.1). Depletion of energy reserves is only practical with cell suspensions, but where possible this will increase the yield of free Ub in the homogenate.

2.3. Purification Procedure

A wide variety of purification schemes have been applied to the preparation of Ub. They have in common, however, many of the principles discussed below. The order of steps is often dictated by the most efficient route from large volumes of starting material but appears to make little difference in the purity of the final product. I discuss our purification from erythrocytes[24] and make reference to other preparations such as those from yeast and thymus when appropriate.

2.3.1. General Considerations

Its molecular weight of 8565 makes Ub one of the smallest major proteins in most cells. It has a neutral isoelectric point $(6.7)^{39}$ and little appreciable charge below pH 9.0. It is extremely resistant to thermal denaturation, showing a reversible denaturation transition at about $85°C.^{40}$ It is also reversibly denatured by urea,[40] guanidinium hydrochloride,[41,42] and alcohol.[43] There are no cysteine or tryptophan residues in the molecule and mild oxidation of the methionine residue with hydrogen peroxide does not affect the activity of the protein.[44] The hardiness of this protein and the ease of attaining the native folded conformation means that quite severe conditions can be applied to the purification of Ub without adverse results.

2.3.2. Heat Treatment

Because of the excellent thermal stability of Ub, one of the earliest steps that should be applied is heat treatment of homogenates. The homogenate is heated to 90°C and held for 5–10 min. Ub is partially denatured by these conditions, but the molecule easily refolds. When large amounts of material are to be processed, filtration can be used to remove the precipitated and aggregated protein. In erythrocyte lysates, this step serves to remove the vast majority of hemoglobin. In other tissues, it may be possible to use lower temperatures to avoid losses of Ub. Losses owing to denaturation and/or aggregation of Ub have been reported to vary from 25% in erythrocyte lysates[24] to 75% in yeast homogenates.[22]

2.3.3. TCA Precipitation

The concentration of large volumes of heat-treated homogenates is conveniently attained by TCA precipitation of the proteins (5% w/v). If the protein concentration is high enough, the precipitate is flocculant and sediments to the bottom of the container at unit gravity. The majority of the supernate can then be removed by means of a siphon. If necessary, the protein can be pelleted by centrifugation, taking care to use the minimum force required. If the pellets are compacted, it is necessary to use a homogenizer to resuspend the pellet. The slurry or the resuspended pellet is neutralized with NaOH, and undissolved protein is removed by centrifugation.

2.3.4. Ammonium Sulfate Fractionation

We next subject the neutralized TCA concentrate to ammonium sulfate precipitation. The 45–80% ammonium sulfate pellet contains the Ub, and this is resuspended in a minimal volume of buffer, followed by dialysis to remove salt and exchange the buffer for one appropriate for the subsequent chromatography steps. It is prudent to assay for Ub in the 45% pellet since, at high concentrations, some Ub can be precipitated by these conditions.

2.3.5. Ion Exchange Chromatography

Since Ub has a neutral isoelectric point, both cation and anion exchange chromatography are useful in its preparation. We routinely chromatograph the dialyzed ammonium sulfate pellet on DEAE cellulose equilibrated with 20 mM ammonium bicarbonate, pH 9.0. Under these conditions, Ub is slightly retained and elutes at about three column volumes. Because of this, the sample volume loaded should not exceed the column volume to obtain the best results. This step affords a significant purification and removes the chromophoric proteins.

The pooled fractions containing Ub from DEAE chromatography can then be adjusted to pH 4.5 with acetic acid and loaded directly onto a carboxymethyl cellulose column equilibrated with 50 mM ammonium acetate, pH 4.5. Ub is tightly bound under these conditions and can be eluted as a broad peak by 50 mM ammonium acetate, pH 5.5.[24]

2.3.6. Other Useful Procedures

Although the above procedures are sufficient to yield homogeneous Ub from erythrocytes,[24] from other tissues it is sometimes necessary to add another step. With yeast Ub, we applied chromatography on QAE Sephadex A25.[22] The protein is applied in 75 mM ethanolamine acetate, pH 9.4 and eluted with 50 mM sodium acetate in the same buffer. Others have used gel filtration profitably because of the small size of Ub, often in the presence of denaturants.[31,39] It has been noted[26] that at low ionic strength and pH, Ub is slightly adsorbed to Biogel P10 resin, and this may be useful as a final purification step. Chromatography on Amberlite CG-50 at neutral pH,[31] chromatography on hydroxylapatite,[36] and crystallization from ammonium sulfate[23] have also been utilized in various preparations.

3. PHYSICAL CHARACTERISTICS OF UBIQUITIN

3.1. Amino Acid Sequence

The amino acid sequences of Ub isolated from a number of sources are summarized in Figure 1. All animal ubiquitins show an identical sequence. This protein lacks cysteine and tryptophan and has only one methionine (amino terminal), one histidine, and one tyrosine. There are 11 acidic and 11 basic residues in addition to the single histidine, giving the protein an isoelectric point of 6.7. The only unusual feature of the primary structure is the presence of a Pro-Pro sequence at residues 37 and 38.

The tryptic peptides from *Tetrahymena* Ub have been purified and subjected to amino acid analysis.[35] Eight of these peptides show compositions that are identical to the corresponding peptides from bovine Ub. One peptide (34–42) was not analyzed, but it chromatographs with the same retention time and peak shape characteristic of the bovine peptide[50] and is probably identical. One peptide (12–27) shows differences in the amino acid composition and elution position[50] from the corresponding bovine peptide. Thus, there are apparently sequence differences in the same region of the molecule as in yeast and oat. Interestingly, Ub from *Tetrahymena* appears to be phosphorylated at serine and possibly ADP-ribosylated[34] at unknown sites. There have been no reports of similar post-translational modification in other organisms. One plant Ub has been sequenced by Vierstra *et al.*,[37] that from oat. This Ub has three amino acid substitutions. These are Ser-19 for Pro, Asp-24 for Glu, and Ala-57 for Ser. Yeast Ub[22,38] also has three amino acid substitutions compared to animal Ub; Ser-19 for Pro, Asp-24 for Glu, and Ser-28 for Ala.

It is interesting to consider the possible mechanisms by which this sequence identity is maintained. Of course, the cellular functions of Ub exert the ultimate selective pressure and these aspects are considered in Section 5.3. The gene sequence has been reported for yeast,[38] *Xenopus laevis,*[33] chicken,[9] and human[46,47] Ub. In each case, the gene consists of tandem repeats of the coding sequence with no intervening sequences, or as a fusion gene with Ub at the amino terminus of a longer coding sequence. How does the cell prevent mutations of one of the copies of Ub within these genes? Such tandem repeats may be a proofreading mechanism that serves to maintain the similarity of the multiple copies of the gene. This does not, however, assure that heteromorphism does not occur in the protein. A possible additional mechanism to maintain sequence conservation is suggested by the fact that the polyprotein translation product must be cleaved by a processing enzyme to release the intact 76 amino acid sequence. It is tempting to speculate that if mutations in one copy of the gene arose, it could change the conformation of the polyubiquitin or the fusion protein sufficiently to prevent its processing. This may be a lethal mutation since a mutation in one copy would be sufficient to prevent release of sufficient amounts of intact Ub. Thus, conservation of the sequence may be attained by the combined effects of the gene structure, specificity of the processing enzyme, and functional selective pressure.

3.2. Crystal Structure

Vijay-Kumar *et al.*[48] have determined the crystal structure of Ub and subsequently refined it to 1.8 Å resolution.[49] The protein has a tightly packed globular conformation (Figure 2)* with a pronounced hydrophobic core. Upon refinement, two sections of helical structure are apparent, three and one-half turns of alpha helix from residue 23–34 and a short section of 3,10 helix at residues 56–59. The other prominent feature is a five-strand beta sheet wrapped around one face of the long helix. These elements of secondary structure are connected by seven reverse turns. About 90% of the polypeptide chain is involved in hydrogen-bonded secondary structure. This extensive bonding and the hydrophobic core probably account for the pronounced thermal stability of this protein.

Several features of the surface of the folded molecule deserve comment. First, the carboxyl terminus protrudes from the core of the protein as it must in order to participate in formation of conjugates with other

---

* The space-filling representations and the ribbon drawings of the ubiquitin structure were obtained at the University of Alabama, Birmingham, Center for Macromolecular Crystallography with the kind assistance of Drs. M. Carson, C. Bugg, W. Cook, and S. Vijay-Kumar.

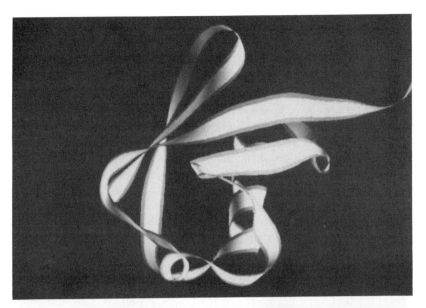

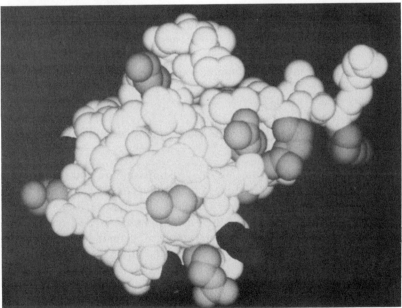

Figure 2. The basic face of Ub. (*Top panel*) Ribbon drawing of the peptide backbone of Ub. Note the carboxyl terminus protruding to the upper right. (*Bottom panel*) Space-filling representation of Ub in the same orientation as top panel. Acidic side chains are dark grey, basic side chains are light grey, and all other atoms are white. Hydrogen atoms have been omitted for clarity. Seven of the 11 basic residues are clearly visible from this orientation.

proteins. The last hydrogen bond to the body of the protein is at Arg-72. Second, as shown in Figure 2, most of the lysine and arginine residues (except Arg-74) are buried, involved in salt-bridges or hydrogen bonds, or sterically blocked to some extent. Three lysines (6,33,63) have been described as being exposed to solvent. The lack of tryptic cleavage at these positions may be related to the presence of large or acidic side chains flanking these lysines[50] (Figure 3). These observations explain the stability of the folded molecule to tryptic cleavage. As shown in Figures 2 and 3, there is a marked partition of the acidic residues to one face of the molecule and the basic residues to another. Finally, Figure 4 shows that there is also a patch of hydrophobic residues on the surface of the protein near the carboxyl terminus. These "landmarks" must be recognized by the various enzymes that bind to and metabolize Ub and its derivatives.

Recently, the crystal structures of yeast and oat Ub were determined.[50] These proteins have a tertiary structure that is identical to that of the bovine protein. This explains the similarity in immunological and functional properties of these three proteins. It is interesting to note that the residues that vary between species (19,24,27, and 57) are clustered in one region of the tertiary structure opposite to the carboxyl terminus (Figure 5). Since these changes have little effect on the activities of Ub (see Section 4.1), it is tempting to speculate that this region of Ub is involved in a nonproteolytic function that varies between yeast, plants, and animals. This may be the immunological function of T-cell homing[10] or mitogenesis.[13]

3.3. Spectroscopic Properties of the Native and Alcohol-Induced Conformations

There is every indication that the crystal structure and the solution structure of Ub are identical. Most of this evidence comes from a variety of spectroscopic studies, which are discussed in this section, and from chemical modification studies, which are discussed in Section 4.2. Recently, we found that modest concentrations of alcohol can cause a conformational change in Ub.[43] Such conformational changes appear to be important in the functioning of Ub. Iodinated derivatives of Ub, which have decreased conformational stability, are actually more active in the stimulation of ubiquitin-dependent protein degradation.[44] Most of our

Figure 3. The acidic face of Ub. (*Top panel*) Ribbon drawing of the peptide backbone. (*Bottom panel*) Space-filling representation. This view is obtained by a 150° rotation of the orientation shown in Figure 2 and the same coloring scheme is used for the atoms. Eight of the 11 acidic residues are clearly visible from this face of the molecule.

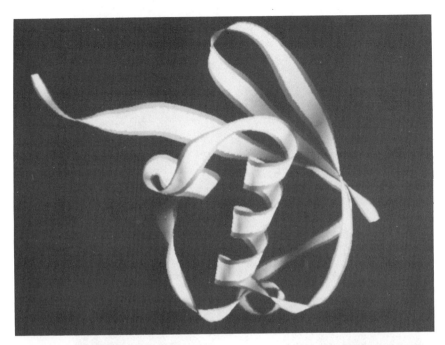

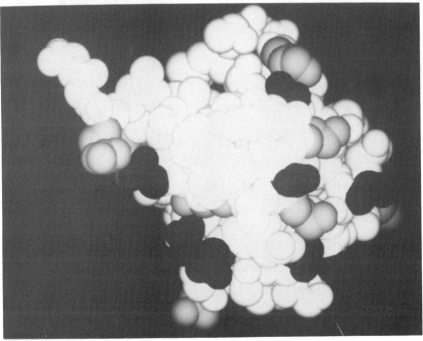

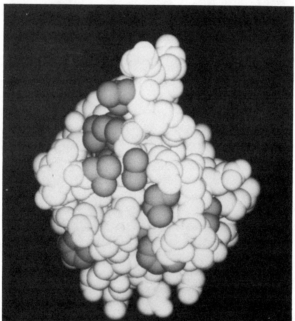

Figure 4. The hydrophobic face of Ub. (*Top panel*) Ribbon drawing of the peptide backbone. (*Bottom panel*) Space-filling representation. The side-chain atoms of hydrophobic residues are colored grey and all others are white. The basic face of the molecule lies to the right of this hydrophobic patch and the acidic face to the left.

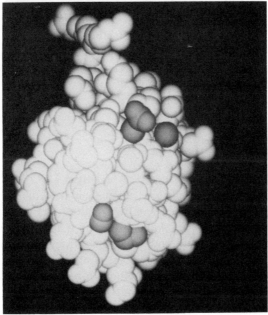

Figure 5. Position of altered residues in oat and yeast Ub. (*Top panel*) Ribbon drawing of the peptide backbone. (*Bottom panel*) Space-filling representation. The side-chain atoms of residues that differ from animal Ub (19,24,28,57) are colored grey and all other atoms white. This face of the molecule is almost directly opposite the carboxyl-terminal strand of beta-sheet.

structural information about this conformational change is obtained by spectroscopic studies discussed below.

3.3.1. NMR

The NMR spectrum of Ub[40,41] contains a number of features that indicate a tightly folded structure is adopted in aqueous solution. One early study[41] demonstrated that Ub is stable to extremes of pH, temperature, and high concentrations of guanidine hydrochloride. A later study[40] investigated more thoroughly the spectral characteristics of this protein. There are a number of hard-to-exchange amide protons, again indicating a tightly folded structure. The pH dependence of the chemical shifts indicates that all the Lys, Asp, and Glu residues exhibit normal pK values. The pK of the histidine is 5.8 while that of the tyrosine is 11.2 with evidence for a ionizable carboxyl group in close proximity. This is most likely Glu-51 as judged by its location in the crystal structure.[48,49] Several unassigned methyl groups were observed to be near the aromatic residues as judged by both titration behavior and NOE measurements. The quality of the NMR spectra obtainable and the availability of the crystal structure make this protein an attractive one for further NMR studies. In particular, two-dimensional NMR experiments could confirm the similarity of the solution and the crystal structure. In addition, we are applying NMR techniques to defining the nature of the alcohol-induced conformational changes.*

3.3.2. Circular Dichroism

The circular dichroism spectrum of native Ub[31,40,42,43] indicates a low amount of helical structure, in agreement with the crystallographic data. There have been no reports of the spectrum in the aromatic region. Upon addition of alcohol, however, there is a marked increase in the helical content of the protein, indicating the formation of a structure with almost 50% helix at a dielectric constant of about 50.[43] The formation of this conformation is also accompanied by an increased solvent exposure of Met-1 and Tyr-59. Oxidation of Met-1 has no effect on the CD signal at neutral or acidic pH, even though oxidation causes a conformational change at acidic pH.[26,44]

---

* These studies are currently in progress (C. Post, K. Wilkinson, and J. Markley).

3.3.3. UV Absorption

The absorption spectrum of Ub is quite weak, reflecting the presence of the single tyrosine and two phenylalanines. The absorbance of a 1 mg/ml solution at 275 nm is $0.17^{20,42}$ and 0.16 at 280 nm.[24] A somewhat higher value for the absorbance at 280 nm has been reported,[23] but this does not agree with the other reported values or that expected for this protein. Spectroscopic measurements of the tyrosine ionization confirm the pK value of $11.2,^{42}$ and indicate that this pK is lowered to 10.5 upon unfolding of the protein.[45] This suggests that the proximity of Glu-51 is responsible for the elevated pK of this tyrosine.

3.3.4. Fluorescence

The fluorescence of Ub is a sensitive probe for the conformation of the molecule around Tyr-59. It has been observed that the fluorescence of the single tyrosine is extremely quenched at neutral pH.[42] Protonation of an acidic group with a pK of 3.9 results in an increase of the fluorescence from about 7 to 50% of that expected for tyrosine. Denaturation of Ub also enhances the fluorescence. Thus, the ionized Glu-51 quenches the tyrosine fluorescence and perturbs the ionization of the tyrosine. We have also found that the Tyr fluorescence is increased upon undergoing the alcohol-induced conformational change, indicating that it becomes exposed to solvent.[43]

3.4. Digestion by Proteases

The susceptibility of Ub to digestion by proteases, particularly trypsin, has been the subject of some disagreement. It appears generally agreed that Ub is digested by pepsin^{4,32} and *Staphylococcus* V-8 protease.^{30,37} Some groups have reported significant cleavage to peptides catalyzed by trypsin,^{23,31,52} while others report no cleavage or release of the carboxyl terminal glycylglycine.^{1,22,30,36,37,45,51} We have investigated this by characterizing and quantitating the digestions using HPLC. With homogeneous Ub and trypsin that has been treated with TPCK and affinity purified on a column of soybean trypsin inhibitor, we see no evidence for significant cleavage of the core of the molecule. Digestion of ubiquitin with 5% (w/w) trypsin at pH 8.2 in the presence of 10 mM calcium chloride results in the rapid cleavage of the diglycine peptide (1 min) followed by the very slow cleavage at Arg-72 to release leucylarginine with a half-time of about 10 hr. In the absence of the stabilizing metal ion, there is less than 20% release of this dipeptide in 24 hr. We have found that it is necessary to partially denature Ub with 6 M urea in order to obtain good yields of peptides on digestion with highly purified trypsin.[51] It is possible that the cleavage of native Ub sometimes observed is due to contamination of commercial preparations of trypsin with an endoprotease activity that is removed upon affinity purification of trypsin.

3.5. Hydrodynamic Properties

The globular conformation of ubiquitin observed in the crystal structure appears to apply in solution also. Its sedimentation coefficient is 1.2 S,[42] and its apparent molecular weight has been reported to be approximately 8500 by gel fitration.[1,20] The electrophoretic mobility on highly cross-linked SDS gels is abnormally high, giving apparent molecular weight of 5500.[20,36] This may be due to the binding of an abnormally high amount of SDS or incomplete denaturation. On 12.5% gels, we and others[26,30] find a more normal apparent molecular weight suggesting the latter, but these molecular weight estimates are less accurate. The migration of Ub in a variety of other electrophoretic systems has also been reported.[1,3,27,29,34]

3.6. Binding and Catalytic Activities of Ubiquitin

In addition to the binding of Ub to the enzymes that metabolize it, there is some evidence for the noncovalent association of Ub to proteins. It has been reported that association of contaminating proteins with insect Ub results in a higher apparent molecular weight on gel filtration and protection from tryptic digestion.[31] Similarly, a complex between Ub and another peptide was isolated from thymus and this complexation was suggested to protect Ub from digestion by trypsin.[52] Gel filtration in the presence of denaturants removed these contaminating proteins but association was not shown to be reversible. In the nucleus, Ub is bound noncovalently in the linker region of chromatin and released by micrococcal nuclease treatment,[30] implying the presence of a binding site. Finally, free Ub is an activator of histone deacetylase activity.[53] Thus, it would appear likely that a number of noncovalent binding interactions may be possible in the cell. It will be of obvious interest to characterize any Ub binding proteins, especially in light of the recent findings that Ub is part of the cell surface architecture.

It has also been reported that Ub may possess catalytic activities: hydrolysis of p-nitrophenyl acetate,[54] p-nitrophenyl phosphate,[55] and hydration of carbon dioxide.[23,53,54] During the course of the esterase assay, Ub is progressively acetylated[53] at lysine residues.[56] Examination of these data reveals that there is no evidence for a catalytic release of p-nitrophenol. The turnover numbers measured range from 0.004 to 0.1 per minute,[54,55] and the amount of p-nitrophenol released can be accounted for by the irreversible acetylation of the lysine residues. The results do suggest that one of the lysine residues (Lys-6) is more reactive at pH 7 since a rapid burst of p-nitrophenol release is observed. The carbonic anhydrase activity of these preparations could be due to an exceedingly small contamination with the very active carbonic anhydrase, but the possibility that this is a true activity of Ub has not been ruled out.

4. STRUCTURE–FUNCTION RELATIONS

Perhaps one of the more interesting structural questions involves the determination of which regions of Ub are necessary for interaction with the various enzymes involved in its physiological functions. The fact that the sequence is so highly conserved suggests that most or all portions of the protein are required for one or more of its functions. The several enzymes that must interact with Ub could exert selective pressure to maintain a fixed conformation over much of the protein's surface. It is less clear why no substitutions are allowed in the core of the molecule. It is possible that these areas are involved in the conformational changes[43–45] that may be required for proper function. In order to examine these questions, a variety of Ub derivatives have been isolated or synthesized and tested with the enzymes of the protein degradation system (Table I). When assays for other Ub functions are available, similar studies could further define the selective pressures exerted on this highly conserved protein.

We have divided the reactions of the protein degradation system into three catagories: activation of the carboxyl terminus catalyzed by E1, conjugation of activated Ub to proteins catalyzed by the ligase system, and proteolysis of the conjugates. A complete characterization of the interaction of Ub derivatives with these enzymes would involve the assay of all these enzymes. At the time these studies were initiated, however, only E1 was available in a purified state. Thus, we have measured the activity of derivatives in stimulating ATP:pyrophosphate exchange and overall proteolysis and, where possible, the steady-state level of conjugates. A kinetic description of the effect of altering specific residues is given below and summarized in Table I.

Table I. Derivatives of Ubiquitin Tested for Activity in the ATP-Dependent Degradation System (References Are Given in the Text)

		Activity measured		
Reagent	AA modified[a]	Activation[b]	Conjugation	Proteolysis[b]
Ubiquitin from oat	P-19, E-24, S-57	100	$++^c$	nd[d]
Ubiquitin from yeast	P-19, E-24, A-28	100	$++$	100
Hydrogen peroxide	M-1	100	nd	100
Methyl iodide	M-1	nd	nd	20
TNM[e]	Y-59	55	nd	45
TNM then dithionite	Y-59	55	nd	45
Iodine	Y-59	100	nd	100
Diethylpyrocarbonate	H-68	100	nd	170
OAPA[f]	R-42	15	nd	0
OAPA	R-72	25	nd	0
OAPA	R-42, R-72	15	nd	0
OAPA	R-74	60	$++$	0
Chloramine-t + I⁻	Y-59, H-68, M-1	100	$++$	180

[a] Identified by tryptic peptide mapping.
[b] Maximal activity (% of native) at saturating concentrations of derivative.[44]
[c] Conjugate formation was detected.
[d] Not determined.
[e] Tetranitromethane.
[f] 4(Oxyacetyl)phenoxy acetic acid.

4.1. Natural Variant Sequences

4.1.1. Oat Ubiquitin

Oat Ub contains conservative substitutions at residues 19,24, and 57.[37] Its tertiary structure is identical to that of animal Ub[50] and there is strong cross-reactivity with anti-human Ub antibodies. This protein is fully active in stimulating ATP:pyrophosphate exchange as shown by the corespondence between its activity and the level of intact Ub detected by radioimmunoassay of several preparations.[36] This Ub also forms a significant level of ubiquitin–protein conjugates, indicating that a functional conjugation system is present in plants. The stimulation of overall protein degradation has not been reported, but the results suggest that oat Ub is active in protein degradation.

4.1.2. Yeast Ubiquitin

Yeast Ub also has three conservative amino acid substitutions: residues 19,24, and 28.[22,38] The crystal structure of yeast Ub is identical to that of animal and oat Ub.[50] Its solution properties, including sensitivity to trypsin, are identical to the animal Ub. The yeast protein shows identical stimulation of ATP:pyrophosphate exchange catalyzed by the reticulocyte E1 and of overall proteolysis catalyzed by reticulocyte fraction II.[22]

4.2. Chemically Modified Ubiquitin

Another way in which we can vary Ub structure to probe its function is by chemically modifying the protein. These studies yield information about the importance of residues in other parts of the molecule. Perhaps most interesting is the fact that alterations of certain residues can modify, but not totally abolish, the activity of Ub. This means that we can use these modifications as probes of the interactions of Ub with the different enzymes of the system. In all cases, the derivatives discussed below have been fully characterized by tryptic peptide mapping[51] and chemical analysis to verify the nature of the alterations introduced upon chemical modification.

4.2.1. Methionine-1

The amino terminal Met is buried in the native structure with the sulfur hydrogen-bonded to the backbone nitrogen of Lys-63. This residue cannot be alkylated in the native structure, but it is exposed to solvent on formation of the alcohol-induced conformation.[43] Oxidation of this residue to the sulfoxides and the sulfone has no influence on its activity in proteolysis, its CD spectrum, or its susceptibility to tryptic digestion.[44,57,58] It is reasonable that this modification would not greatly interfere with the ability of this side chain to hydrogen bond. Some preparations of Ub from stored human blood contain Met-1 in the oxidized form as isolated, and the properties of this natural oxidation product are identical.[43] It has been reported[26] that oxidation of testis Ub results in the appearance of three new bands in acidic urea gels. This oxidation was not reversible upon reduction, leading the authors to suggest that it was not due to oxidation of Met. Breslow *et al.*,[44] however, found that oxidation of Met-1 leads to a derivative with an altered conformation at acidic (but not neutral) pH and point out that oxidation to the sulfone would not be reversed by thiols. Thus, at neutral pH the structure and activities of Ub oxidized at Met-1 are very similar to native Ub.

In contrast, if the methionine is alkylated with methyl iodide,[43] the resultant derivative is inactive in overall proteolysis. This modification introduces a positive charge and would be expected to interfere with the native hydrogen-bonding interactions. Thus, the conformation of Ub in this region of the molecule is important for the proteolytic functions of Ub.

4.2.2. Tyrosine-59

Tyr-59 is located in a turn structure at the end of the short run of helical structure, and the phenolic proton is hydrogen-bonded to the backbone nitrogen of Glu-51. The side chain of Glu-51 electrostatically elevates the Tyr pK to 11.2. When Tyr-59 is nitrated,[42,57,58] a derivative is obtained that is less active in both ATP:pyrophosphate exchange (55%) and in supporting overall proteolysis (45%). Nitration of tyrosine is known to lower the phenolate pK from 11.2 to 6.85.[42] We considered it possible that the introduction of an additional negative charge on the phenolate would interfere with hydrogen-bonding and electrostatically repulse Glu-51, resulting in an alteration of the structure in this region of the molecule.

To test this hypothesis, the nitrated tyrosine was reduced to amino tyrosine using dithionite.[57,58] This restores the pK of the phenolate to a value similar to tyrosine. When this derivative is assayed, it shows the same levels of activity as the nitrated derivative. Thus, the alteration of activities associated with tyrosine modification are not due to electrostatic interactions but must be due to the introduction of steric bulk to the tyrosine ring or alteration of hydrogen bonding in this region of Ub.

If Ub is monoiodinated at tyrosine, it is indistinguishable from native

Ub in the ATP:pyrophosphate exchange assay and in the proteolysis assay.[45] It would be anticipated that the iodination of tyrosine would introduce steric bulk similar to that observed with the amino- or nitrotyrosine derivatives. The fact that this derivative is fully active in these assays suggests that it is primarily hydrogen-bonding interactions of the tyrosine that are important for maintaining the native conformation about this residue. These results are consistent with the hypothesis that although the conformation around Tyr-59 is important for interaction with the E1, it has little effect on the interactions with the rest of the enzymes of the proteolysis system.

4.2.3. Histidine-68

His-68 is exposed to solvent in the crystal structure, although NMR data suggest that it is shielded to some extent.[40,41] The pK of this residue is 5.8,[40] also suggesting some hydrophobic character to its environment. When this residue is modified using ethoxyformic anhydride,[57,58] the derivative has full activity in supporting ATP:pyrophosphate exchange but almost twice the activity in supporting overall protein degradation. This surprising result suggests that the region around His-68 is critical for interaction with one or more of the enzymes downstream from the E1 and that alterations in this region can enhance the interactions with the enzyme(s). There is good evidence (see Section 4.2.5) that the primary effect of modification of histidine is to increase the binding of Ub and Ub conjugates with the proteases of the system.

4.2.4. Arginine-42, -72, and -74

Arg-42, Arg-72, and Arg-74 are located at the surface of the molecule in a region near the carboxyl terminus of Ub. Four different derivatives are formed when Ub is reacted with an anionic arginine reagent, 4-(oxyacetyl)phenoxyacetic acid.[59] Arg-54 is located near the opposite end of the molecule and does not react with this reagent, perhaps because of two close carboxylate groups. Two of these derivatives, one modified at Arg-42 and one doubly modified at Arg-42 and Arg-72, are much less active in stimulation of ATP:pyrophosphate exchange and completely inactive in stimulation of proteolysis. A third derivative modified at Arg-72 shows about 25% activity in stimulation of ATP:pyrophosphate exchange and no activity in stimulation of proteolysis. These results suggest that the region around Arg-42 and Arg-72 is important for the interaction of the E1 with Ub.

The fourth derivative, modified at Arg-74, is also inactive in stimulating proteolysis but shows 60% of the activity of native Ub in stimulating ATP:pyrophosphate exchange. This result is unexpected because of the inactivity of the derivative of the nearby Arg-72 and because of the proximity of this residue to the carboxyl terminus that becomes adenylated. Apparently, the interactions of Ub with the binding site of the E1 terminate near residue 72 or 73, and the bulky side chain introduced at residue 74 in this derivative can be accommodated by the active site. An alternative possibility is suggested by examination of the crystal structure. Arg-42 and Arg-72 are clustered on one face of the molecule and the side chain of Arg-72 protrudes from the opposite side of the exposed carboxyl-terminal peptide (Figure 6). Thus, it may be that E1 recognizes the face containing Arg-42 and Arg-72, without making contact with the molecule in the region of the Arg-74 side chain. We have also shown that Ub and this Ub derivative can be conjugated with similar efficiency to the same populations of endogenous proteins.[59] Thus, modification of Arg-74 with this bulky, anionic moiety has little effect on the activation or the conjugation of Ub. It does, however, abolish the activity in stimulating overall protein degradation. Thus, Arg-74 is important for the interaction of Ub with one or more of the enzymes involved in the proteolytic steps.

4.2.5. Iodinated Ubiquitin

We have also investigated the possible synergistic effects of modification of multiple residues by iodination of the protein. When Ub is iodinated using chloramine-T, two derivatives are produced containing oxidized Met, diiodinated histidine, and either mono- or diiodinated tyrosine.[45] Both of these derivatives have been shown to be conformationally less stable than the native protein, being more sensitive to tryptic digestion and alcohol-induced helix formation. These derivatives are unaltered in their ability to support ATP:pyrophosphate exchange catalyzed by E1, but they are nearly twice as active as native Ub in the stimulation of proteolysis. In light of the results with the His-68 derivative discussed in Section 4.2.3, it appears that iodination of the histidine is responsible for this increased activity. An additional kinetic feature of the extensively iodinated derivatives is the fact that excess derivative present in the assay is an inhibitor of proteolysis. The site of inhibition is at the proteolytic step since the steady-state level of conjugates actually increases over the concentration range where the rate of proteolysis is decreasing.[45]

4.3. Conclusions

These results have led us to postulate that the conformation of Ub, perhaps in the region of the histidine, is important for interaction of Ub

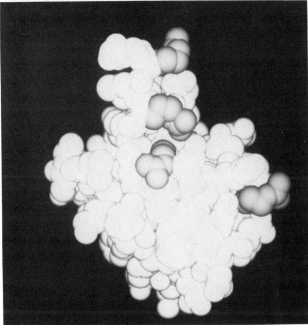

Figure 6. Location of the Arg residues in Ub. (*Top panel*) Ribbon drawing. (*Bottom panel*) Space-filling representation. The side-chain atoms of Arg residues are colored grey and all others white. Note that Arg-74 at the top of the molecule points into the page. Arg-54 is unreactive with anionic reagents, probably because of the two nearby acidic side chains.

conjugates with the proteases of the system. The evidence suggests that this conformational change can be induced by modest increases in the hydrophobicity of the environment, perhaps upon conjugation to denatured proteins. Thus, conformational changes in the Ub portion of the conjugate structure could have an effect on the partition of these conjugates between their various metabolic fates. We have suggested that this system of protein–protein interactions could be a proofreading mechanism that assures that abnormal or denatured proteins will be targeted for proteolysis by this system.[45] Conjugates with Ub in an aqueous conformation would be disassembled by the action of the isopeptidase and would not be degraded by this system.

These results also suggest that the enzymes of the system recognize much of the Ub surface. E1 appears most sensitive to alterations around Tyr-59, Arg-42, and Arg-72. This corresponds to the right half of the molecule as depicted in Figure 6. The protease components of this system appear to recognize the surface of the molecule around His-68 and Arg-74. This corresponds to the left half of the molecule as depicted in Figure 6. No derivatives have been found that are perturbed in the ligase reaction, suggesting that other regions of Ub are involved in these specific binding interactions.

Finally, the enzymes of this system appear to be indifferent to conservative changes around residues 19,24,28, and 57 as occur in the yeast and oat proteins. This region of the molecule is most likely involved in some of the other functions of Ub.

5. FUTURE DIRECTIONS OF STRUCTURAL CHEMISTRY

It is always important to ask what the future directions in a research area should be. We have gained much information from the approach of studying the structural chemistry of Ub. There are, however, many questions left to be answered by further studies in this area. In this section I try to suggest some of these approaches and describe initial attempts in this area.

5.1. Site-Directed Mutagenesis

The chemical modification studies described above have been important in beginning to define the regions of Ub that interact with the various enzymes of the protein degradation system. It can be anticipated that similar studies will define more enzymes that interact with Ub and therefore yield insight into the other functions of Ub. There is a serious disadvantage to these types of study in that only a small subset of changes can be made to the Ub structure, and many regions of the molecule cannot be altered. Site-directed mutagenesis of Ub appears to be a very promising approach to overcoming this shortcoming. Our studies with chemical modification make it clear that we can change the structure of Ub and retain at least some activity, a prerequisite for these types of study. Recently, the gene has been synthesized by Ecker et al.,[60] using the cassette method and inserted into yeast under the control of the inducible metallothionine promoter. This construct strongly expresses the altered protein and, along with other genetic manipulations such as deletion of the endogenous genome, these constructs could be very useful in defining the role of Ub in yeast. The gene has also been inserted into E. coli, which does not contain Ub. Under the control of the heat-shock promoter, Ub is expressed at levels up to 20% of the cellular protein.[60] This system will serve as a facile source of mutant Ub structures to use in the structure–function studies described in Section 4. Using these mutants, we should be able to fully map the interactions between Ub and the various enzymes of the system.

5.2. Carboxyl-Terminal Derivatives of Ubiquitin

The studies discussed above have concentrated on the structure and chemistry of free Ub. Many of the enzymes of the proteolytic system, as well as those concerned with the other roles of Ub, act on carboxyl-terminal derivatives of Ub. Preparations of these derivatives are necessary to begin to understand the downstream enzymes and the metabolism of Ub derivatives. Small amounts of derivatives such as carboxyl-terminal amides,[61] thiol esters,[61] and conjugates[62,63] have been produced enzymatically and used in studies on these enzymes. There are, however, a number of disadvantages to this approach such as the low yield and limited types of conjugate that can be produced.[64] We have taken a more general approach of chemically activating the carboxyl terminus of Ub for the synthesis of defined carboxyl-terminal derivatives.

The central problem with this approach is to activate just the carboxyl terminus of Ub without modifying other residues. We have used the enzyme trypsin to accomplish this.[65] In aqueous solution and in the absence of metal ions, the only bond cleaved by trypsin is that following Arg-74, resulting in the release of Gly-Gly. In the presence of high concentrations of glycylglycine ethyl ester, trypsin will catalyze the formation of the carboxyl-terminal ethyl ester of Ub. This is essentially a transpeptidation reaction and approximately 50% of the starting Ub can be converted to the ethyl ester. This ester has been purified and characterized to demonstrate the specificity of this reaction. Two applications of this chemistry are discussed next.

5.2.1. Substrates for Ubiquitin Isopeptidase and Hydrolase

The ethyl ester has been shown to be a good substrate for Ub carboxyl-terminal hydrolase and an HPLC assay has been described.[65] Recently, we have used this assay to monitor the purification of other activities that hydrolyze this ester. In addition to UCH, three Ub esterase activities have been resolved from calf thymus,[66] and they are currently being purified and characterized. All four of the activities detected have a specific binding site for Ub and appear to be thiol proteases with specificity for hydrolyzing Ub derivatives at the Gly-76 terminus of Ub. This substrate will be of value in the purification, characterization, and comparison of the enzymes that cleave carboxyl-terminal derivatives of Ub. Such enzymes include UCH,[61] which may be responsible for the cleavage of limit-digest peptides from Ub protein conjugates; A24 lyase,[67] a 30 kDa enzyme that removes the Ub from uH2a; the isopeptidase,[62,63] a 200 kDa enzyme activity that removes Ub from a wide variety of Ub protein conjugates; and a putative processing enzyme[9,33,38,46,47] that is responsible for the cleavage of the polyprotein translation product of the Ub gene and the fusion proteins that contain Ub at the amino terminus.

5.2.2. Synthesis of Ubiquitin–Protein Conjugates

The carboxyl-terminal ester of Ub is a key intermediate in our recently developed route to the synthesis of Ub–protein conjugates. This ester can be converted to the acyl hydrazide by treatment with hydrazine, and subsequently to the acyl azide by the action of nitrous acid. In the presence of a nucleophile, the acyl azide undergoes nucleophilic attack yielding carboxyl-terminal derivatives of Ub. The amino groups of peptides and proteins are sufficiently reactive and we have shown that this reaction leads to synthetic Ub–protein conjugates.[68] To date, cytochrome c and three different calmodulins have been conjugated to Ub using this chemistry. We are currently characterizing the sites of conjugation and the metabolism of these conjugates by the enzymes of the proteolytic system. It is clear that these pure conjugates, available in milligram amounts, will be important in defining the specificity of the enzymes of Ub metabolism as well as in clarifying the nature of the conformation of Ub and the target protein in such conjugates.

5.3. Approaches to Defining "Other Functions"

The proteolytic role of Ub, if not its consequences, is well understood. Less clear are a number of other functions of Ub, such as its involvement in the immune response, chromatin structure, and receptor action. In this section, I suggest some potential approaches to defining these roles that utilize our knowledge of the structural chemistry of the Ub system.

5.3.1. Specific Inhibitors of Ubiquitin Metabolism

One way to examine specific functions of Ub in the cell is through the use of specific inhibitors of Ub metabolism. Several nonspecific inhibitors of proteolysis[69-72] have shed light on that function but suffer from the criticism that they are nonspecific. Perhaps the most specific inhibitor would be one based on the Ub structure itself. Rose has reported the preparation of the carboxyl-terminal aldehyde of Ub, which is a good inhibitor of the carboxyl-terminal hydrolase and the isopeptidase.[73] It has not, however, been synthesized in amounts necessary for *in vivo* studies. We have recently made large amounts of this aldehyde using chemistry similar to that used in synthesizing conjugates.[66] The carboxyl-terminal ethyl ester of Des-Gly Ub has been made by trypsin-catalyzed transpeptidation with glycine ethyl ester. Activation of this to the acyl azide and condensation with aminoacetaldehyde diethylacetal yields the protected Ub aldehyde, which can be deprotected by mild acid hydrolysis. We have demonstrated that this aldehyde is a tightly bound inhibitor of all four Ub esterase activities from thymus.[66] Microinjection of this inhibitor into cells would be expected to interfere with a number of enzymes that catalyze reactions requiring the formation of a tetrahedral intermediate at the carboxyl terminus of Ub. Similar chemistry can be applied to make a series of irreversible inhibitors of enzymes that bind Ub. This class of ubiquitin-based inhibitors will be of great value in specifically interfering with Ub metabolism and defining its role(s) in the cell.

5.3.2. Deubiquitination of Cell Surface Receptors

The recent finding that cell surface receptors are ubiquitinated suggests a whole new set of questions. What is the role of this modification? One way to address this question is to examine the binding properties and the recycling of these receptors under conditions where they are and are not ubiquitinated. The isopeptidase or A24 lyase may be useful in such a study. If one or both of these enzymes can release Ub from these receptors, an experimental system is available to study the effects of ubiquitination on receptor structure and function. This is analogous to the use of endoglycosidases to study the role of carbohydrates in cell surface proteins or limited proteolysis to remove cell surface proteins.

5.3.3. Genetic Approaches

Finally, the powerful approaches of molecular biology should be applied to this problem. If a synthetic Ub gene were introduced into a mammalian cell line and the endogenous copies either deleted or inactivated, the cell would be forced to use the synthetic gene product. If the yeast sequence or one of the site-directed mutants still functional in protein degradation were introduced, it is possible that some nonproteolytic function of Ub would be perturbed. This is anticipated, since all portions of Ub appear to be under intense selective pressure and all portions of the protein appear to be critical for some function in the cell. One potential approach to inactivating the endogenous genes would be to make use of anti-sense RNA complimentary to the promoter regions or the junction sequences of the polyubiquitin gene and the Ub fusion genes. This could be designed so as not to interfere with the synthetic gene but would prevent translation of the endogenous genes. Another approach to the same end would be to identify and delete the gene for the Ub processing enzyme. In the absence of this enzyme, no free Ub would be produced, and the synthetic gene product could be expressed.

6. CONCLUDING REMARKS

The purpose of this chapter has been to introduce Ub to the reader and to point out how studies on the structural chemistry of Ub have contributed to our understanding of the importance of this fascinating molecule. The following chapters present an up-to-date treatment of our understanding of the enzymology and the cell biology of this system. The multidisciplinary approaches exemplified in this book will be important in deciphering the role of this regulator molecule that participates in all facets of protein metabolism.

ACKNOWLEDGMENTS. Some of the work described here was supported by grants from the NIH (GM30308) and the Emory University Research Council. I would like to thank the following individuals for furnishing manuscripts to me in advance of their publication: C. E. Bugg, T. Butt,

W. J. Cook, S. T. Crooke, D. J. Ecker, M. I. Khan, J. Marsh, I. A. Rose, J. V. B. Warms, and S. Vijay-Kumar.

REFERENCES

1. Schlesinger, D. H., Goldstein, G., and Niall, H. D., 1975, The complete amino acid sequence of ubiquitin, an adenylate cyclase stimulating polypeptide probably universal in living cells, *Biochemistry* 14: 2214–2218.
2. Ciechanover, A., Elias, S., Heller, H., Ferber, S., and Hershko, A., 1980, Characterization of the heat-stable polypeptide of the ATP-dependent proteolytic system from reticulocytes, *J. Biol. Chem.* 255: 7525–7528.
3. Wilkinson, K. D., Urban, M. K., and Haas, A. L., 1980, Ubiquitin in the ATP-dependent proteolysis factor I of rabbit reticulocytes, *J. Biol. Chem.* 255: 7529–7532.
4. Wilkinson, K. D., and Audhya, T. K., 1981, Stimulation of ATP-dependent proteolysis requires ubiquitin with the COOH-terminal sequence Arg-Gly-Gly, *J. Biol. Chem.* 256: 9235–9241.
5. Matsui, S., Seon, B. K., and Sandberg, A. A., 1979, Disappearance of a structural chromatin protein A24 in mitosis: Implications for molecular basis of chromatin condensation, *Proc. Natl. Acad. Sci. U.S.A.* 76: 6386–6390.
6. Levinger, L., and Varshavsky, A., 1982, Selective arrangement of ubiquitinated and D1 protein-containing nucleosomes within the *Drosophila* genome, *Cell* 28: 375–385.
7. Mueller, R. D., Yasuda, H., Hatch, C. L., Bonner, W. M., and Bradbury, E. M., 1985, Identification of ubiquitinated histones 2A and 2B in *Physarum polycephalum*, *J. Biol. Chem.* 260: 5147–5153.
8. Finley, D., Ciechanover, A., and Varshavsky, A., 1984, Thermolability of ubiquitin-activating enzyme from the mammalian cell cycle mutant ts85, *Cell* 37: 43–55.
9. Bond, U., and Schlesinger, M. J., 1985, Ubiquitin is a heat shock protein in chicken embryo fibroblasts, *Mol. Cell. Biol.* 5: 949–956.
10. Siegelman, M., Bond, M. W., Gallatin, W. M., St. John, T., Smith, H. T., Fried, V. A., and Weissman, I. L., 1986, Cell surface molecule associated with lymphocyte homing is a ubiquitinated branched-chain glycoprotein, *Science* 231: 823–829.
11. Yarden, Y., Escobedo, J. A., Kuang, W.-J., Yang-Feng, T. L., Daniel, T. O., Tremble, P. M., Chen, E. Y., Ando, M. E., Harkins, R. N., Francke, U., Fried, V. A., Ullrich, A., and Williams, L. T., 1986, Structure of the receptor for platelet-derived growth factor helps define a family of closely related growth factor receptors, *Nature* 323: 226–232.
12. Meyer, E. M., West, C. M., and Chau, V., 1986, Antibodies directed against ubiquitin inhibit high affinity [³H]choline uptake in rat cerebral cortical synaptosomes, *J. Biol. Chem.* 261: 14365–14368.
13. Scheid, M. P., Goldstein, G., and Boyse, E. A., 1978, The generation and regulation of lymphocyte populations, *J. Exp. Med.* 147: 1727–1743.
14. Haas, A. L., Murphy, K. E., and Bright, P. M., 1985, The inactivation of ubiquitin accounts for the inability to demonstrate ATP-dependent proteolysis in liver extracts, *J. Biol. Chem.* 260: 4694–4703.
15. Hershko, A., Eytan, E., Ciechanover, A., and Haas, A. L., 1982, Immunochemical analysis of the turnover of ubiquitin–protein conjugates in intact cells, *J. Biol. Chem.* 257: 13964–13970.

16. Haas, A. L., and Bright, P. M., 1985, The immunochemical detection and quantitation of intracellular ubiquitin–protein conjugates, *J. Biol. Chem.* **260:** 12464–12473.

17. Ciechanover, A., Elias, S., Heller, H., and Hershko, A., 1982,"'Covalent-affinity" purification of ubiquitin-activating enzyme, *J. Biol. Chem.* **257:** 2537–2542.

18. Haas, A. L., and Rose, I. A., 1982, The mechanism of ubiquitin activating enzyme, *J. Biol. Chem.* **257:** 10329–10337.

19. Rose, I. A., and Warms, J. V. B., 1987, A specific endpoint assay for ubiquitin, *Proc. Natl. Acad. Sci. U.S.A.* **84:** 1477–1481.

20. Ciechanover, A., Hod, Y., and Hershko, A., 1978, A heat-stable polypeptide component of an ATP-dependent proteolytic system from reticulocytes, *Biochem. Biophys. Res. Commun.* **81:** 1100–1104.

21. Evans, A. C., Jr., and Wilkinson, K. D., 1985, Ubiquitin-dependent proteolysis of native and alkylated bovine serum albumin: Effects of protein structure and ATP concentration on selectivity, *Biochemistry* **24:** 2915–2923.

22. Wilkinson, K. D., Cox, M. J., O'Connor, L. B., and Shapira, R., 1986, Structure and activities of a variant ubiquitin sequence from Bakers' yeast, *Biochemistry* **25:** 4999–5004.

23. Jabusch, J. R., and Deutsch, H. F., 1983, Isolation and crystallization of ubiquitin from mature erythrocytes, *Prep. Biochem.* **13:** 261–273.

24. Haas, A. L., and Wilkinson, K. D., 1985, The large scale purification of ubiquitin from human erythrocytes, *Prep. Biochem.* **15:** 49–60.

25. Goldstein, G., Scheid, M., Hammerling, U., Boyse, E. A., Schlesinger, D. H., and Niall, H. D., 1975, Isolation of a polypeptide that has lymphocyte-differentiating properties and is probably represented universally in living cells, *Proc. Natl. Acad. Sci. U.S.A.* **72:** 11–15.

26. Loir, M., Caraty, A., Lanneau, M., Menezy, Y., Muh, J. P., and Sautiere, P., 1984, Purification and characterization of ubiquitin from mammalian testis, *FEBS Lett.* **169:** 199–204.

27. Seidah, N. G., Crine, P., Benjannet, S., Scherrer, H., and Chretien, M., 1978, Isolation and partial characterization of a biosynthetic N-terminal methionyl peptide of bovine pars intermedia: Relationship to ubiquitin, *Biochem. Biophys. Res. Commun.* **80:** 600–608.

28. Hamilton, J. W., and Rouse, J. B., 1980, The biosynthesis of ubiquitin by parathyroid gland, *Biochem. Biophys. Res. Commun.* **96:** 114–120.

29. Scherrer, H., Seidah, N. G., Benjannet, S., Crine, P., Lis, M., and Chretian, M., 1978, Biosynthesis of a ubiquitin-related peptide in rat brain and in human and mouse pituitary tumors, *Biochem. Biophys. Res. Commun.* **84:** 874–885.

30. Watson, D. C., Levy-Wilson, B., Gordon, W., and Dixon, G. H., 1978, Free ubiquitin is a non-histone protein of trout testis chromatin, *Nature* **276:** 196–198.

31. Gavilanes, J. G., de Buitrago, G. G., Perez-Castells, R., and Rodriguez, R., 1982, Isolation, characterization and amino acid sequence of a ubiquitin-like protein from insect eggs, *J. Biol. Chem.* **257:** 10267–10270.

32. Levenbrook, L., Bauer, A. C., and Chou, J. Y., 1986, Ubiquitin in the blowfly *Calliphora vicina, Insect Biochem.* **16:** 509–515.

33. Dworkin-Rastl, E., Shrutkowski, A., and Dworkin, M. B., 1984, Multiple ubiquitin mRNAs during *Xenopus laevis* development contain tandem repeats of the 76 amino acid coding sequence, *Cell* **39:** 321–325.

34. Levy-Wilson, B., Denker, M. S., and Ito, E., 1983, Isolation, characterization, and postsynthetic modifications of *Tetrahymena* high mobility group proteins, *Biochemistry* **22:** 1715–1721.

35. Fusauchi, Y., and Iwai, K., 1985, *Tetrahymena* ubiquitin–histone conjugate uH2A. Isolation and structural analysis, *J. Biochem.* **97:** 1467–1476.

36. Vierstra, R. D., Langan, S. M., and Haas, A. L., 1985, Purification and initial characterization of ubiquitin from the higher plant, *Avena sativa, J. Biol. Chem.* **260:** 12015–12021.

37. Vierstra, R. D., Langan, S. M., and Schaller, G. E., 1986, Complete amino acid sequence of ubiquitin from the higher plant *Avena sativa, Biochemistry* **25:** 3105–3108.

38. Ozkaynak, E., Finley, D., and Varshavsky, A., 1984, The yeast ubiquitin gene: Head-to-tail repeats encoding a polyubiquitin precursor protein, *Nature* **312:** 663–666.

39. Low, T. L. K., Thurman, G. B., McAdoo, M., McClure, J., Rossio, J. L., Naylor, P. H., and Goldstein, A. L., 1979, The chemistry and biology of thymosin, *J. Biol. Chem.* **254:** 981–986.

40. Cary, P. D., King, D. S., Crane-Robinson, C., Bradbury, M., Rabbani, A., Goodwin, G. H., and Johns, E. W., 1980, Structural studies on two high-mobility-group proteins from calf thymus, HMG-14 and HMG-20 (ubiquitin), and their interaction with DNA, *Eur. J. Biochem.* **112:** 557–580.

41. Lenkinski, R. E., Chen, D. M., Glickson, J. D., and Goldstein, G., 1977, Nuclear magnetic resonance studies of the denaturation of ubiquitin, *Biochim. Biophys. Acta* **494:** 126–130.

42. Jenson, J., Goldstein, G., and Breslow, E., 1980, Physical–chemical properties of ubiquitin, *Biochim. Biophys. Acta* **624:** 378–385.

43. Wilkinson, K. D., and Mayer, A. N., 1986, Alcohol-induced conformational changes of ubiquitin, *Arch. Biochem. Biophys.* **250:** 390–399.

44. Breslow, E., Chauhan, Y., Daniel, R., and Tate, S., 1986, Role of methionine-1 in ubiquitin conformation and activity, *Biochem. Biophys. Res. Commun.* **138:** 437–444.

45. Cox, M. J., Haas, A. L., and Wilkinson, K. D., 1986, Role of ubiquitin conformations in the specificity of protein degradation: Iodinated derivatives with altered conformations and activities, *Arch. Biochem. Biophys.* **250:** 400–409.

46. Lund, P. K., Moats-Staats, B. M., Simmons, J. G., Hoyt, E., D'Ercole, A. J., Martin, F., and Van Wyk, J. J., 1985, Nucleotide sequence analysis of a cDNA encoding human ubiquitin reveals that ubiquitin is synthesized as a precursor, *J. Biol. Chem.* **260:** 7609–7613.

47. Wiborg, O., Pedersen, M. S., Wind, A., Berglund, L. E., Marcker, K. A., and Vuust, J., 1985, The human multigene family: Some genes contain multiple directly repeated ubiquitin coding sequences, *EMBO J.* **4:** 755–759.

48. Vijay-Kumar, S., Bugg, C. E., Wilkinson, K. D., and Cook, W. J., 1985, Three-dimensional structure of ubiquitin at 2.8 Å resolution, *Proc. Natl. Acad. Sci. U.S.A.* **82:** 3582–3585.

49. Vijay-Kumar, S., Bugg, C. E., and Cook, W. J., 1987, Structure of ubiquitin refined at 1.8 Å resolution, *J. Mol. Biol.* **194:** 531–544.

50. Vijay-Kumar, S., Bugg, C. E., Wilkinson, K. D., Vierstra, R. D., and Cook, W. J., 1987, Comparison of three-dimensional structures of yeast and oat ubiquitin with human ubiquitin, *J. Biol. Chem.* **262:** 6396–6399.

51. Cox, M. J., Shapira, R., and Wilkinson, K. D., 1986, Tryptic peptide mapping of ubiquitin and derivatives using reverse-phase high performance liquid chromatography, *Anal. Biochem.* **154:** 345–352.

52. Low, T. L. K., and Goldstein, A. L., 1979, The chemistry and biology of thymosin, *J. Biol. Chem.* **254:** 987–995.

53. Mezquita, Z., Chiva, M., Vidal, S., and Mezquita, C., 1982, Effects of high mobility group nonhistone proteins HMG-20 (ubiquitin) and HMG-17 on histone deacetylase activity assayed *in vitro, Nucleic Acids Res.* **10:** 1781–1797.

54. Matsumoto, H., Taniguchi, N., and Deutsch, H. F., 1984, Isolation, characterization, and esterase and CO hydration activities of ubiquitin from bovine erythrocytes, *Arch. Biochem. Biophys.* **234:** 426–433.
55. Taniguchi, N., and Matsumoto, H., 1985, The *p*-nitrophenyl phosphatase activity of ubiquitin from bovine erythrocytes, *Comp. Biochem. Physiol.* **81B:** 587–590.
56. Jabusch, J. R., and Deutsch, H. F., 1985, Localization of lysine acetylated in ubiquitin reacted with *p*-nitrophenyl acetate, *Arch. Biochem. Biophys.* **238:** 170–177.
57. Cox, M. J., 1986, Chemical modification of ubiquitin, M.S. Thesis, Emory University.
58. Wilkinson, K. D., 1987, Protein ubiquitinization: A regulatory post-translational modification, *Anti-Cancer Drug Des.* **2:** 211–229.
59. Duerksen-Hughes, P. J., Xu, X., and Wilkinson, K. D., 1987, The ubiquitin binding sites of the activating enzyme and proteases: Evidence for differential interactions around Arg-74 of ubiquitin, *Biochemistry* **26:** 6980–6987.
60. Ecker, D. J., Khan, M. I., Marsh, J., Butt, T., and Crooke, S. T., 1987, Chemical synthesis and expression of a cassette adapted ubiquitin gene, *J. Biol. Chem.* **262:** 3524–3527.
61. Pickart, C. M., and Rose, I. A., 1985, Ubiquitin carboxyl-terminal hydrolase acts on ubiquitin carboxyl-terminal amides, *J. Biol. Chem.* **260:** 7903–7910.
62. Hough, R., and Rechsteiner, M., 1986, Ubiquitin–lysozyme conjugates: Purification and susceptibility to proteolysis, *J. Biol. Chem.* **261:** 2391–2399.
63. Hershko, A., Leshinsky, E., Ganoth, D., and Heller, H., 1984, ATP-dependent degradation of ubiquitin–protein conjugates, *Proc. Natl. Acad. Sci. U.S.A.* **81:** 1619–1623.
64. Lee, P. L., Midelfort, C. F., Murakami, K., and Hatcher, V. B., 1986, Multiple forms of ubiquitin–protein ligase. Binding of activated ubiquitin to protein substrates, *Biochemistry* **25:** 3134–3138.
65. Wilkinson, K. D., Cox, M. J., Mayer, A. N., and Frey, T., 1986, Synthesis and characterization of ubiquitin ethyl ester, a new substrate for ubiquitin carboxyl-terminal hydrolase, *Biochemistry* **25:** 6644–6649.
66. Mayer, A. N., 1986, Resolution of ubiquitin carboxyl-terminal hydrolases using ubiquitin ethyl ester as the substrate, M.S. Thesis, Emory University.
67. Kanda, F., Sykes, D. E., Yasuda, H., Sandberg, A. A., and Matsui, S.-I., 1986, Substrate recognition of isopeptidase: Specific cleavage of the (α-glycyl)lysine linkage of ubiquitin protein conjugates, *Biochim. Biophys. Acta* **870:** 64–75.
68. Wilkinson, K. D., Marriott, D., and Chau, V., 1988, Non-enzymatic synthesis of ubiquitin–calmodulin conjugates: A general synthetic route to prepare ubiquitin conjugates (manuscript submitted).
69. Haas, A. L., and Rose, I. A., 1981, Hemin inhibits ATP-dependent proteolysis: Role of hemin in regulating conjugate degradation, *Proc. Natl. Acad. Sci. U.S.A.* **78:** 6845–6848.
70. Tanaka, K., Waxman, L., and Goldberg, A. L., 1984, Vanadate inhibits the ATP-dependent degradation of proteins in reticulocytes without affecting ubiquitin conjugation, *J. Biol. Chem.* **259:** 2803–2809.
71. Breslow, E., Daniel, R., Ohba, R., and Tate, S., 1986, Inhibition of ubiquitin-dependent proteolysis by non-ubiquitinable proteins, *J. Biol. Chem.* **261:** 6530–6535.
72. Hough, R., Pratt, G., and Rechsteiner, M., 1986, Ubiquitin–lysozyme conjugates, identification and characterization of an ATP-dependent protease from rabbit reticulocyte lysates, *J. Biol. Chem.* **261:** 2400–2408.
73. Pickart, C. M., and Rose, I. A., 1987, Mechanism of ubiquitin carboxyl-terminal hydrolase: Borohydride and hydroxylamine inactivate in the presence of ubiquitin, *J. Biol. Chem.* **261:** 10210–10217.

Chapter 2

Molecular Genetics of the Ubiquitin System

Daniel Finley, Engin Özkaynak, Stefan Jentsch,
John P. McGrath, Bonnie Bartel, Michael Pazin,
Robert M. Snapka, and Alexander Varshavsky

1. INTRODUCTION AND SUMMARY

The ubiquitin (Ub) system was first approached experimentally using purely biochemical methods.[1] Genetic techniques provide not only an alternative way to address previously posed questions in this field[2,3] but also a strategy for the identification and dissection of the physiological functions of Ub. The genetic approach has turned up new components of the Ub system, such as polyubiquitin[4-12] and the Ub–hybrid proteins,[12-17]

DANIEL FINLEY, JOHN P. MCGRATH, BONNIE BARTEL, and ALEXANDER VARSHAVSKY • Department of Biology, Massachusetts Institute of Technology, Cambridge, Massachusetts 02139.　ENGIN ÖZKAYNAK • Department of Biology, Massachusetts Institute of Technology, Cambridge, Massachusetts 02139. *Present address:* Creative Biomolecules, Hopkinton, Massachusetts 01748.　STEFAN JENTSCH • Department of Biology, Massachusetts Institute of Technology, Cambridge, Massachusetts 02139. *Present address:* Friedrich Miescher Institute, Max Planck Society, 7400 Tübingen, Federal Republic of Germany.　MICHAEL PAZIN • Department of Biology, Massachusetts Institute of Technology, Cambridge, Massachusetts 02139. *Present address:* Department of Biochemistry and Biophysics, School of Medicine, University of California, San Francisco, California 94143.　ROBERT M. SNAPKA • Department of Biology, Massachusetts Institute of Technology, Cambridge, Massachusetts 02139. *Present address:* The Ohio State University, Division of Radiobiology, Department of Radiology, Columbus, Ohio 43210-1214.

and more recently has led to identification of the product of the yeast DNA repair gene *RAD6* as a Ub-conjugating enzyme.[18] Our current studies are focused primarily on yeast, whose Ub system, as described in Section 3, closely resembles that of mammals. Thus, insights gained from molecular genetic analysis of the yeast system should be relevant to other eukaryotes as well. While Ub genes have now been cloned from a number of organisms (reviewed in refs. 19 and 20), mutational analysis of the Ub system has been restricted to yeast[21] and the mammalian cell line ts85.[2,3,22]

Achieving a detailed understanding of a biochemical pathway ultimately requires means to perturb the pathway specifically *in vivo*. For example, the physiological substrates of the Ub-dependent proteolytic pathway can be inferred by determining which proteins are stabilized when ubiquitination is selectively inhibited *in vivo*. As described in Section 2, the conditional ubiquitination phenotype of the murine cell line ts85, which results from a thermolabile Ub-activating enzyme, has been used to show that, at least in higher eukaryotes, the bulk of short-lived intracellular proteins is degraded through a Ub-dependent pathway.[2,3] Moreover, amino acid analog-containing proteins,[3] puromycyl peptides,[3] and heat-damaged[22] proteins are similarly degraded by a Ub-dependent pathway in ts85 cells, apparently unlike the bulk of long-lived intracellular proteins.[22]

For a deeper analysis of the Ub system, we have chosen the yeast *Saccharomyces cerevisiae,* an organism for which powerful genetic approaches have been developed.[23] Yeast Ub was found to be derived from precursor proteins encoded by a family of natural gene fusions.[4,16] The genes *UBI1, UBI2,* and *UBI3* encode hybrid proteins in which Ub is fused to unrelated ("tail") amino acid sequences, whereas *UBI4* contains five consecutive Ub-coding repeats in a spacerless head-to-tail arrangement and thus encodes polyubiquitin.[4] The basic structural features of polyubiquitin and of the Ub-hybrid proteins have been highly conserved in the evolution of eukaryotes.[4-17] The analysis of yeast strains with an engineered deletion of the polyubiquitin gene (*UBI4*) has shown that the essential function of this gene is to provide Ub to cells under stress, and also that Ub is an essential component of the stress response system.[21] The more recent analysis of deletions of the *UBI1–UBI3* genes has provided initial insight into their functions and has also uncovered antagonistic genetic interactions among them. Our recent studies on the yeast Ub–protein ligase system have led to the cloning of genes for several of its components and to the discovery[18] that one of the Ub-conjugating enzymes of this system is encoded by the DNA repair gene *RAD6*. A different genetically based approach, in which plasmids are used to encode well-defined proteolytic substrates, has recently led to an advance in understanding of the rules of targeting in selective protein turnover.[24] These latter findings are discussed in Chapter 11 of this book.

2. CONDITIONAL UBIQUITINATION IN A MAMMALIAN CELL CYCLE MUTANT

2.1. Temperature Sensitivity of Ubiquitin-Activating Enzyme from ts85 Cells

ts85 is a temperature-sensitive derivative of FM3A, a cell line established from a spontaneous mouse mammary carcinoma.[25] ts85 cells have a complex but well-defined phenotype: shift-up of ts85 cultures to a nonpermissive temperature (39.0–39.5°C) rapidly and completely inhibits the generation of new mitotic cells,[25,26] and the initially unsynchronized ts85 cultures evolve a stable distribution of cell cycle positions composed of early G_2 and (fewer) late S phase cells.[25,27] The rate of histone H1 phosphorylation *in vivo* was measured in synchronized G_2 and G_1/S ts85 cultures and was shown[26,27] in both instances to be abnormally low at 39°C. However, both cytoplasmic and nuclear protein kinase activities from ts85 cells, tested *in vitro* with histone H1 as a substrate, behaved as wild type in thermoinactivation experiments,[27] suggesting that the reduction in H1 phosphorylation is secondary to some other defect.

Remarkably, a temperature-sensitive modification of yet another histone was identified. The Ub–H2A semihistone, uH2A, disappears from ts85 chromatin at 39°C with a half-life of approximately 3 hr.[28,29] A temperature-resistant growth revertant, ts85R-MN3, is phenotypically wild type for both H1 phosphorylation and H2A ubiquitination, strongly suggesting that the apparently single mutation responsible for cell cycle arrest is also responsible for the defects in histone modification.[27]

In vivo pulse-chase experiments indicated that the disappearance of uH2A at 39°C was due to a reduced rate of Ub–H2A conjugation in ts85 cells.[2] We therefore tested whether the reduced synthesis of uH2A at 39°C was directly due to thermolability of some component of the Ub–protein ligase system. The conjugation of radioactively labeled Ub to protein substrates was indeed temperature sensitive in crude extracts prepared from ts85 cells as compared to those from wild-type or growth-revertant cells.[2] Mixed (mutant plus wild-type) extracts had an intermediate thermolability, indicating that temperature sensitivity of *in vitro* ubiquitination is not due to catalytic enzyme modification (e.g., phosphorylation, proteolysis) by an exchangeable factor peculiar to ts85 cells. After affinity purification of the Ub–protein ligase enzymes on Ub–Se-

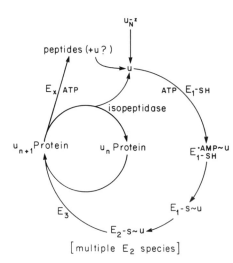

Figure 1. Pathways of the Ub system. Ub precursors ($u_N - z$), which may contain one or more ($N = 1, 2, 3 \ldots$) Ub repeats, and C-terminal non-Ub sequences (z) are processed into mature Ub (u). Ub conjugated through high-energy bonds ($\sim$) is represented to the right of the acceptor molecule; for low-energy (amide bond) conjugates, Ub is at the left, followed by a subscript ($n = 0, 1, 2, \ldots$) denoting the multiplicity of Ub moieties covalently bound to the acceptor. The question mark (top, left) denotes the current uncertainty as to whether a significant proportion of Ub in Ub–protein conjugates is regenerated upon degradation of acceptor proteins. (Modified from ref. 2, with permission. Copyright 1984, Cell Press.)

pharose (see Figure 1 for a description of the Ub cycle), the ligase from ts85 cells grown at the permissive temperature was found, unlike the wild-type ligase, to conjugate Ub efficiently to a specific endogeneous substrate, which appeared to be Ub-activating enzyme (E1), a component of the ligase system. Since many defective proteins are substrates for ubiquitination, these results[2] suggested that E1 may be the temperature-sensitive component of the Ub–protein ligase system in ts85 cells. To test this possibility, we directly assayed the formation of Ub–E1 thiolester (E1-S~u), an intermediate in the transfer of activated Ub to acceptor proteins[30] (see Figure 1). E1 enzymes were purified from ts85 and FM3A cells, then assayed in the presence of [^{125}I]ubiquitin and ATP. Figure 2 shows that a 40°C preincubation rapidly and specifically inactivates the E1 enzyme from ts85 cells. These and other experiments[2] suggest that the mutation conferring conditional cell cycle arrest in ts85 cells lies within the structural gene for E1.

2.2. Temperature Sensitivity of Selective Protein Degradation in ts85 Cells

The ts85 cells provide a well-defined experimental system for testing whether intracellular protein degradation is dependent on Ub–protein conjugation. To assay protein degradation *in vivo,* amino acid analogs were initially used. When supplied to cells, these can substitute for cognate amino acids in translation, producing "abnormal" proteins, most of

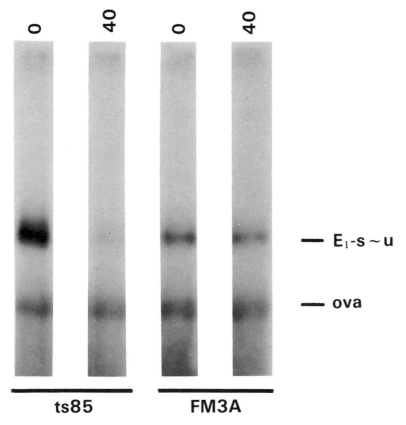

Figure 2. Heat inactivation of purified Ub-activating enzyme (E1) from mouse ts85 cells. E1 enzymes, purified from ts85 and wild-type (FM3A) cells, were preincubated at 40°C for either 0 or 40 min, then assayed at 30°C by the addition of [^{125}I]ubiquitin. After 10 min at 30°C, SDS-containing sample buffer lacking 2-mercaptoethanol was added, followed by electrophoresis in a 7.5% polyacrylamide–SDS gel. E1-S~u denotes a derivative of E1 coupled to Ub *via* a thiolester bond; ova denotes a band of ovalbumin (used as a carrier) that is associated with a small fraction of [^{125}I]ubiquitin in a noncovalent, nonspecific, and E1-independent manner. See ref. 2 for details. (Modified from ref. 2, with permission. Copyright 1984, Cell Press.)

which are rapidly degraded *in vivo*.[31] The latter effect can be used to monitor the selective component of intracellular protein turnover.

Wild-type (FM3A), mutant (ts85), and revertant (ts85R-MN3) cells were incubated with the amino acid analogs L-threo-α-amino-β-chlorobutyric acid (a valine analog) and 2-aminoethyl-L-cysteine (thialysine, a lysine analog), followed by pulse labeling of newly synthesized proteins with [^{35}S]methionine in the continued presence of analogs. The degradation of ^{35}S-labeled proteins was then followed during a cold chase incubation. At the permissive temperature, abnormal intracellular proteins decayed at approximately equal rates in the three cell lines examined, with more than 70% of the ^{35}S-labeled proteins degraded within 4 hr (Figure 3A). In striking contrast, at 39.5°C less than 15% of the ^{35}S-labeled proteins were degraded in ts85 cells during the same period (Figure 3B). Neither the wild-type cells (FM3A) nor the growth revertant cells (ts85R-MN3) displayed the temperature sensitivity of protein degradation that was clearly seen with ts85 cells (Figure 3B).[3]

The ts85 degradation defect identified in Figure 3B was shown to apply more generally to selectively degraded proteins.[3] One indication of this is shown in Figure 3C, in which the degradation of pulse-labeled, truncated polypeptide chains is followed during a cold chase incubation. Prematurely terminated translation products, released from ribosomes as a consequence of puromycin treatment, generally undergo rapid degradation in both prokaryotic and eukaryotic cells.[32] The degradation of such proteins is Ub dependent in ts85 cells: upon shift-up to the nonpermissive temperature, the truncated polypeptides are stabilized in the ts85 mutant but not in the wild-type (FM3A) or revertant (ts85R-MN3) cells (Figure 3C).[3] Similar experiments have shown that not only abnormal proteins but apparently also the bulk of normal, relatively short-lived proteins fail to be degraded efficiently in ts85 cells at nonpermissive temperature.[3] However, specific examples of otherwise undamaged, normally short-lived proteins that are stabilized upon shift-up to nonpermissive temperature in ts85 cells remain to be identified.

Closer examination of the substrate specificity of the degradative pathway that is temperature sensitive in ts85 cells is shown in Figure 4, in which the *in vivo* degradation of abnormal (amino acid analog-containing) proteins is followed by SDS–PAGE during a cold chase incubation.[3] The degradation of more than 20 discrete protein species can be inspected individually in Figure 4. The existence of a minor Ub-independent selective degradative system would be suggested in particular if the rate of degradation of specific protein species in Figure 4 were unaffected by the temperature shift-up of ts85 cells; however, Figure 4D shows that in ts85 cells each of the electrophoretically distinguishable protein species is stabilized by shift-up. The electrophoretic bands formed by these otherwise unstable proteins retain their sharpness over the 4 hr of chase incubation at nonpermissive temperature in ts85 cells. Thus, in the absence of a functional Ub system, the (nonlysosomal) intracellular environment appears to be substantially free from activities of partial or limited proteolysis, at least with regard to amino acid analog-containing proteins.

These results[2,3] identify protein ubiquitination as a required step in

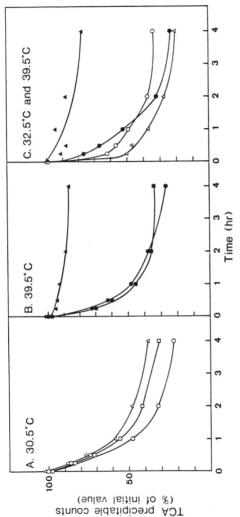

Figure 3. Protein turnover in ts85 cells. (A) ts85, ts85R-MN3, and FM3A cells grown at 32.5°C were incubated for 15 min at 32.5°C with amino acid analogs followed by a 6 min pulse with [^{35}S]methionine, and a chase with unlabeled methionine in the absence of amino acid analogs. Intracellular ^{35}S radioactivity precipitable with hot TCA was measured at different times after the pulse. △-△, ts85 cells; □-□, ts85R-MN3 cells; ○-○, FM3A cells. (B) Same as (A) but the cells were shifted to 39.5°C for 6 hr before incubation with analogs and [^{35}S]methionine at the same temperature (39.5°C). Symbols as above, but closed. (C) Instead of amino acid analogs, the cells (ts85 and FM3A) were incubated with puromycin before the 6 min pulse with [^{35}S]methionine. Open and closed symbols correspond to experiments carried out at 32.5 and 39.5°C, respectively. See ref. 3 for details. (Modified from ref. 3, with permission. Copyright 1984, Cell Press.)

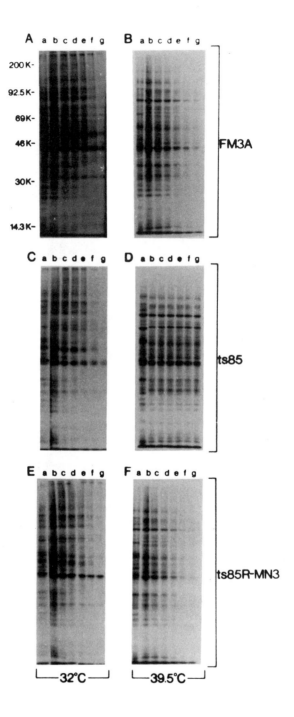

the turnover of most, though not necessarily all, selectively degraded intracellular proteins. Indeed, a slow but detectable protein degradation occurs in ts85 cells even at nonpermissive temperature (Figure 3). Whether the slight leakiness of the protein degradation defect in ts85 cells is due entirely to the observed leakiness in the protein ubiquitination defect in these cells[2] or whether a minor proportion of short-lived intracellular proteins can be degraded via Ub-independent pathways remains to be determined. Unlike relatively short-lived intracellular proteins, the bulk of long-lived proteins is not stabilized at nonpermissive temperature in ts85 cells,[22] suggesting that turnover of this latter class of proteins is largely independent of Ub–protein conjugation.

2.3. Proteins Induced in ts85 Cells at Nonpermissive Temperature

The patterns of pulse-labeled proteins synthesized by ts85 and FM3A cells at permissive temperature are indistinguishable by one-dimensional polyacrylamide gel electrophoresis (Figure 5, lanes A and B). However, 4 hr after shift-up to nonpermissive temperature (39°C), the relative rates of labeling are significantly altered for several proteins in ts85 but not wild-type cells (Figure 4, lanes C and D). In particular, two of the major protein species synthesized in ts85 cells at nonpermissive temperature, "p65" and "p45," are undetectable in control patterns and are thus subject to rapid and strong induction upon shift-up (Figure 5).[33]

When ts85 cells are subjected to an acute heat stress (43°C for 15 min) and thereafter returned to permissive temperature (31°C) for 6 hr, p65 and p45 species are undetectable by pulse labeling (data not shown). Furthermore, the above acute heat stress of ts85 cells, when followed by a 39°C incubation, while resulting in the induction of p65 and p45 proteins, does not significantly intensify these effects over that in controls that have not been acutely heat stressed (Figure 5, lanes H and C). Moreover, the

←─────────────────────────────

Figure 4. SDS–PAGE patterns of ^{35}S-labeled, amino acid analog-containing proteins in pulse-chase *in vivo* experiments. FM3A, ts85, and ts85R-MN3 cells were incubated with amino acid analogs, pulse-labeled with [^{35}S]methionine, and chased as described in the legend to Figure 3. At the end of each chase period, aliquots were removed from cell suspensions and centrifuged. The cell pellets were dissolved in SDS sample buffer and electrophoresed in a 10% polyacrylamide–SDS gel followed by a fluorographic detection of ^{35}S-labeled proteins. Pulse-chase experiments were carried out with FM3A cells (A,B), ts85 cells (C,D), and ts85R-MN3 cells (E,F) at either 32.5°C (A,C,E) or 39.5°C (B,D,F). Lane a in all panels is the pattern of proteins pulse labeled for 6 min in the absence of amino acid analogs; lane b, the same but the pulse labeling was in the presence of amino acid analogs (lanes b were all loaded with equal amounts of total ^{35}S radioactivity); lanes c–g, 15, 30, 60, 120, and 240 min, respectively, of chase with unlabeled methionine in the absence of amino acid analogs. (From ref. 3, with permission. Copyright 1984, Cell Press.)

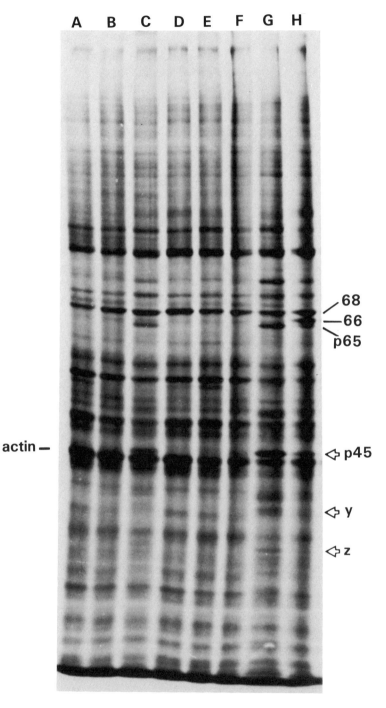

identical treatment (43°C for 15 min followed by a 39°C incubation) does not induce p65 or p45 in the wild-type (FM3A) cells (Figure 5, lane E). Taken together, these data suggest that expression of the temperature-sensitive defect of ts85 cells is sufficient for the induction of p65 and p45 proteins, and that effects of heat stress *per se* are not relevant to this ts85-specific response.[33]

The ts85-specific induction of a third protein, HSP66, shows quite different characteristics. As found for p65 and p45, acute heat stress does not result in a significant induction of HSP66 if followed by a 30°C incubation (Figure 5, lane F). However, when ts85 cells are shifted to nonpermissive temperature (39°C), HSP66 is still not significantly induced (Figure 5, lane C) unless a more severe heat stress is first administered (43°C for 15 min; Figure 5, lane H). Moreover, as the 39°C incubation is extended, the synthesis of HSP66 falls once again to undetectable levels (Figure 5, lane G). Subjeck *et al.*[34] have shown, using ts85 cells, that HSP66 is a member of the ~70 kDa family of heat-shock proteins (hsps). Although the identity of p45 has not been definitively established as yet, the existing data, described below, strongly suggest that p45 is polyubiquitin. First, the synthesis of Ub itself is strongly induced in ts85 cells at nonpermissive temperature.[19] Second, the strong induction of p45 (Figure 5, lane G, and data not shown) is associated with the appearance of two additional, less intensely labeled, smaller protein species (designated y and z in Figure 5), consistent with their being generated from p45 by proteolytic processing.[33] Third, p45, y, and z are all metabolically unstable proteins.[33] Fourth, both p45 and Ub have neutral isoelectric points (ref. 34 and P. Swerdlow, personal communication).

The above results suggest that the impact of a failure of ubiquitination on gene expression is selective and focused primarily on stress-inducible genes such as those encoding polyubiquitin (see Section 3) and specific members of the *HSP70* gene family. These and related findings have led

←————————————————————————————————

Figure 5. Patterns of protein synthesis in ts85 cells after temperature shifts. Exponential 31°C cultures of ts85 and FM3A cells were heat shocked at 43°C for 15 min (lanes E–H), then incubated at either 31 or 39°C. Controls (nonshocked cultures) are shown in lanes A–D. Cells were pulse labeled with [^{35}S]methionine for 1 hr under the following conditions: (A) FM3A cells, 31°C; (B) ts85 cells, 31°C; (C) ts85 cells, 39°C, labeled at 4 hr after shift-up; (D) FM3A cells, 39°C: (E) FM3A cells, 39°C, labeled at 18 hr after shift-up; (F) ts85 cells, 31°C, labeled at 18 hr after shift-up; (G) ts85 cells, as lane E; (H) ts85 cells, as lane E, but labeling was at 6 hr after shift-up. After pulse labeling, cells were pelleted, and the pellets were dissolved in SDS sample buffer, followed by electrophoresis in a 10% polyacrylamide–SDS gel and fluorography. "68," HSP68, the major constitutively expressed member of the 70 kDa family of stress (heat-shock) proteins. "66," HSP 66, a different member of the 70 kDa family of stress proteins. For additional explanations of the band designations, see main text.

us to suggest[2] that the induction of stress proteins may be generally triggered by either overloading or inactivation of the Ub system. Overloading of the Ub system as a consequence of heat stress could result from the thermal denaturation and aggregation of a variety of proteins *in vivo*.[2] In mammalian cells, proteins damaged by the incorporation of amino acid analogs are degraded via the Ub pathway[3]; this apparently holds true in yeast cells as well[21] (see Section 3.2). Moreover, it appears that, if not degraded, such damaged proteins can be extremely toxic.[21] This result supports the suggestion[2] that protein damage may be a causative agent of the toxicity and lethality of heat stress and other conditions that induce stress proteins. Furthermore, it has recently been shown that microinjection of a denatured protein into living cells is sufficient to induce the expression of stress response genes.[35] The mechanism of induction, in this as well as other cases, may be through competitive inhibition of either the degradation or ubiquitination of endogenous positive regulators of stress-specific gene expression.

2.4. Other Phenotypes of ts85 Cells

2.4.1. Temperature Sensitivity of Polyoma Virus Replication in ts85 Cells

While most ts85 cells arrest in the G_2 phase of the cell cycle at nonpermissive temperature, a subpopulation arrests in S phase.[27] To address this apparent DNA replication defect, we have followed the lytic cycle of polyoma virus in ts85 cells. The inhibition of polyoma DNA replication in ts85 (but not in FM3A) cells upon shift-up to nonpermissive temperature is both rapid (~3 hr after shift-up) and nonleaky (R. M. Snapka, D. Finley, E. Özkaynak, and A. Varshavsky, unpublished results). Further analysis of this system may uncover a specific step in the control of DNA replication for which ubiquitination is required.

2.4.2. Nucleoside Transport Defect in ts85 Cells

In the course of analysis of the polyoma DNA replication in ts85 cells, we noted that, unlike the wild-type (FM3A) and revertant (ts85R-MN3) cells, ts85 cells did not incorporate exogenously added [³H]thymidine efficiently into DNA even under conditions of permissive temperature and exponential growth. A more detailed study of this phenomenon (M. Pazin, S. Jentsch, R. M. Snapka, D. Finley, and A. Varshavsky, unpublished results) has produced the following results:

1. When the rate of incorporation of [³H]thymidine into DNA was

measured in a 1 hr assay, the rate of incorporation with ts85 cells (at 30°C) was approximately tenfold lower than that with FM3A cells, provided that the initial concentration of the added [^{3}H]thymidine was significantly below 0.5 μM. The difference in DNA incorporation rates progressively decreased as the concentration of extracellular [^{3}H]thymidine was increased, and disappeared above ~2 μM [^{3}H]thymidine.

2. When the rate of [^{3}H]thymidine transport across the plasma membrane was measured at the extracellular [^{3}H]thymidine concentration of 0.2 μM in a 1–70 sec assay, FM3A cells displayed kinetics of transport characteristic of a facilitated diffusion mechanism, with the intracellular concentration of [^{3}H]thymidine reaching its extracellular concentration within 30 sec after time zero. In striking contrast, ts85 cells transported [^{3}H]thymidine much faster than the wild-type (FM3A) cells, and, moreover, within 20 sec accumulated [^{3}H]thymidine inside the cells to a concentration that was approximately fivefold higher than the extracellular concentration of [^{3}H]thymidine. This burst of the active transport of thymidine was followed by a net efflux of thymidine, which began approximately 20 sec after time zero and lasted throughout the rest of the experiment. When this type of assay was carried out with ATP-depleted ts85 and FM3A cells, transport characteristics differed little between these cells and were those of a typical facilitated diffusion process.

The experiments described above were carried out at permissive temperature, and in all of them the revertant (ts85R-MN3) cells behaved similarly to the wild-type (FM3A) cells. Thus, the assumption that the specific and complex nucleoside transport phenotype of ts85 cells is due to the same (putative) mutation in the gene for Ub-activating enzyme that accounts for the other phenotypes of ts85 cells is supported by the transport data with the revertant (ts85R-MN3) cells. Although nucleoside transporters in mammalian cells are largely of the passive (facilitated diffusion) type, active transport of nucleosides is also known to occur in some specialized somatic cells.[36-38] However, the combination of an active influx of a nucleoside followed by its rapid efflux, as seen in ts85 cells, is to our knowledge an unprecedented pattern of abnormal transport.

One possible explanation of these results is that the (putative) mutational lesion in E1 from ts85 cells, while conferring thermolability onto the Ub activation function of E1, at the same time nonconditionally perturbs the ability of E1 to provide activated Ub for the modification of either nucleoside transporters themselves or regulators thereof. Further analysis of the transport phenotype of ts85 cells is under way.

3. THE UBIQUITIN SYSTEM OF *SACCHAROMYCES CEREVISIAE*

3.1. The Yeast Ubiquitin Genes: A Family of Natural Gene Fusions

In yeast, as in other eukaryotes, Ub is encoded by a complex multigene family.[16] One striking feature of these genes is that while all of them contain Ub-coding sequences, none of them encodes mature Ub (Figure 6). Instead, the four yeast Ub genes all encode hybrid proteins in which Ub is fused at its carboxyl terminus either to itself, as in polyubiquitin, or to unrelated ("tail") amino acid sequences.[16] Thus, in *S. cerevisiae*, Ub is invariably a product of post-translational processing of precursor proteins. The peptide bond that joins the carboxyl-terminal Gly of Ub to the α-amino group of the next residue of the fusion protein is analogous to the isopeptide bond in post-translationally formed, "branched" Ub–protein conjugates.

Although the Ub-coding elements of the *UBI1–UBI4* genes (Figure 6) differ significantly at the nucleotide sequence level, they encode identical amino acid sequences. Low-stringency Southern hybridization of genomic DNA to probes spanning each of the four Ub-coding loci does not reveal any cross-hybridizing sequences that cannot be accounted for

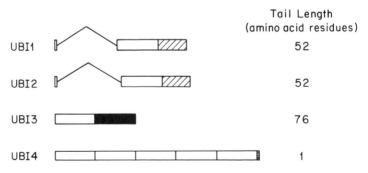

Figure 6. Organization of yeast *UBI1–UBI4* genes. The relative sizes of coding elements and introns in *UBI1* and *UBI2* are drawn approximately to scale. Open blocks denote the 228 bp Ub-coding elements. Two striped blocks and a dark block denote tail-coding elements in *UBI1*, *UBI2*, and *UBI3*, respectively. The tail amino acid sequence in the polyubiquitin product of *UBI4* consists of a single residue, Asn. See ref. 16 for details. (From ref. 16, with permission. Copyright 1987, IRL Press.)

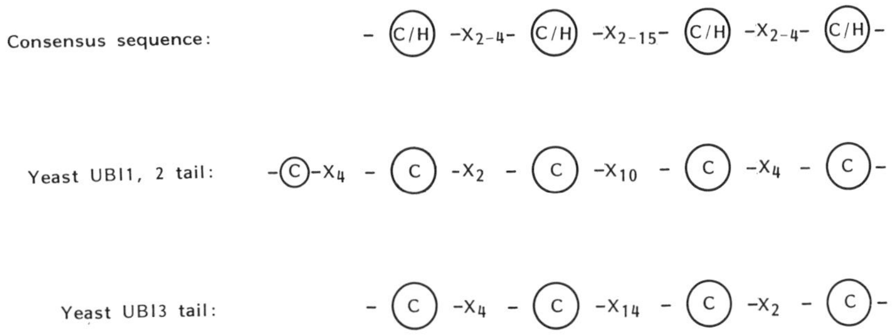

Consensus sequence: $- \boxed{C/H} -X_{2-4}- \boxed{C/H} -X_{2-15}- \boxed{C/H} -X_{2-4}- \boxed{C/H} -$

Yeast UBI1, 2 tail: $- \boxed{C} -X_4 - \boxed{C} -X_2 - \boxed{C} -X_{10} - \boxed{C} -X_4 - \boxed{C} -$

Yeast UBI3 tail: $- \boxed{C} -X_4 - \boxed{C} -X_{14} - \boxed{C} -X_2 - \boxed{C} -$

Figure 7. Putative metal-binding, nucleic acid-binding domains in UBI1–UBI3 proteins. The consensus sequence shown, in which C and H stands for Cys and His, respectively, is a generalized version of the motif originally found in the transcription factor TFIIIA of *Xenopus* and proposed to represent a Zn^{2+}-binding domain that interacts with nucleic acids.[40–42] Homologous motifs have since been detected in many other nucleic acid-binding and metal-binding proteins (reviewed in ref. 42). (From ref. 16, with permission. Copyright 1987, IRL Press.)

by the known Ub genes.[16] Thus, it is likely that we have identified all the Ub-coding genes in *S. cerevisiae*.

The fusion protein encoded by the *UBI1* gene is 128 residues long and consists of Ub followed by an unrelated 52 residue tail sequence.[16] The tail of the UBI1 fusion protein is basic, containing ~31% Lys and Arg residues. A highly basic stretch of seven residues at the carboxyl terminus of the UBI1 tail strongly resembles a sequence motif shown previously to be required for protein localization to the nucleus (reviewed in ref. 39). Thus, either the intact fusion protein or its 52 residue tail may function as a nucleic acid binding protein. This possibility is strengthened by the presence within the tail of a cysteine-containing sequence motif (Figure 7)[16] originally identified within the *Xenopus* transcription factor TFIIIA, where it was found repeated nine times.[40] The corresponding nine domains of TFIIIA each contain a coordinated Zn^{2+} ion and have been proposed to recognize approximately 5 bp of DNA per domain.[41] TFIIIA-like, potential metal-binding domains have been more recently identified in a number of proteins, many of which have been implicated in nucleic acid binding.[42]

The fusion protein encoded by the *UBI2* gene is identical in sequence to that of *UBI1*, despite a ~15% divergence at the nucleotide sequence level between their coding regions.[16] Although the 367 bp *UBI2* intron interrupts the Ub-coding sequence of *UBI2* at exactly the same position as in *UBI1*, the two introns differ in size and are not obviously similar except for the essential sequence elements that are generally conserved among yeast introns.[16] The flanking regions of the *UBI1* and *UBI2* genes are also not obviously similar, suggesting that *UBI1* and *UBI2*, while encoding identical proteins, may be differentially regulated *in vivo*.

The amino acid sequence of the 76 residue UBI3 tail bears little similarity to that of the 52 residue tail of the UBI1 and UBI2 proteins.[16] Nonetheless, the UBI1, UBI2, and UBI3 tails share a number of general structural features. The tails are all basic and contain putative nuclear localization signals, although in the UBI3 protein this sequence is present at the beginning of the tail rather than at its end as in the UBI1 and UBI2 proteins. Moreover, a TFIIIA-like, putative nucleic acid-binding motif present in the UBI1 and UBI2 tails is also found in the UBI3 tail (Figure 7).[16]

The amino acid sequence of Ub is identical in mammals, frogs, fish, and insects and differs in only three of 76 residues from the sequence of yeast Ub[4,16,43] (Figure 8). The tails of the UBI1–UBI3 proteins are also highly conserved in evolution.[16] Specifically, the deduced amino acid sequence of the tail of yeast UBI1/UBI2 protein is ~83% similar to the deduced sequence of its putative human counterpart[17] (Figure 8). Furthermore, the deduced amino acid sequence of the tail of the UBI3 protein is ~67% similar to the deduced sequence of its putative human counterpart[13] (Figure 8). The cDNA clones that encode Ub-hybrid proteins highly similar to the UBI1 and UBI2 proteins of yeast have also been isolated from the slime mold *Dictyostelium discoideum*[15] and the parasitic protozoan *Trypanosoma cruzi*.[12] The tail amino acid sequences are thus conserved to a high degree over great evolutionary distances. A search for similarities between the sequences of the UBI1–UBI3 tails and known proteins, using the National Biomedical Research Foundation database and the algorithm of Lipman and Pearson,[44] has not revealed statistically significant homologies (data not shown).

The expression of polyubiquitin and Ub-hybrid genes is under developmental and metabolic control in various organisms.[5–7,9,10,13,15,16,47] The yeast genes *UBI1, UBI2,* and *UBI3* are all expressed in exponentially growing cells, while *UBI1* and *UBI2* are turned off when cells are heat stressed or starved.[16] *UBI3*, unlike most yeast genes, is apparently expressed under all these conditions.[16] A third pattern of regulation is shown by the *UBI4* gene, which is expressed at low levels in exponentially growing cells but is strongly induced by heat shock, starvation, and other stresses.[16,21]

The *UBI4* gene (Figure 6) encodes a ~43 kDa primary translation product composed of five identical repeats of the Ub amino acid sequence joined head-to-tail, without spacers.[4,16] To generate mature Ub from this precursor, a processing protease would have to cleave at Gly-Met junc-

Yeast ubiquitin :: MQIFVKTLTGKTITLEVESSDTIDNVKSKIQDKEGIPPDQQRLIFAGKQLEDGRTLSDYNIQKESTLHLVLRLRGG

Mammalian ubiquitin :: ——————————————————————————P———E———A———————————————————————————————

Yeast UBI1,2 tail :: •••IIEPSLKALASKYNCDKSVCRKCYARLPPRATNCRKRKCGHTNQLRPKKKLK

Human :: ——————RQ——Q————MI——————H——V——K————N————V—

Yeast UBI3 tail :: •••GKKRKKKVYTTPKKIKHKHKKVKLAVLSYYKVDAEGKVTKLRRECSNPTCGAGVFLANHKDRᴸYCGKCHSVYKVNA

Human :: •••A————S————N——R————K————EN——ISRᴸ————PSDE————RM—S—F——H————CLT—CF—KPEDK

Figure 8. Conservation of the deduced amino acid sequences of the tails of UBI1–UBI3 proteins between yeast and mammals. The deduced amino acid sequences of yeast Ub and of the tails encoded by *UBI1–UBI3* are taken from refs. 4 and 16. The deduced amino acid sequence of human homologs of the UBI1/UBI2 and UBI3 proteins are taken from ref. 17 and 13, respectively. (Modified from ref. 16, with permission. Copyright 1987, IRL Press.)

tions between the adjacent repeats. Although the polyubiquitin-processing protease remains to be characterized biochemically, a yeast protease that deubiquitinates Ub–β-galactosidase fusion proteins *in vivo* has recently been identified in this laboratory[24] (see Chapter 11). The substrate requirements of this enzyme[24] are those expected of a protease that processes both polyubiquitin and the other natural Ub fusions encoded by the *UBI1, UBI2,* and *UBI3* genes.

The number of repeats per polyubiquitin locus varies considerably among and also apparently within species,[4–12] suggesting that these loci frequently engage in unequal crossover events.[45] Such variation has also been found within the Ub-hybrid genes: although typically the Ub genes that contain "tail" sequences, such as *UBI1, UBI2,* and *UBI3,* have only a single Ub-coding element, a gene recently isolated from *Trypanosoma cruzi* contains a tail sequence similar to that of *UBI1* and *UBI2,* joined to a stretch of ~50 contiguous Ub-coding elements.[12] In yeast polyubiquitin (Figure 6), the last of the five Ub repeats is followed by a single Asn residue.[4,16] Most polyubiquitin variants in higher eukaryotes also contain a single extra amino acid residue at the end of their last Ub repeat, the extra residue being different in polyubiquitin variants from different species.[4–12] By blocking the carboxyl terminus of the preceding Gly residue, the extra carboxyl-terminal residue could serve to prevent participation of unprocessed polyubiquitin in Ub–protein conjugation. The properties of a recently characterized Ub-specific hydrolase from mammalian cells that cleaves small adducts off the carboxyl terminus of Ub[46] are compatible with its involvement in removing this extra residue.

3.2. The Polyubiquitin Gene Is Essential for Stress Resistance

A precise, oligonucleotide-directed deletion of *UBI4* was constructed *in vitro* and substituted in the yeast genome in place of the wild-type allele.[21] The *ubi4* deletion mutants grow vegetatively at rates comparable to those of wild-type strains, at least between 23 and 36°C, and the steady-state levels of free Ub in *ubi4* cells are indistinguishable from those in wild-type cells. Thus, *UBI4* plays little if any role in the physiology of exponentially growing cells. However, the major (~1.5 kb) *UBI4* transcript is strongly induced either by heat stress or in stationary-phase cultures.[16,21] The implication that *UBI4* expression may be under the control of the stress response regulatory network is strengthened by the identification within the *UBI4* promoter of strong homologies to the consensus "heat-shock box" nucleotide sequence required in *cis* for stress inducibility of other eukaryotic genes.[16] These features of the polyubiquitin gene appear to be highly conserved in evolution, since a chicken polyubiquitin gene is also heat stress inducible and contains a heat-shock box within its promoter region.[6,47] These considerations led us to test whether *UBI4* is required for survival under the diverse conditions known to induce stress proteins.

Cultures of *ubi4* cells grow at rates indistinguishable from those of wild type at temperatures up to approximately 37°C, whereas at significantly higher temperatures neither culture can grow indefinitely. To test whether *ubi4* mutants might be hypersensitive to chronic heat stress, wild-type and isogenic *ubi4* cells were spread onto YPD plates, incubated for various periods of time at 38.5°C, and then shifted down to 23°C to allow surviving cells to form colonies. That *ubi4* mutants are hypersensitive to heat stress of this type is shown in Figure 9. This phenotype can be complemented by a *UBI4* gene carried on a single-copy, centromere-containing (*CEN*) plasmid.[21] Moreover, complementation is also observed using a *CEN* plasmid carrying a *UBI4* "minigene" in which four of the five Ub repeats of the natural *UBI4* gene have been deleted *in vitro*.[21] These results indicate that Ub is an essential component of cellular thermotolerance mechanisms.

The *ubi4*-complementing ability of the *in vitro* constructed *UBI4* "minigene," which contains flanking sequences of *UBI4* but only a single

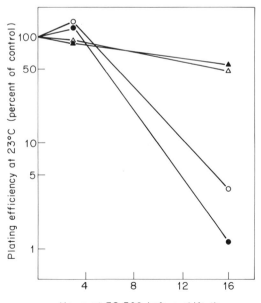

Figure 9. *ubi4* mutants are hypersensitive to incubation at 38.5°C. Cultures growing exponentially at 30°C in rich (YPD) medium were diluted appropriately, spread onto YPD plates, and shifted to 38.5°C. After the times indicated, the plates were returned to 23°C to allow colony formation by viable cells. Open circles, SUB60 (*ubi4*); closed circles, SUB63 (*ubi4*); open triangles, SUB61 (*UBI4*); closed triangles, SUB62 (*UBI4*). See ref. 21 for details. (From ref. 21, with permission. Copyright 1987, Cell Press.)

Plating efficiency at 23°C (percent of control)

Hours at 38.5°C before shift-down

Ub-coding repeat, indicates that the required function of *UBI4* is to provide mature Ub monomers rather than polyubiquitin per se. Since the polyubiquitin gene functions specifically during stress, its repetitive structure may provide an additional selective advantage in reducing the metabolic cost of Ub synthesis under conditions in which metabolic activities are severely compromised. In addition, formation of the spacerless polyubiquitin gene (presumably from a monoubiquitin gene) in the course of evolution may have been greatly facilitated by the prior existence of Ub carboxyl-terminal proteases.[21] Such proteases, having evolved presumably to recycle Ub from post-translationally formed Ub–protein conjugates, were thus preadapted to process polyubiquitin (as well as the UBI1, UBI2, and UBI3 gene products). The (still to be characterized) protease that processes polyubiquitin may thus have at least one more function *in vivo*, namely, the deubiquitination of a subset of post-translationally formed Ub–protein conjugates.

The *ubi4* mutants are also hypersensitive to either carbon or nitrogen starvation.[21] When yeast cells are starved for essential nutrients, they respond by repressing the initiation of new rounds of cell division and entering into stationary (G_0) phase of the cell cycle. Several previously described starvation-sensitive yeast mutants fail to enter stationary phase upon nutrient withdrawal. This class of mutants includes *bcyl* (ref. 48), *RAS2*^{val19} (ref. 49), and *ardl* (ref. 50). However, starved *ubi4* mutants appear to be capable of entering stationary phase, since they display at least some characteristics of stationary-phase cells, such as accumulation of glycogen and enhanced resistance to acute heat stress.[21] Thus, *UBI4* does not appear to be required for cells to enter stationary phase but rather may be required for their survival once in stationary phase.

Another striking phenotype of the *ubi4* mutants is that, although *ubi4/UBI4* diploids can sporulate normally, *UBI4* is subsequently required to maintain the viability (ability to germinate) of haploid spores.[21] This is apparently a novel phenotype in yeast and is unexpected because yeast spores appear to be metabolically dormant.[51] We note that other genes whose mutations result in the phenotype of spore-specific, time-dependent loss of viability might have been mistaken for essential genes if only standard times of sporulation had been used in tetrad analyses.

Since *UBI4* is necessary for the maintenance of spore viability,[21] the Ub system may be active even in mature yeast spores. An alternative explanation assumes that the rate of accumulation of damaged proteins in both metabolically dormant spores and growing cells may be significant even in the absence of explicit environmental stress. However, the metabolic dormancy of the spore would prevent it from degrading or otherwise metabolizing damaged proteins that would therefore increase in concentration with time. Upon germination of the spore, the accumulated damaged proteins would interfere with the resumption of growth unless they were degraded or otherwise processed by the Ub system. The longer the time interval between sporulation and germination, the higher the load of damaged proteins in the spore, with a resulting increased demand for Ub upon germination of older spores. In wild-type spores, but not in $ubi4$ spores, increased amounts of Ub would either be stored within the spore or produced upon germination via the $UBI4$ gene, thus possibly accounting for the time-dependent loss of viability of $ubi4$ spores.[21]

Stress proteins can be induced not only by heat and starvation but also by a variety of toxic compounds. In particular, the exposure of cells to amino acid analogs results in the production of structurally abnormal proteins and efficiently induces the stress response.[52,53] Remarkably, the plating efficiency of $ubi4$ strains on amino acid analog-containing plates is 1000-fold or more below that of wild-type strains.[21] Similar effects were observed using either canavanine, an arginine analog, or L-threo-α-amino-β-chlorobutyric acid (TACB), a valine analog. As described in Section 2.2, amino acid analog-containing proteins are degraded via a Ub-dependent pathway in the mammalian cell cycle mutant ts85. The striking hypersensitivity of $ubi4$ mutants to amino acid analogs[21] strongly suggests that the degradation of abnormal proteins is similarly Ub dependent in yeast.

The above results indicate that the essential function of $UBI4$ is to provide Ub under conditions of stress. In a test of this interpretation, the levels of free (unconjugated) Ub in wild-type and $ubi4$ cells were determined by immunoblotting before and after shifting the cells to 39°C (Figure 10).[21] As previously noted, exponentially growing wild-type and $ubi4$ cells have comparable levels of free Ub. However, 30 min after the shift to 39°C, the amount of free Ub in $ubi4$ cells is greatly reduced, whereas the wild-type cells show a slight increase in free Ub (Figure 10, lanes b). Upon further incubation at 39°C, the amount of free Ub in wild-type cells returns to a level comparable to that in unstressed cells, while in $ubi4$ cells the level of free Ub continues to decline (Figure 10, lanes c,d). These results[21] suggest that the stress-specific loss of viability of $ubi4$ cells reflects their incapacity to maintain required levels of free Ub during stress.

Why is the provision of Ub from genes other than $UBI4$ apparently sufficient during exponential growth but not during stress? A variety of evidence indicates that certain forms of cellular stress, such as heat stress and the accumulation of amino acid analog-containing proteins, result in an increased demand for Ub.[21,22,54,55] In particular, it has been shown, using ts85 cells, that a moderate heat stress results in a burst of Ub-dependent protein degradation, possibly because additional proteolytic

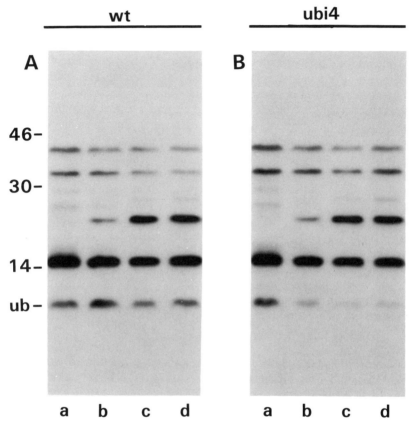

Figure 10. Effect of heat stress on free Ub levels in *ubi4* mutant cells. Immunoblot analysis of proteins from either wild-type (A) or *ubi4* mutant (B) cells. Proteins in SDS extracts of whole cells were electrophoresed in an SDS–polyacrylamide gel, transferred to nitrocellulose, and probed with an antibody to Ub. Extracts were prepared from cultures growing exponentially in YPD at 23°C (lanes a) and from cultures shifted to 39°C for either 30 min (lanes b), 1 hr (lanes c), or 3 hr (lanes d). Designations at the left of (A) indicate the molecular weights (in kDa) of electrophoretic standards and the position of free Ub. Higher molecular weight bands reacting with the antibody include Ub–protein conjugates in the extract. Each lane was loaded with extract from 3×10^7 cells. Relative plating efficiencies were essentially constant throughout the experiment. See ref. 21 for details. (From ref. 21, with permission. Copyright 1987, Cell Press.)

substrates are generated through heat damage.[22] (A relatively severe heat stress inhibits the turnover of at least some proteins, perhaps as a result of overloading, direct heat inactivation, or metabolic regulation of the degradative system.[22,54,56]) Thus, the role of *UBI4* in heat-stressed cells may be not to increase the levels of free Ub over those of unstressed cells but to maintain free-Ub levels against an increased rate at which free Ub is depleted through the formation of Ub–protein conjugates.

Ub appears to be unusual among stress proteins (SPs) in that heat stress does not necessarily induce dramatic increases in the cellular levels of free Ub. However, an apparently general property of the well-characterized SPs is that they are distributed *in vivo* between free and tightly protein-bound forms. Proteins damaged by heat or other stresses may increase the pool of SP-binding substrates and thus decrease the levels of free SPs, including that of Ub. Thus, the increase in total levels of stress proteins after stress may simply reflect a regulatory mechanism that acts to maintain required levels of free stress proteins.

The consideration of Ub as a stress protein underscores previously proposed common features in the mechanism of action of stress proteins.[2] Although there are clearly significant distinctions between the detailed mechanisms of action of such stress proteins as Ub, HSP90 (refs. 57 and 58), HSP70 (refs. 59–62), GRP78 (an HSP70 analog),[63,64] and protease La (a product of the *lon* gene of *E. coli*),[65] these proteins share the following functional characteristics: (1) they appear to function through binding, often very stably, to other proteins; (2) they appear to recognize specifically and bind more than one, and possibly a large number of, specific proteins *in vivo*; and (3) their binding specificity may include a preference for structurally abnormal proteins, in particular misfolded,[2,3] unassembled,[63] or misassembled[59] proteins. The stress-inducible proteins and their constitutively produced cognates thus appear to comprise an integrated metabolic system that regulates the biogenesis, conformation, and selective destruction of a variety of specific, though as yet poorly defined, classes of protein structures.

3.3. Analysis of the Ubiquitin-Hybrid Genes

We have recently constructed precise deletions of the Ub-hybrid genes, *UBI1, UBI2,* and *UBI3* (D. Finley and A. Varshavsky, unpublished results). Deletions of the Ub-hybrid genes, unlike the deletion of *UBI4*, each confer a slow-growth phenotype (Table I). The strongest phenotype is seen in the *ubi3* mutant, which has a doubling time of 6.8 hr as compared to 1.6 hr for the wild-type cells under the same conditions. *ubi3* cells are abnormally large, and the slow growth of their cultures is due to a prolonged cell cycle period rather than to a frequent generation of inviable cells. Moreover, exponentially growing *ubi3* cultures have an extremely high (~65%) proportion of unbudded cells, whereas only ~28% of wild-type cells are unbudded under exponential growth conditions. These results suggest that the approximately fourfold lower growth rate of the *ubi3*

Table I. Growth Rates of Ubiquitin Mutants

Strain	Doubling time^a (hr)
WT	1.6
ubi1	2.4
ubi2	2.0
ubi3	6.8
ubi4	1.6
ubi1 ubi2	(nonviable)
ubi1 ubi3	5.3
ubi1 ubi4	2.4
ubi2 ubi3	5.6
ubi2 ubi4	2.0
ubi3 ubi4	7.3

^{a} Doubling times were measured by monitoring optical densities (at 600 nm) of cultures growing exponentially in rich (YPD) medium at 30°C.

mutant (Table I) is due largely to an approximately tenfold prolongation of the G1 phase in exponentially growing *ubi3* cultures as compared to their wild-type counterparts.

The *ubi3* phenotype is highly pleiotropic. For example, *UBI3* is required not only for efficient growth but also, like *UBI4*, for the survival of cells starved for nitrogen. Moreover, upon nitrogen starvation, the *ubi3 ubi4* double mutant loses viability much more rapidly than either the *ubi3* or *ubi4* single mutants, indicating that in starved wild-type cells the function of *UBI3* cannot be confined to the positive control of UBI4 expression.

The pleiotropic phenotype of the *ubi3* mutant is not due to a Ub deficiency. This is indicated by the fact that the deletion of *UBI3* can be largely if not completely complemented by a derivative of the *UBI3* gene in which the Ub-coding element has been deleted, that is, by a *UBI3* derivative that encodes the tail of the UBI3 protein (Figure 6). Thus, the slow-growth phenotype of *ubi3* mutants is largely due to a deficiency in UBI3 tail function. This inference is supported by the observation that the levels of free Ub are scarcely reduced in *ubi3* cells as compared to their wild-type counterparts and that a plasmid that constitutively expresses free Ub[66] does not complement the *ubi3* deletion.

The *ubi1* and *ubi2* deletions confer milder but significant growth inhibition, while the *ubi1 ubi2* double mutant is nonviable (Table I). Thus, *UBI1* and *UBI2* (which encode identical proteins) together constitute an essential subfamily of the Ub genes. Since a single extra copy of the *UBI2* gene can fully complement the growth defect of *ubi1* cells, the phenotypes of the *ubi1* and *ubi2* single mutants may be viewed as essentially gene dosage effects. If so, the cellular growth rates must be extremely sensitive to small perturbations in the levels of the UBI1 and UBI2 gene products.

Remarkably, both the *ubi1 ubi3* and *ubi2 ubi3* double mutants grow *faster* than the *ubi3* single mutant (Table I). This is particularly striking since the *ubi1* and *ubi2* single mutants grow more slowly than wild-type cells (Table I). Thus, the *ubi1* and *ubi2* deletions are both weak suppressors of the *ubi3* deletion. The suppressive effect of introducing either *ubi1* or *ubi2* deletions into the *ubi3* genetic background suggests that, to a greater or lesser degree, the slow-growth phenotype of the *ubi3* cells may be due to the UBI1 and UBI2 gene products having a deleterious effect in the *ubi3* genetic background but not in that of the wild type. These apparent regulatory effects of the UBI1–UBI3 gene products may be mediated indirectly, through changes in the expression of genes encoding other proteins of the Ub system, as suggested by the structural similarity between the tails of the UBI1–UBI3 fusion proteins and known DNA binding proteins (Figure 7). Alternatively, the UBI1, UBI2, and UBI3 gene products may interact directly with enzymatic components of the Ub-dependent proteolytic pathway, for instance, with a protease that degrades Ub–protein conjugates (enzyme E_x in Figure 1; see also Chapter 4). This possibility[16] is consistent with the fact that, being Ub–protein fusions, the UBI1–UBI3 proteins resemble post-translationally formed Ub–protein conjugates, which are intermediates in Ub-dependent protein degradation.

We are currently testing whether the essential function of *UBI1* and *UBI2* requires cotranslational synthesis of Ub and the UBI1/UBI2 tails. Since the UBI3 tail can carry out its known functions in the absence of an amino-terminal Ub moiety (see above), the hybrid nature of the *UBI1– UBI3* genes may not be strictly essential for their function. Similarly, the tandemly repeated structure of the *UBI4* gene, while presumably beneficial, is not strictly essential, as discussed in Section 3.2. It is possible that cotranslational synthesis of the Ub and the tail proteins serves to hard-wire stoichiometric synthesis and coordinate regulation between the corresponding protein species. In this view, the Ub-hybrid genes may be functionally analogous to prokaryotic operons.

3.4. The Yeast Ubiquitin–Protein Ligase System

All known functions of Ub are mediated by its covalent attachment to target proteins. The enzymatic reactions involved in the formation of Ub–protein conjugates have been extensively studied with enzymes isolated from mammalian reticulocytes[67,68] (see Figure 1). In an initial, ATP-requiring step, the carboxyl-terminal glycine residue of Ub is joined, through a high-energy thiolester bond, to a thiol group (presumably of a cysteine residue) in the Ub-activating enzyme, E1. The activated Ub is then transferred to thiol groups of several smaller proteins, collectively referred to as Ub carrier proteins or E2s[69,70] (see Chapter 3). These proteins then serve to donate Ub to specific protein substrates, forming branched conjugates in which the carboxyl terminus of Ub is joined via an isopeptide bond to ε-amino groups of lysine residues in acceptor proteins. Ubiquitination by some of the E2 enzymes requires the presence of another distinct protein, E3. The E3-dependent Ub conjugation appears to involve specifically those protein substrates that are destined for entry into the Ub-dependent proteolytic pathway[71] (see also Chapters 3, 11, and 12).

We have recently purified several enzymatic components of the *S. cerevisiae* Ub–protein ligase system, using Ub affinity chromatography in the presence of ATP. Under these conditions,[18,69,70] both E1 and E2 enzymes became covalently joined to the immobilized Ub via thiolester bonds and could be selectively eluted from the Ub column with dithiothreitol (DTT) (Figure 11a). The electrophoretic pattern of these yeast proteins (Figure 11a) closely parallels that of the mammalian E1 and E2 enzymes (ref. 18 and data not shown). This similarity is not entirely unexpected, given the exceptional conservation of the Ub amino acid sequence between yeast and higher eukaryotes (Figure 8).[4,16,43] The protein species in the DTT eluate (Figure 11a) were identified as Ub-activating enzyme (E1) and several Ub carrier proteins (E2s) by carrying out assays for their activities as has been described previously for the mammalian E1 and E2 enzymes.[18] Gel exclusion chromatography of the DTT eluate (data not shown) separated E1 and several E2 species, thereby allowing examination of substrate specificities of the individual Ub carrier proteins. One of the enzymes, a ~20 kDa species ($E2_{20k}$ in Figure 11a), catalyzed the transfer of Ub to histones H2A and H2B in the presence of ATP and purified Ub-activating enzyme. Additional evidence that the Ub–histone conjugating activity was intrinsic to the ~20 kDa protein species was obtained by recovering this activity in the protein renatured after elution from an SDS–polyacrylamide gel slice (Figure 11b). The $E2_{20k}$ Ub-conjugating enzyme produces a monoubiquitinated derivative of either histone H2A (data not shown) or histone H2B in an E3-independent *in vitro* reaction (Figure 11b).[18] In mammalian cells, an identical catalytic activity has been assigned to a similarly sized, ~20 kDa protein species present in the DTT eluate of the Ub affinity column.[70]

To begin the molecular genetic analysis of the yeast Ub–protein ligase system, we undertook the cloning of genes encoding its protein compo-

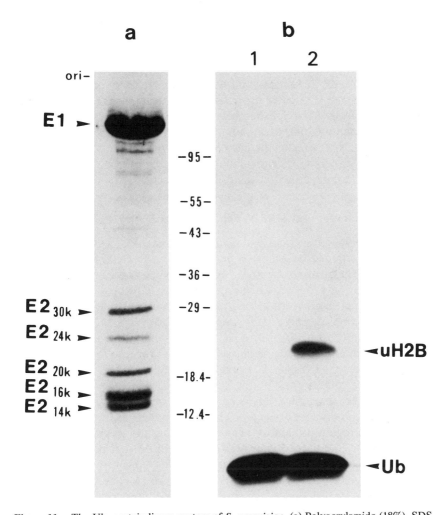

Figure 11. The Ub–protein ligase system of *S. cerevisiae*. (a) Polyacrylamide (18%)–SDS electrophoretic pattern of the dithiothreitol (DTT) eluate from Ub affinity column. Coomassie-stained bands corresponding to the Ub-activating enzyme (E1) and five Ub-conjugating enzymes (E2s) are indicated. Sizes of protein standards (in kDa) are shown on the right. Minor protein bands immediately below the E1 band are degradation products of E1 (data not shown). (b) After electrophoretic fractionation of an (E2$_{20k}$)-enriched chromatographic fraction, proteins were eluted either from a gel slice containing the putative E2$_{20k}$ enzyme (lane 2) or from a slice containing the E2$_{30k}$ enzyme (lane 1). The eluted material was incubated in the presence of ATP, [^{125}I]ubiquitin, histone H2B, and the purified E1 enzyme. Products of the reaction were electrophoresed in an 18% polyacrylamide–SDS gel and detected by autoradiography. Designations: *Ub*, ubiquitin; *uH2B*, monoubiquitinated H2B. See ref. 18 for details. (From ref. 18, with permission. Copyright 1987, Macmillan Journals Ltd.)

nents. Two cloning strategies were employed. Initially, antibodies against more abundant proteins in the DTT eluate (Figure 11a) were generated and used to screen a yeast genomic library carried in the λgt11 expression vector. This approach yielded the genes for the Ub-activating enzyme, E1 (J. P. McGrath, S. Jentsch, and A. Varshavsky, unpublished results), and for one of the E2 carrier proteins, $E2_{30k}$ (S. Jentsch, J. P. McGrath, and A. Varshavsky, unpublished results). The alternative approach consisted of determining partial amino acid sequences of the purified E2 enzymes and the synthesis of the corresponding oligonucleotide probes for use in library screening. The results of applying this strategy to the carrier protein $E2_{20k}$ are described in the following section.

3.5. The DNA Repair Gene *RAD6* Encodes a Ubiquitin Carrier Protein

A 14 residue amino acid sequence of the yeast $E2_{20k}$ enzyme (Figure 11a) was determined by microsequencing cyanogen bromide fragments derived from the purified enzyme.[18] A search for similarities between this sequence and the sequences of known proteins collected in the Institut Pasteur Protein Sequence database revealed that the protein encoded by the *RAD6* gene of *S. cerevisiae*[72] contains an identical 14 residue amino acid sequence (Figure 12). Moreover, the calculated molecular mass of the RAD6 protein, 19.7 kDa, is very close to the apparent molecular mass of the $E2_{20k}$ enzyme (Figures 11a and 12).

To verify directly that the $E2_{20k}$ Ub carrier protein is the product of the *RAD6* gene, we cloned this gene using an oligonucleotide probe corresponding to a nucleotide sequence of *RAD6* (refs. 72 and 73) outside the observed region of identity between the two proteins (see the legend to Figure 12). The cloned gene was verified to be *RAD6* by restriction mapping. A ~600 bp *EcoRI* fragment containing the entire *RAD6* coding sequence was subcloned into an *E. coli* expression vector. Extracts from induced cultures of *E. coli* expressing the *RAD6* gene and control extracts from cells harboring the same expression vector lacking the *RAD6* insert were then assayed for the presence of a Ub-conjugating activity. When supplemented with ATP and purified yeast Ub-activating enzyme, only extracts from *RAD6*-expressing cells were capable of mediating the covalent conjugation of [^{125}I]ubiquitin to histone H2B (Figure 13). The presence of a distinct Ub–histone conjugating activity in bacterial cells expressing the yeast *RAD6* gene product directly identifies the RAD6 protein as a Ub carrier protein. This result, together with the close similarity of the molecular sizes and the originally revealed region of sequence identity between these two proteins, RAD6 and $E2_{20k}$, confirms that the $E2_{20k}$ Ub carrier protein is the product of the *RAD6* gene.

5'- GAATTCCAAAGATTATTTTTAGGCAGACAGACTAAAAGATAAAGCGTC ATG↓TCC ACA
Met Ser Thr   3

CCA GCT AGA AGA AGG TTG ATG↓AGA GAT TTT AAA CGT ATG↓AAG GAA GAT GCC CCA
Pro Ala Arg Arg Arg Leu Met Arg Asp Phe Lys Arg Met Lys Glu Asp Ala Pro   21

CCG GGT GTA TCT GCT TCA CCA TTA CCT GAT AAC GTC ATG↓GTA TGG AAC GCC ATG↓
Pro Gly Val Ser Ala Ser Pro Leu Pro Asp Asn Val Met Val Trp Asn Ala Met   39

ATT ATC GGG CCA GCC GAT ACT CCA TAT GAA GAC GGA ACT TTT AGG TTA TTG TTG
Ile Ile Gly Pro Ala Asp Thr Pro Tyr Glu Asp Gly Thr Phe Arg Leu Leu Leu   57

GAG TTT GAT GAA GAA TAT CCC AAT AAG CCA CCG CAT GTC AAA TTT TTG AGT GAA
Glu Phe Asp Glu Glu Tyr Pro Asn Lys Pro Pro His Val Lys Phe Leu Ser Glu   75

ATG↓TTT CAT CCC AAT GTC TAT GCA AAT GGT GAA ATT TGT TTG GAT ATT TTG CAG
Met | Phe His Pro Asn Val Tyr Ala Asn Gly Glu Ile |Cys| Leu Asp Ile | Leu Gln   93
Phe His Pro Asn Val Tyr Ala Asn Gly Glu Ile --- Leu Asp Ile

AAC AGA TGG ACT CCA ACA TAT GAT GTC GCA TCC ATA TTG ACA TCC ATT CAA AGT
Asn Arg Trp Thr Pro Thr Tyr Asp Val Ala Ser Ile Leu Thr Ser Ile Gln Ser   111

TTA TTC AAC GAT CCA AAT CCA GCT TCG CCA GCA AAC GTT GAA GCT GCA ACA TTA
Leu Phe Asn Asp Pro Asn Pro Ala Ser Pro Ala Asn Val Glu Ala Ala Thr Leu   129

                             3' - CAA TTC CTC TGC CAT CTC TTT
                                ||| ||| ||| ||| ||| ||| |||
TTC AAA GAT CAT AAA TCA CAG TAC GTC AAA AGA GTT AAG GAG ACG GTA GAG AAA
Phe Lys Asp His Lys Ser Gln Tyr Val Lys Arg Val Lys Glu Thr Val Glu Lys   147

AGA ACC CTC - 5'
||| ||| |||
TCT TGG GAG GAT GAT ATG↓GAC GAT ATG↓GAC GAT GAT GAT GAT GAT GAT GAC GAC
Ser Trp Glu Asp Asp Met Asp Asp Met Asp Asp Asp Asp Asp Asp Asp Asp Asp   165

GAC GAC GAC GAC GAA GCA GAC TGA GAAAAATCAAAAGAATCTTAATGATGAATGCCGAGCCG
Asp Asp Asp Asp Glu Ala Asp STOP   172

ATATTATGAATTC -3'

Figure 12. Region of identity between the *RAD6* gene product and the sequenced portion of E2$_{20k}$ enzyme. Shown are the nucleotide sequence and the deduced amino acid sequence of the previously cloned[72,73] *RAD6* gene of *S. cerevisiae*. The boxed region indicates a directly determined 14 residue amino acid sequence from the purified E2$_{20k}$ enzyme (see main text) that is identical to the deduced sequence of the RAD6 protein between residues 77 and 91. A single cysteine residue of the RAD6 protein that is presumably essential for E2 activity is present within this stretch. Under the conditions of the Edman sequencing procedure used, unambiguous identification of Cys residues was not possible, hence the gap in the sequenced region of the E2$_{20k}$ enzyme. Arrows indicate sites of cyanogen bromide cleavage. Shown also is the position within *RAD6* gene of a 30 nucleotide sequence that has been used in the present work to isolate the *RAD6* gene from a EMBL3A-based yeast genomic library by screening with the corresponding synthetic oligonucleotide. *EcoRI* sites used for subcloning are underlined. See ref. 18 for details. (From ref. 18, with permission. Copyright 1987, Macmillan Journals Ltd.)

These findings[18] identify the *RAD6* gene as a member of a family of closely related genes, which we propose to denote as the *UBC* genes (for *Ub-c*onjugating enzymes). The *UBC* gene family encodes at least five distinct E2 enzymes in yeast (Figure 11a and our unpublished data). The gene encoding the $E2_{30k}$ enzyme (Figure 11a) was cloned previously and was designated *UBC1* (S. Jentsch, J. P. McGrath, and A. Varshavsky, unpublished results). In this terminology, the *RAD6* locus corresponds to the *UBC2* gene, which encodes the $E2_{20k}$ Ub carrier protein.

In *S. cerevisiae*, the *RAD6(UBC2)* gene plays a central role in one of the three epistasis groups of genes involved in DNA repair.[74,75] Phenotypes of *rad6* mutants are remarkably pleiotropic and include slow growth, extreme sensitivity to UV light, x-rays, and chemical mutagens, hypersensitivity to the antifolate drug trimethoprim, and severe defects in induced mutagenesis.[74-76] In addition, diploids homozygous for some *rad6* alleles are defective in sporulation.[77,78] The observation that a highly conserved counterpart of the $E2_{20k}$ enzyme exists in mammals (ref. 70 and our unpublished results) implies that *RAD6*-dependent functions are relevant to all eukaryotic cells. Moreover, these results[18] raise the possibility that some of the human hereditary diseases that involve defects in DNA repair may in fact be due to mutations in genes for specific Ub-conjugating enzymes. For example, in certain types of the disease *xeroderma pigmentosum,* the cells are apparently defective in an early "preincision" event believed to regulate the initiation of repair processes.[79,80]

The observed enzymatic activity of the *RAD6* gene product[18] suggests that at least some of the multiple functions attributed to this gene are mediated by ubiquitination of specific target proteins. The $E2_{20k}$ enzyme, the product of the *RAD6(UBC2)* gene, while ubiquitinating histones such as H2A and H2B *in vitro*, does not appear to ubiquitinate endogenous proteins in the *E. coli* extract (Figure 13). The substrate specificity of the RAD6 protein *in vitro* suggests that its *in vivo* substrates may be few in number and may include basic proteins such as chromosomal histones. The latter possibility is also consistent with the presence of 14 consecutive negatively charged residues at the carboxyl-terminal region of the RAD6 protein[72] (Figure 12).

In vitro studies utilizing wild-type and mutant mammalian cells defective in DNA repair have emphasized the importance of chromatin structure in regulating the accessibility of lesions in the DNA to the repair enzymes.[79-82] One possibility is that ubiquitination of specific proteins mediated by the RAD6(UBC2) protein results in the selective degradation of the target proteins. Alternatively, RAD6(UBC2)-mediated ubiquitination of target proteins could directly induce structural alterations in chromatin, thereby allowing access by enzymes of the DNA repair and mu-

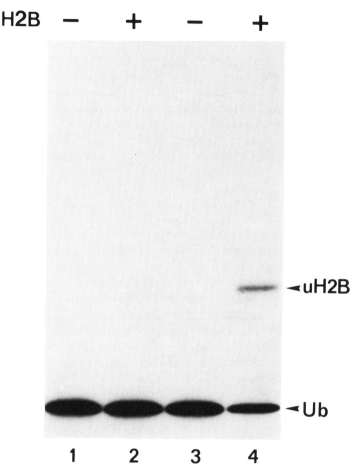

Figure 13. The *RAD6* gene product is a Ub-conjugating enzyme. *E. coli* cells harboring expression vectors that either contained or lacked the *RAD6* gene (see Section 3.5) were induced and lysed. Extracts of cells containing the control vector (lanes 1 and 2) and the vector containing the *RAD6* gene (lanes 3 and 4) were assayed for Ub–histone conjugation. Products of the reaction were electrophoresed in an 18% polyacrylamide–SDS gel and detected by autoradiography. Designations are as in Figure 11. See ref. 18 for details. (From ref. 18, with permission. Copyright 1987, Macmillan Journals Ltd.)

tagenesis pathways to the underlying DNA lesions. Targeting of ubiquitination events to the sites of lesions in chromosomes may be due to selective exposure of ubiquitination sites in the RAD6(UBC2) substrates after localized damage to either DNA or other components of the chromosome. It remains to be determined whether transient ubiquitination events are required not only for DNA repair but also for DNA transcription, recombination, and replication, and if so, whether specific E2 enzymes are involved.

Sporulation, another distinct pathway defective in certain *rad6* mutants,[77,78] is accompanied by drastic changes in the protein composition of the forming ascospore and also by a more compact packing of the chromosomal DNA. This latter process may require changes in the protein composition of the chromatin, possibly analogous to those that occur during spermatogenesis in higher eukaryotes. The requirement for the *RAD6(UBC2)* function in this pathway may thus be due to an involvement of ubiquitination of chromosomal proteins in the process of chromatin remodeling during sporulation.

The discovery that the *RAD6* gene product is a Ub carrier protein[18] underscores the striking functional diversity of the Ub system, which is now known to be involved in selective protein turnover, DNA repair, and a variety of other processes that have stress-related functions.

Progress through the cell cycle may also involve specific ubiquitination events, as suggested by work on the mammalian cell line ts85.[2,3] Moreover, the yeast cell division cycle gene *CDC34* encodes a protein whose amino acid sequence is highly similar to that of the RAD6(UBC2) protein (B. Byers and M. G. Goebl, personal communication). The CDC34 protein is also highly similar to Ub carrier proteins other than RAD6(UBC2) (S. Jentsch, J. P. McGrath, and A. Varshavsky, unpublished results), suggesting that CDC34 is yet another member of the family of Ub carrier proteins.

ACKNOWLEDGMENTS.   We thank Mary Jalenak for technical assistance, Tauseef Butt (Smith Kline and French Laboratories) for the gift of the Yep46 plasmid,[66] Mark Hochstrasser and Pamela Larsen (MIT) for comments on the manuscript, and Barbara Doran for secretarial assistance. This work was supported by grants to A.V. from the National Institutes of Health (GM31530 and CA43309). E.Ö. was a Fellow of the Charles A. King Trust and the Medical Foundation. S.J. was a Fellow of the Deutsche Forschungsgemeinschaft. B.B. was supported by a predoctoral fellowship from the National Science Foundation.

REFERENCES

1. Hershko, A., 1983, Ubiquitin: Roles in protein modification and breakdown, *Cell* **34:** 11–12.

2. Finley, D., Ciechanover, A., and Varshavsky, A., 1984, Thermolability of ubiquitin-activating enzyme from the mammalian cell cycle mutant ts85, *Cell* **37:** 43–55.
3. Ciechanover, A., Finley, D., and Varshavsky, A., 1984, Ubiquitin dependence of selective protein degradation demonstrated in the mammalian cell cycle mutant ts85, *Cell* **37:** 57–66.
4. Özkaynak, E., Finley, D., and Varshavsky, A., 1984, The yeast ubiquitin gene: Head-to-tail repeats encoding a polyubiquitin precursor protein, *Nature* **312:** 663–666.
5. Dworkin-Rastl, E., Shrutkowski, A., and Dworkin, M. B., 1984, Multiple ubiquitin mRNAs during *Xenopus laevis* development contain tandem repeats of the 76 amino acid coding sequence, *Cell* **39:** 321–325.
6. Bond, U., and Schlesinger, M. J., 1985, Ubiquitin is a heat shock protein in chicken embryo fibroblasts, *Mol. Cell. Biol.* **5:** 949–956.
7. Wiborg, O., Pedersen, M. S., Wind, A., Berglund, L. E., Marcker, K. A., and Vuust, J., 1985, The human ubiquitin multigene family; some genes contain multiple directly repeated ubiquitin coding sequences, *EMBO J.* **4:** 755–759.
8. Arribas, C., Sampedro, J., and Izquierdo, M., 1986, The ubiquitin genes in *D. melanogaster:* Transcription and polymorphism, *Biochim. Biophys. Acta* **868:** 119–127.
9. Gausing, K., and Barkardottir, R., 1986, Structure and expression of ubiquitin genes in higher plants, *Eur. J. Biochem.* **158:** 57–62.
10. Giorda, R., and Ennis, H. L., 1987, Structure of two developmentally regulated *Dictyostelium discoideum* ubiquitin genes, *Mol. Cell. Biol.* **6:** 2097–2103.
11. Baker, R. T., and Board, P. G., 1987, The human ubiquitin gene family: Structure of a gene and pseudogenes from the Ub B subfamily, *Nucleic Acids Res.* **15:** 443–463.
12. Swindle, J., Ajioka, J., Eisen, H., Sanwal, B., Jacquemot, C., Browder, Z., and Buck, G., 1988, The genomic organization and transcription of the ubiquitin genes of *Trypanosoma cruzi, EMBO J.* (in press).
13. Lund, P. K., Moats-Staats, B. M., Simmons, J. G., Hoyt, E., D'Ercole, A. J., Martin, F., and Van Wyk, J. J., 1985, Nucleotide sequence analysis of a cDNA encoding human ubiquitin reveals that ubiquitin is synthesized as a precursor, *J. Biol. Chem.* **260:** 7609–7613.
14. St. John, T., Gallatin, W. M., Siegelman, M., Smith, H. T., Fried, V. A., and Weissman, I. L., 1986, Expression cloning of a lymphocyte homing receptor cDNA: Ubiquitin is the reactive species, *Science* **221:** 845–850.
15. Westphal, M., Muller-Taubenberger, A., Noegel, A., and Gerisch, G., 1986, Transcript regulation and carboxyl terminal extension of ubiquitin in *Dictyostelium discoideum,* *FEBS Lett.* **209:** 92–96.
16. Özkaynak, E., Finley, D., Solomon, M. J., and Varshavsky, A., 1987, The yeast ubiquitin genes: A family of natural gene fusions, *EMBO J.* **6:** 1429–1439.
17. Salvesen, G., Lloyd, C., and Farley, D., 1987, cDNA encoding a human homolog of yeast ubiquitin 1, *Nucleic Acids Res.* **15:** 5485–5486.
18. Jentsch, S., McGrath, J. P., and Varshavsky, A., 1987, The DNA repair gene *RAD6* encodes a ubiquitin-conjugating enzyme, *Nature* **329:** 131–134.
19. Finley, D., and Varshavsky, A., 1985, The ubiquitin system: Functions and mechanisms, *Trends Biochem. Sci.* **47:** 275–284.
20. Schlesinger, M. J., and Bond, U., 1987, Ubiquitin genes, *Oxford Surv. Eukaryotic Genes* **4:** 77–91.
21. Finley, D., Özkaynak, E., and Varshavsky, A., 1987, The yeast polyubiquitin gene is essential for resistance to high temperatures, starvation and other stresses, *Cell* **48:** 1035–1046.
22. Parag, H. A., Raboy, B., and Kulka, R. G., 1987, Effect of heat shock on protein

degradation in mammalian cells: Involvement of the ubiquitin system, *EMBO J.* **6:** 55–61.

23. Struhl, K., 1983, The new yeast genetics, *Nature* **305:** 391–397.
24. Bachmair, A., Finley, D., and Varshavsky, A., 1986, *In vivo* half-life of a protein is a function of its amino-terminal residue, *Science* **234:** 179–186.
25. Mita, S., Yasuda, H., Marunouchi, T., Ishiko, S., and Yamada, M., 1980, A temperature-sensitive mutant of cultured mouse cells defective in chromosome condensation, *Exp. Cell Res.* **126:** 407–416.
26. Matsumoto, Y., Yasuda, H., Mita, S., Marunouchi, T., and Yamada, M., 1980, Evidence for involvement of H1 histone phosphorylation in chromosome condensation, *Nature* **284:** 181–183.
27. Yasuda, H., Matsumoto, Y., Mita, S., Marunouchi, T., and Yamada, M., 1981, A mouse temperature-sensitive mutant defective in H1 histone phosphorylation is defective in deoxyribonucleic acid synthesis and chromosome condensation, *Biochemistry* **20:** 4414–4419.
28. Marunouchi, T., Yasuda, H., Matsumoto, Y., and Yamada, M., 1980, Disappearance of a chromosomal basic protein from cells of a mouse temperature-sensitive mutant defective in histone phosphorylation, *Biochem. Biophys. Res. Commun.* **95:** 126–131.
29. Matsumoto, Y., Yasuda, H., Marunouchi, T., and Yamada, M., 1983, Decrease in uH2A (protein A24) in a mouse temperature-sensitive mutant, *FEBS Lett.* **151:** 139–142.
30. Ciechanover, A., Elias, S., Heller, H., and Hershko, A., 1982, "Covalent" affinity purification of ubiquitin-activating enzyme, *J. Biol. Chem.* **257:** 2537–2542.
31. Rabinovitz, M., and Fisher, J. M., 1964, Characteristics of the inhibition of hemoglobin synthesis in rabbit reticulocytes by threo-α-amino-β-chlorobutyric acid. *Biochim. Biophys. Acta* **91:** 313–322.
32. Goldberg, A. L., and St. John, A. C., 1976, Intracellular protein degradation in mammalian and bacterial cells, *Annu. Rev. Biochem.* **45:** 747–803.
33. Finley, D., 1984, Approaches to molecular genetics of the ubiquitin system, Ph.D. thesis, Massachusetts Institute of Technology.
34. Subjeck, J. R., Sciandra, J. J., and Shyy, T. T., 1985, Analysis of the expression of the two major proteins of the 70 kilodalton mammalian heat shock family, *Int. J. Radiat. Biol.* **47:** 275–284.
35. Anathan, J., Goldberg, A. L., and Voellmy, R., 1986, Abnormal proteins serve as eukaryotic stress signals and trigger the activation of heat shock genes, *Science* **232:** 522–524.
36. Aronow, B., Toll, D., Patrick, J., McCartan, K., and Ullman, B., 1986, Dipyridamole-insensitive nucleoside transport in mutant murine T lymphoma cells, *J. Biol. Chem.* **261:** 14467–14473.
37. Pickard, M. A., Brown, R. R., Paul, B., and Paterson, A. R. P., 1973, Binding of the nucleoside transport inhibitor 4-nitrobenzyl-6-thioinosine to erythrocyte membranes, *Can. J. Biochem.* **51:** 666–672.
38. Plagemann, P. G. W., and Wohlhueter, R. M., 1980, Permeation of nucleosides, nucleic acid bases, and nucleotides in animal cells, *Curr. Top. Membr. Transp.* **14:** 225–330.
39. Dingwall, C., and Laskey, R. A., 1986, Protein import into the cell nucleus, *Annu. Rev. Cell Biol.* **2:** 366–390.
40. Miller, J., McLachlan, A. D., and Klug, A., 1985, Repetitive zinc-binding domains in the protein transcription factor IIIA from *Xenopus* oocytes, *EMBO J.* **4:** 1609–1614.
41. Rhodes, D., and Klug, A., 1986, An underlying repeat in some transcriptional control sequences corresponding to half a double helical turn of DNA, *Cell* **46:** 123–132.
42. Berg, J. M., 1985, Potential metal-binding domains in nucleic acid binding proteins, *Science* **232:** 485–487.

43. Wilkinson, K. D., Cox, M. J., O'Connor, L. B., and Shapira, R., 1986, Structure and activities of a variant ubiquitin sequence from baker's yeast, *Biochemistry* **25:** 4999–5004.
44. Lipman, D. J., and Pearson, W. R., 1985, Rapid and sensitive protein similarity searches, *Science* **227:** 1435–1441.
45. Sharp, P. M., and Li, W., 1987, Ubiquitin genes as a paradigm of concerted evolution of tandem repeats, *J. Mol. Evol.* **25:** 58–64.
46. Pickart, C. M., and Rose, I. A., 1985, Ubiquitin carboxyl-terminal hydrolase acts on ubiquitin carboxyl-terminal amides, *J. Biol. Chem.* **260:** 7903–7910.
47. Bond, U., and Schlesinger, M. J., 1986, The chicken ubiquitin gene contains a heat shock promoter and expresses an unstable mRNA in heat-shocked cells, *Mol. Cell. Biol.* **6:** 4602–4610.
48. Matsumoto, K., Uno, I., and Ishikawa, T., 1983, Control of cell division in *Saccharomyces cerevisiae* mutants defective in adenylate cyclase and cAMP-dependent protein kinase, *Exp. Cell Res.* **146:** 151–161.
49. Toda, T., Uno, I., Ishikawa, T., Powers, S., Kataoka, T., Broek, D., Cameron, S., Broach, J., Matsumoto, K., and Wigler, M., 1985, In yeast, RAS proteins are controlling elements of adenylate cyclase, *Cell* **40:** 27–36.
50. Whiteway, M., and Szostak, J. W., 1985, The *ARD1* gene of yeast functions in the switch between the mitotic cell cycle and alternative developmental pathways, *Cell* **43:** 483–492.
51. Thevelein, J. M., den Hollander, J. A., and Shulman, R. C., 1984, Trehalase, control of dormancy and induction of germination in fungal spores, *Trends Biochem. Sci.* **9:** 495–497.
52. Hightower, L. E., and White, F. P., 1982, Preferential synthesis of rat heat-shock and glucose-regulated proteins in stressed cardiovascular cells, in: *Heat Shock* (M. J. Schlesinger, M. Ashburner, and A. Tissieres, eds.), Cold Spring Harbor Laboratory, Cold Spring Harbor, NY, pp. 369–377.
53. Hall, B., 1983, Yeast thermotolerance does not require protein synthesis, *J. Bacteriol.* **156:** 1363–1365.
54. Carlson, N., Rogers, S., and Rechsteiner, M., 1987, Microinjection of ubiquitin: Changes in protein degradation in HeLa cells subjected to heat-shock, *J. Cell Biol.* **104:** 547–555.
55. Rose, I. A., and Warms, J. V. B., 1987, A specific endpoint assay for ubiquitin, *Proc. Natl. Acad. Sci. U.S.A.* **84:** 1477–1481.
56. Munro, S., and Pelham, H. R. B., 1984, Use of peptide tagging to detect proteins expressed from cloned genes: Deletion mapping functional domains from *Drosophila* hsp70, *EMBO J.* **3:** 3087–3093.
57. Brugge, J., Yonemoto, W., and Darrow, D., 1983, Interaction between the Rous sarcoma virus transforming protein and two cellular phosphoproteins: Analysis of the turnover and distribution of this complex, *Mol. Cell. Biol.* **3:** 9–19.
58. Catelli, M. G., Binart, N., Jung-Testas, I., Renoir, J. M., Baulieu, E. E., Feramisco, J. R., and Welch, W. J., 1985, The common 90-kd protein component of non-transformed "8S" steroid receptors is a heat-shock protein, *EMBO J.* **4:** 3131–3135.
59. Lewis, M. J., and Pelham, H. R. B., 1985, Involvement of ATP in the nuclear and nucleolar functions of the 70 kd heat shock protein, *EMBO J.* **4:** 3137–3143.
60. Ungewickell, E., 1985, The 70-kd mammalian heat shock proteins are structurally and functionally related to the uncoating protein that releases clathrin triskelia from coated vesicles, *EMBO J.* **4:** 3385–3391.
61. Chappell, T. G., Welch, W. J., Schlossman, D. M., Palter, K. B., Schlesinger, M. J.,

and Rothman, J. E., 1986, Uncoating ATPase is a member of the 70 kilodalton family of stress proteins, *Cell* **45:** 3–13.

62. Pelham, H. R. B., 1985, Speculations on the functions of the major heat shock and glucose-regulated proteins, *Cell* **46:** 959–961.

63. Munro, S., and Pelham, H. R. B., 1986, An hsp70-like protein in the ER: Identity with the 78 kd glucose-regulated protein and immunoglobulin heavy chain binding protein, *Cell* **46:** 291–300.

64. Gething, M. M., McCammon, K., and Sambrook, J., 1986, Expression of wild-type and mutant forms of influenza hemagglutinin: The role of folding in intracellular transport, *Cell* **46:** 939–950.

65. Goff, S. A., and Goldberg, A. L., 1985, Production of abnormal proteins in *E. coli* stimulates transcription of *lon* and other heat shock genes, *Cell* **41:** 587–595.

66. Ecker, D. J., Khan, M. I., Marsh, J., Butt, T., and Crooke, S. T., 1987, Chemical synthesis and expression of a cassette adapted ubiquitin gene, *J. Biol. Chem.* **262:** 3524–3527.

67. Hershko, A., and Cienchanover, A., 1986, The ubiquitin pathway for the degradation of intracellular proteins, *Prog. Nucleic Acids Res. Mol. Biol.* **33:** 19–56.

68. Rechsteiner, M., 1987, Ubiquitin-mediated pathways for intracellular proteolysis, *Annu. Rev. Cell. Biol.* **3:** 1–30.

69. Hershko, A., Heller, H., Elias, S., and Ciechanover, A., 1983, Components of ubiquitin–protein ligase system. *J. Biol. Chem.* **258:** 8206–8214.

70. Pickart, C. M., and Rose, I. A., 1985, Functional heterogeneity of ubiquitin carrier proteins, *J. Biol. Chem.* **260:** 1573–1581.

71. Hershko, A., Heller, H., Eytan, E., and Reiss, Y., 1986, The protein substrate binding site of the ubiquitin–protein ligase system, *J. Biol. Chem.* **261:** 11992–11999.

72. Reynolds, P., Weber, S., and Prakash, L., 1985, *RAD6* gene of *Saccharomyces cerevisiae* encodes a protein containing a tract of 13 consecutive aspartates, *Proc. Natl. Acad. Sci. U.S.A.* **82:** 168–172.

73. Prakash, L., Polakowski, R., Reynolds, P., and Weber, S., 1983, Molecular cloning and preliminary characterization of the *RAD6* gene of the yeast *Saccharomyces cerevisiae*, in: *Cellular Responses to DNA Damage* (E. C. Friedberg and B. A. Bridges, eds.), Alan R. Liss, New York, pp. 559–568.

74. Lawrence, C. W., 1982, Mutagenesis in *Saccharomyces cerevisiae, Adv. Genet.* **21:** 173–254.

75. Haynes, R. H., and Kunz, B. A., 1981, DNA repair and mutagenesis in yeast, in: *The Molecular Biology of the Yeast Saccharomyces cerevisiae: Life Cycle and Inheritance* (J. Strathern, E. Jones, and J. Broach, eds.), Cold Spring Harbor Laboratory, Cold Spring Harbor, NY., pp. 371–414.

76. Tuite, M. F., and Cox, B. S., 1981, *RAD6* gene of *Saccharomyces cerevisiae* codes for two mutationally separable deoxyribonucleic acid repair functions, *Mol. Cell. Biol.* **1:** 153–157.

77. Crame, J. C., and Mortimer, R. K., 1974, A genetic study of X-ray sensitive mutants in yeast, *Mutat. Res.* **24:** 281–292.

78. Montelone, B. A., Prakash, S., and Prakash, L., 1981, Recombination and mutagenesis in *rad6* mutants of *Saccharomyces cerevisiae*: Evidence for multiple functions of the *RAD6* gene, *Mol. Gen. Genet.* **184:** 410–415.

79. Friedberg, E. C., 1983, in: *DNA Repair*, W. H. Freeman, San Francisco, CA, pp. 506–525.

80. Fujiwara, Y., and Kano, Y., 1983, Characteristics of thymine dimer excision from *xe-*

roderma pigmentosum chromatin, in: *Cellular Responses to DNA Damage* (E. C. Friedberg and B. A. Bridges, eds.), Alan R. Liss, New York, pp. 215–224.

81. Mellon, I., Bohr, V. A., Smith, C. A., and Hanawalt, P. C., 1986, Preferential DNA repair of an active gene in human cells, *Proc. Natl. Acad. Sci. U.S.A.* **83:** 8878–8882.

82. Hanawalt, P. C., and Sarasin, A., 1976, Cancer-prone hereditary diseases with DNA processing abnormalities, *Trends Genet.* **2:** 124–129.

Chapter 3

Ubiquitin Activation and Ligation

Cecile M. Pickart

1. INTRODUCTION

The energy dependence of intracellular protein turnover has been recognized for several decades. In the 1950s, Simpson[1] and Schwieger *et al.*[2] showed that respiratory inhibitors decrease the rate of release of amino acids from cells. Therefore, these compounds, which decrease intracellular ATP concentration, inhibit intracellular protein turnover. Since peptide bond hydrolysis is a thermodynamically favorable process, why should protein break down require ATP? And how might ATP hydrolysis be coupled to peptide bond hydrolysis? Our understanding of the mechanistic basis for the ATP requirement has increased substantially in recent years. There are now several known ATP-dependent proteases. One of these is part of the multienzyme pathway that requires the small protein ubiquitin (Ub) as a cofactor. Although Ub-dependent protein breakdown is not the only ATP-dependent protein breakdown in higher eukaryotes,[3] it is clear that the Ub-dependent pathway is quantitatively important. Turnover of at least 90% of the short-lived proteins in a mouse mammary carcinoma cell line is Ub dependent.[4]

Ub-dependent proteolysis requires ATP at two distinct steps. ATP

CECILE M. PICKART • Department of Biochemistry, SUNY/Buffalo, Buffalo, New York 14214.

is hydrolyzed during formation of a covalent bond between Ub and the proteolytic substrate. There is convincing evidence that formation of this Ub–target protein conjugate is a necessary event in ATP- and Ub-dependent turnover. The actual proteolytic step, for which the conjugate is the substrate, is also ATP dependent.

The existence of Ub conjugates in the nucleus[5] and on the cell surface,[6,7] sites where a proteolytic function for Ub seems unlikely, raises the possibility that conjugation of Ub to proteins may sometimes serve regulatory functions. Conjugates can undergo nondegradative disassembly, providing a mechanism whereby regulatory modification by Ub could be reversed. Understanding completely the biological purposes served by Ub conjugation to proteins requires thorough knowledge of mechanisms and specificities of the enzymes responsible for conjugate formation.

2. SIGNIFICANCE OF UBIQUITIN–PROTEIN CONJUGATES IN UBIQUITIN-DEPENDENT PROTEIN TURNOVER

All known or postulated roles of Ub involve its covalent conjugation to cellular proteins, and it is likely that specificity in conjugate formation contributes to specificity in most processes involving Ub. Certainly this is true for Ub-dependent protein turnover (Section 3.4), where the function of Ub in proteolysis is best understood. Prior to considering the actual conjugating enzymes, it is useful to review briefly some of the evidence concerning the role of Ub–protein conjugates in Ub-dependent protein turnover.

2.1. Historical Background

Early work by Goldberg, Hershko, and their co-workers established that crude soluble lysates from rabbit reticulocytes degraded alkali-denatured and amino acid analog-containing globin molecules in a Mg^{2+}- and ATP-dependent fashion.[8,9] Subsequent fractionation of the cytoplasmic ATP-dependent proteolytic activity by anion exchange chromatography generated two fractions.[10] Unadsorbed neutral or basic proteins constituted fraction I, and fraction II contained adsorbed (acidic) proteins that eluted in 0.5 M KCl. Recombination of the two fractions was necessary for ATP-dependent proteolytic activity. Further characterization of the active species in fraction I showed that it was a small (about 8 kDa), heat-stable protein. It was named ATP-dependent proteolysis factor 1 (APF-1) and was easily purified as a result of its unusual stability.[10,11] In contrast, fraction II appeared to contain multiple components that were required for ATP-dependent proteolysis in the presence of APF-1.

APF-1 was subsequently shown to be identical to Ub, a highly conserved protein of previously unknown function.[12] Ubiquitin had been identified in nuclei, both free and covalently linked to certain histones.[5] The structure of the nuclear Ub–histone adduct had direct bearing on the hypothesis that was developed concerning the function of Ub in protein degradation. The key experiment leading to our present understanding of the role of Ub in ATP-dependent protein turnover demonstrated that purified, radiolabeled Ub, when added with ATP to fraction II, formed covalent adducts with many proteins.[13] These adducts were much larger than free Ub and were easily detected by exclusion chromatography or SDS gel electrophoresis and autoradiography. Their heterogeneous sizes suggested that they were conjugates of "substrate" proteins in fraction II, rather than catalytic intermediates involving Ub and specific enzymes. The adducts, termed conjugates, were not indefinitely stable. Upon enzymatic removal of ATP from fraction II, all radiolabel eventually returned to a species with the molecular weight of Ub. That this species was indeed intact Ub was shown by the fact that another round of conjugation could occur on addition of more ATP.[13]

Even more suggestive evidence that Ub–substrate conjugation is a central event in ATP-dependent protein turnover was the observation that Ub is conjugated to added substrates when they are being degraded. For example, during the degradation of radioiodinated lysozyme, higher molecular weight forms of lysozyme were observed.[14] Their sizes on SDS–PAGE were approximately consistent with the attachment of one to five molecules of Ub to each molecule of lysozyme. If labeled Ub was used with unlabeled lysozyme, the same labeled bands were observed (here, however, against the large background of adducts of endogenous proteins in fraction II). These results suggested that the higher molecular weight forms of lysozyme were Ub adducts containing variable numbers of Ub molecules. Based on the chemical properties of the Ub–protein bond, Hershko and co-workers proposed that Ub was linked to proteins via amide bonds, most likely involving lysine ε-amino groups of the target protein and carboxyl groups of Ub.[13]

Based on these and other results, Hershko and co-workers proposed a two-step mechanism for Ub-dependent protein breakdown (Figure 1). Step one is ATP-dependent attachment of Ub to the protein substrate. In a subsequent step, a protease recognizes the covalent Ub–protein conjugate and hydrolyzes the substrate portion of the adduct. Covalent attachment of Ub to a cellular protein is thus a signal that commits the protein to degradation. Ub acts catalytically in this process, by virtue of

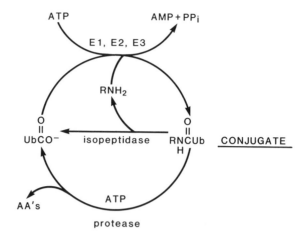

Figure 1. Schematic representation of Ub-dependent proteolytic system. The Ub C terminus is activated through formation of a thiol ester with an activating enzyme (E1) and then transferred to a thiol group on a carrier protein (E2). Transfer to substrate protein amino groups is catalyzed by a ligase (E3). The Ub–protein conjugate is degraded by an ATP-dependent protease, and intact Ub is regenerated through the action of an isopeptidase. See also Figure 3 and Table I.

its regeneration by an "isopeptidase" activity at some point in the degradative process (Figure 1). The original model underwent some revision as more information became available, but it remains the basic hypothesis for the role of Ub in protein turnover.

2.2. Conjugate Function

It is possible that conjugation of Ub to cellular proteins sometimes has nonproteolytic functions, although it is not yet clear what such functions might be. But it is clear that conjugation plays a critical role in Ub-dependent proteolysis. Several distinct lines of evidence indicate that conjugation is a necessary event in Ub-dependent breakdown and that it renders the protein susceptible to breakdown by the protease of the pathway.

2.2.1. Immunochemical Evidence

Incorporation of amino acid analogs into cellular proteins results in abnormal proteins that are very rapidly degraded.[9,15] Immunoprecipitation of conjugates from lysates of cells that have synthesized proteins in the presence of amino acid analogs shows that a large fraction of cellular conjugates are derived from these abnormal proteins.[15] This result suggests that the Ub-dependent proteolytic pathway is involved in turnover of abnormal proteins and that conjugation is necessary for such turnover.

2.2.2. Genetic Evidence

Finley et al.[16] studied the properties of a mutant mouse mammary carcinoma cell line (ts85) that exhibited cell cycle arrest at high temperature (39°C). The only apparent lesion in the cell was a temperature-sensitive mutation in Ub activating enzyme.[16] The ts85 cells were unable to conjugate Ub to cellular proteins at the nonpermissive temperature,[16] and turnover of both abnormal and short-lived proteins was profoundly inhibited at the nonpermissive temperature.[4] These results establish a clear correlation between conjugation of Ub to cellular proteins and degradation of certain proteins. They also confirm other observations[15] that suggest that degradation of abnormal proteins is Ub dependent.

2.2.3. Evolutionary Evidence

Ub is the most highly conserved protein known in eukaryotes.[17] Data obtained from ts85 cells,[4,16] reticulocytes,[15,18] ascites cells,[15] and HeLa cells[19] indicate that Ub has a proteolytic function in all these higher eukaryotic cells. In each case, data support the hypothesis that Ub conjugation to proteins must occur before proteins can be degraded by the protease of the system. Ub conjugation also occurs in yeast lysates[17] and is probably involved in yeast protein turnover.

2.2.4. Biochemical Evidence

Hershko and co-workers[20] and Hough and Rechsteiner[21] purified Ub conjugates of radiolabeled lysozyme. Both groups find that the lysozyme in lysozyme–Ub conjugates is degraded about ten times faster than is free lysozyme. Degradation of conjugates does not require the presence of free Ub[20,21] or the enzymes that catalyze conjugate formation.[20,22] Furthermore, the rate of degradation of the substrate increases as the size of the conjugate increases.[19–21] Presumably, size reflects increasing numbers of attached Ub molecules.

Taken together, these points strongly suggest that conjugation is a *necessary* event in the Ub-dependent degradation. In particular, the biochemical evidence shows that conjugation of Ub to a protein creates susceptibility to a protease. Conjugation therefore precedes peptide bond hydrolysis, and the necessity for conjugation can account for the Ub dependence of protein breakdown. But this is not to say that conjugation of Ub to a protein is *sufficient* to ensure degradation of that protein. Lysozyme conjugates can be hydrolyzed by isopeptidases, thereby regenerating intact lysozyme and intact Ub (Figure 1). Indeed, isopeptidase and protease activities appear to compete for the same conjugate pool.[20,21]

2.3. Conjugate Structure

Since conjugates are the product of the enzymatic reactions to be discussed below, some discussion of conjugate structure is warranted. At present, it is not clear that Ub transfer to a protein results in a single predominant product. Conjugates of most substrates appear to be highly and variably ubiquitinated, probably in part reflecting a processive or cooperative mechanism of Ub transfer. Product heterogeneity has hindered the structural characterization of conjugates. The following evidence, however, suggests that the Ub–protein linkages in conjugates are frequently amide (isopeptide) bonds between Ub carboxyl termini and lysine ϵ-amino groups on the target protein.

2.3.1. Chemical Stability of Ubiquitin–Protein Bonds

The Ub–protein bond is stable to strong denaturants that would dissociate a noncovalent complex.[14] It is also stable to treatments, such as alkali, mild acid, high concentrations of hydroxylamine, or mercaptoethanol, that would cleave esters or thiol esters.[13,14] These chemical properties are those expected for a peptide bond.

2.3.2. Involvement of the Ubiquitin C Terminus in the Ubiquitin–Protein Bond

Ub lacking its C-terminal dipeptide (Gly-Gly) is unable to form conjugates and does not support protein breakdown.[23] The C-terminal glycine residue of Ub is activated during the first step in conjugate formation,[24] and enzymatic digestion of [^{3}H]lysozyme conjugates (formed from Ub labeled biosynthetically with [^{3}H]glycine) yields [^{3}H]Gly-Lys.[25] Thus, the Ub C terminus contributes the carbonyl group to the isopeptide bond, and the protein must contribute the amino group.

2.3.3. Involvement of Protein Substrate Amino Groups in Ubiquitin–Protein Bonds

As shown originally by Hershko *et al.*,[14] more than one Ub molecule can be conjugated to a single protein. This implicates ϵ-amino groups of

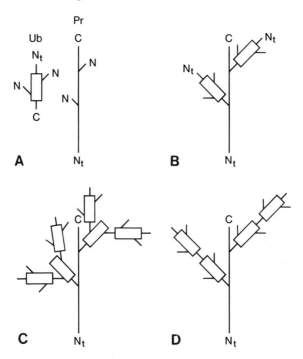

Figure 2. Possible conjugate structures. (A) Components and their functional groups: Ub, ubiquitin; Pr, protein substrate; N_t, amino terminus; C, carboxyl terminus; N, ε-amino group of an internal lysine residue. (B) Conjugate containing one molecule of Ub per protein substrate lysine residue. (C) Conjugate containing Ub molecules conjugated to Ub lysine residues. (D) Conjugate containing Ub molecules conjugated in a head-to-tail arrangement (poly-Ub conjugation).

lysine residues on the protein as other partners in the (iso)peptide bonds (Figure 2B). The issue of possible Ub conjugation to the N terminus of the protein is discussed in Section 3.4. The isolation of [^{3}H]Gly-Lys after digestion of conjugates formed from [^{3}H]Ub^{25} is direct evidence for isopeptide bonds between protein lysines and Ub C termini. Other observations consistent with this conclusion are: (1) polylysine can serve as a substrate for Ub conjugation^{14} and (2) blocking lysine amino groups by reductive methylation abolishes a protein's ability to be conjugated to Ub and to be degraded in a Ub-dependent fashion.^{3,26}

2.3.4. Known Conjugate Structures

In the naturally occurring Ub–H2A conjugate, the Ub C terminus is involved in an isopeptide bond to the ε-amino group of lysine 119 of the histone.[5] *D. discoideum* calmodulin is a substrate for Ub-dependent turnover in reticulocyte fraction II,[27] and calmodulin antibody precipitates a calmodulin–Ub conjugate in which Ub is linked to Lys-115 of calmodulin.[27] These two well-characterized Ub conjugates provide evidence that some of the Ub–protein bonds in conjugates are isopeptide bonds involving substrate protein lysine amino groups and carboxyl termini of Ub.

Certain observations, however, are inconsistent with the hypothesis that all conjugates have the structure shown in Figure 2B. In the models shown there, the molecular weights of conjugates of a given protein reflect variation in the number of lysine residues ubiquitinated. The most serious problem is that some conjugates appear to be far larger than predicted, even assuming that every amino group of the protein can be conjugated to a single Ub molecule. Lysozyme (14 kDa) has six lysine residues. According to the model of Figure 2B, the largest conjugate of lysozyme could have a molecular weight of 65 or 74 kDa, depending on whether or not the amino terminus is also conjugated. Yet a large fraction of the lysozyme conjugates produced using purified enzymes or fraction II, under conditions in which degradation of conjugates is inhibited, have molecular weights between 100 and 250 kDa.[20,21]

These large conjugates may still be accommodated by the general model of Figure 2B if conjugation of additional molecules of Ub to lysine residues of primary, substrate-conjugated Ub molecules is allowed (Figure 2C). Such secondary conjugation may occur, based on the observation that substitution of reductively methylated Ub for normal Ub abolishes the formation of very large (>50 kDa) lysozyme conjugates.[25,28] Although reductively methylated Ub cannot participate in Ub–Ub conjugation, it does support lysozyme breakdown, at a somewhat reduced rate.[25,28]

Ub–Ub conjugation involving the N terminus of a previously conjugated Ub (Figure 2D) is suggested by the structure of some Ub genes, which indicate that Ub can be synthesized as a poly-Ub protein of tandem Ub molecules linked C terminus to N terminus.[17] However, cyanogen bromide cleavage of lysozyme conjugates containing radioiodinated Ub does not yield radiolabeled fragments the size of Ub[25]; such fragments would be expected for "head-to-tail" poly-Ub conjugates, since the only methionine in Ub is at its N terminus. So there is some evidence that the model of Figure 2C (but not 2D) might account for some "excessively" large conjugates. However, problems remain. Hough and Rechsteiner[21] report double-labeling data consistent with a Ub/lysozyme ratio of about 10 in the largest (~220 kDa) lysozyme conjugates. This ratio predicts a molecular weight of about 100 kDa. There is some uncertainty in the ratio, but it seems unlikely to be off by a factor of 2. Thus, the sizes of certain conjugates may not be explained entirely by the model of Figure 2B, even as modified in Figure 2C. Possibly, these large conjugates are covalent complexes involving enzymes of the Ub pathway.[21] Present understanding of the mechanism of conjugate formation does not encompass such complexes as necessary intermediates, but that understanding is by no means complete. It will be important to understand how these very large conjugates are formed, because they are the preferred substrates for lysozyme degradation.[20,21] However, based on the results with methylated Ub, their formation does not appear to be absolutely necessary for protein breakdown.

3. ENZYMES OF UBIQUITIN ACTIVATION AND LIGATION

3.1. Introduction

We know most about the Ub activation pathway in reticulocytes, and what follows is based almost entirely on studies involving these cells. Since soluble lysates from liver,[29] yeast,[17] Friend cells,[30] and oat plants[31] are also capable of catalyzing conjugate formation, the reactions studied with reticulocyte components seem likely to apply to many tissues and organisms.

The Ub affinity methodology[32,33] developed by Hershko, Rose, and co-workers has been important in elucidation of the three-step mechanism for conjugate formation described in Section 3.2.1. The method goes beyond use of simple noncovalent affinity and takes advantage, in some cases, of an actual catalytic mechanism. For enzymes of the Ub system, purification has provided more insight into mechanism that is often the case in enzymology.

What might be a biochemically reasonable minimal mechanism for amide bond formation between the Ub C terminus and a protein amino group? Certainly the C terminus of Ub must be activated for nucleophilic attack, so there should be at least one activated form of Ub in the pathway for conjugate formation [Equation (1)]. Ub activation will probably require ATP hydrolysis.

$$\underset{\displaystyle UbCO^-}{\overset{\displaystyle O \atop \|}{}} \xrightarrow{\text{(ATP)}} \underset{\displaystyle UbCXR}{\overset{\displaystyle O \atop \|}{}} \tag{1}$$

In Equation (1), X could be oxygen or sulfur, and R could be derived from ATP or from an enzyme. The final activated Ub species could be an adenylate or acyl phosphate, or an enzyme-bound ester, thiol ester, or anhydride. In fact, there are *three* different forms of activated Ub in the conjugation pathway. Initial activation of Ub by Ub activating enzyme

Table I. Enzymes of Ub Activation and Ligation

| Enzyme | MW (kDa) | | Catalyzed reaction | Approximate concentration^a | Function |
	Subunit	Native			
Ub activating enzyme (E1)	100	200^b	E1–SH + 2ATP + 2Ub → E1–S–Ub·AMP–Ub + 2PP + AMP	0.4	Initial activation of Ub C terminus
Ub carrier protein (E2)^c	14	25	E1–S–Ub + E2–SH → E2–S–Ub + E1–SH	0.1	"Carrier" of activated Ub to protein substrate
Ub–protein ligase (E3)	200(?)	250	protein–NH$_2$ + E2–S–Ub → protein–NH–Ub + E2–SH	Unknown	Binding of specific protein substrate, and catalysis of Ub transfer from E2–S–Ub to substrate amino groups

^{a} In rabbit reticulocytes.^{41,42} Values in nmole/ml cells.
^{b} Dimer seen mainly at high concentration.^{32}
^{c} There are four additional carrier proteins with the following subunit (native) molecular weights (kDa): 16 (~30), 20 (~50), 25 (~40), 30 (~70). However, only the one shown in the table appears to function in most conjugative reactions (ref. 41 and text). The others are present at concentrations comparable to that shown in the table for the 14 kDa subunit species.

(E1) involves formation of an E1–Ub thiol ester, via a Ub adenylate intermediate. The E1–Ub thiol ester then undergoes a transthiolation reaction with a Ub carrier protein (E2), yielding an E2–Ub thiol ester. This latter adduct is the actual donor of Ub to the protein substrate. Ub transfer from this thiol ester to the protein is catalyzed by a specific ligase. These reactions are shown schematically in Figure 3; properties of the enzymes are summarized in Table I. A more thorough description of Ub activation and ligation follows.

3.2. Initial Ubiquitin Activation: Ubiquitin Activating Enzyme (E1)

3.2.1. Mechanism

Ciechanover et $al.$[34] approached the problem of Ub activation by searching for Ub-dependent isotope exchange reactions involving ATP. Using reticulocyte fraction II, they found Ub-dependent exchange of labeled PP_i into ATP, consistent with initial formation of a Ub adenylate (AMP–Ub) intermediate.[34] The exchange reaction was used as an assay to purify the Ub activating enzyme E1. The purified enzyme catalyzes two Ub-dependent isotope exchange reactions, PP_i–ATP exchange and AMP–ATP exchange in the presence of PP_i.[34,35] The proposed mechanism therefore involves two steps: (1) formation of AMP–Ub with release of PP_i and (2) transfer of Ub from AMP–Ub to an enzyme functional group with release of AMP [Equations (2a)–(2c)].[34]

$$\text{E1–SH} + \text{ATP} + \text{UbC}\overset{\text{O}}{\overset{\|}{\text{}}}\text{O}^- \overset{\text{Mg}^{2+}}{\rightleftharpoons} \text{E1–SH·AMP–Ub} + \text{PP}_i \quad (2a)$$

$$\text{E1–SH·AMP–Ub} + \text{ATP} + \text{UbC}\overset{\text{O}}{\overset{\|}{\text{}}}\text{O}^- \rightleftharpoons$$
$$\text{E1–SC}\overset{\text{O}}{\overset{\|}{\text{}}}\text{Ub·AMP–Ub} + \text{AMP} + \text{PP}_i \quad (2b)$$

$$\text{Net:} \quad \text{E1–SH} + 2\text{ATP} + 2\text{UbC}\overset{\text{O}}{\overset{\|}{\text{}}}\text{O}^- \rightleftharpoons$$
$$\text{E1–SC}\overset{\text{O}}{\overset{\|}{\text{}}}\text{Ub·AMP–Ub} + \text{AMP} + 2\text{PP}_i \quad (2c)$$

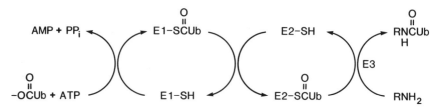

Figure 3. Overall pathway for Ub–protein conjugate formation. For simplicity, the Ub adenylate intermediate has been omitted.

A large body of data now supports this mechanism for E1. Initial AMP–Ub formation requires Mg^{2+}; attack of an enzyme thiol group on bound AMP–Ub to generate an E1–Ub thiol ester [Equation (2b)] is Mg^{2+} independent.[36] Once the thiol ester forms, there is formation of another AMP–Ub [Equation (2b)]. In addition to the observed isotope exchange reactions,[32,34–36] the following facts support this mechanism:

1. The AMP–Ub intermediate, which is formed stoichiometrically with the enzyme, can be isolated following denaturation of the ternary complex. It has the structural properties expected for AMP–Ub.[36]
2. The ternary complex, E1–SC(O)Ub·AMP–Ub, can be isolated by gel filtration. It can be formed either from ATP and Ub[37] or by adding excess AMP–Ub to E1.[36] The ternary complex contains Ub and AMP in the ratio 2:1.
3. The chemical properties of the covalent E1–Ub bond are those expected for a thiol ester, that is, acid stability, lability to alkali, thiols, and mercurials.[34]
4. Free enzyme treated with a thiol reagent, iodoacetamide, catalyzes AMP–Ub formation and PP_i–ATP exchange, but it cannot form a covalent adduct with Ub or catalyze AMP–ATP exchange.[37] These observations support the order of the two reactions presented above. Because the enzyme thiol group is protected in the E1–Ub adduct, treatment of the ternary complex with iodoacetamide has no effect on E1 activity.[33]
5. Reductive cleavage of the E1–Ub thiol ester with tritiated borohydride yields [^{3}H]Ub; acid hydrolysis of the labeled Ub yields [^{3}H]ethanolamine.[24] Therefore, a glycine residue of Ub is involved in the covalent bond to E1, and Ub must be activated at its C terminus.

Results from affinity purification of E1 are also consistent with the mechanism presented in Equations (2a)–(2c). Retention of E1 by a Ub affinity column occurs only in the presence of ATP, and E1 is covalently bound to the resin. E1 can be eluted with AMP and PP_i by reversal of the normal catalytic reaction, or by cleaving the E1–Ub thiol ester with thiols or alkali. The "covalent affinity" method can yield homogeneous E1; the reticulocyte enzyme has a 100 kDa subunit and forms a dimer at high concentrations.[32] The kinetics of the E1 reaction have been studied extensively. All substrates and products have association constants in the micromolar range.[35]

Ub activation by E1 resembles two other biological activation processes. In amino acid activation by aminoacyl tRNA synthetases, the initial step is also adenylate formation [Equation (3a)]. The tightly bound aminoacyl adenylate (AA–AMP) is then attacked by the 3'-OH group

$$E + ATP + AA \rightleftharpoons E{\cdot}AA{-}AMP + PP_i \qquad (3a)$$

$$E{\cdot}AA{-}AMP + tRNA{-}OH \rightleftharpoons E{\cdot}AA{-}OtRNA + AMP \qquad (3b)$$

of the cognate tRNA.[38] In this sequence, adenylate formation serves the general function of activating the amino acid carboxyl group to nucleophilic attack, and the aminoacyl tRNA is analogous to the E1–Ub thiol ester. Specificity in tRNA binding to the synthetase ensures that the amino acid is transferred to the correct tRNA. The E1-catalyzed reaction is even more analogous to the activation of amino acids for peptide bacterial antibiotic synthesis.[39] Here aminoacyl adenylate formation is followed by attack of an enzyme thiol group to form an enzyme–amino acid thiol ester.

3.2.2. Significance

The intermediate E1–SC(O)Ub can be synthesized by adding stoichiometric amounts of Ub to E1 in the presence of ATP and pyrophosphatase (PPase). Although E1 itself does not catalyze Ub transfer to protein amino groups, addition of this intermediate to ATP-depleted fraction II results in the formation of Ub–protein conjugates.[37] On the other hand, addition of the intermediate E1·AMP–Ub (formed from thiol-blocked E1) to ATP- and E1-depleted fraction II does *not* result in the synthesis of conjugates.[37] These results argue that the Ub thiol ester is the terminal intermediate in the E1 reaction. The role of AMP–Ub is to provide the activated Ub necessary for thiol ester synthesis.

Tight binding of AMP–Ub to E1 serves to position AMP–Ub correctly for reaction with the E1 thiol group. Ub in the thiol ester is more reactive with nonspecific nucleophiles than is Ub in enzyme-bound AMP–

Ub,[37] but it is also more accessible to appropriate acceptors. Strong interactions between the E1 portion of the adduct and the "correct" acceptor allow for a specificity in Ub transfer from E1 that would not be possible were free AMP–Ub the product of the reaction. The acceptor, a Ub carrier protein, has no inherent affinity for Ub,[33] so interactions between the E1 portion of the E1–Ub adduct and E2 must be important in mediating this specific transfer (Section 3.3).

E1 appears to be responsible for Ub activation in all cellular conjugation processes, since both cytoplasmic and nuclear conjugates disappear rapidly from ts85 cells at the nonpermissive temperature, and the only apparent defect in these cells is a temperature-sensitive E1.[16]

3.3. Ubiquitin Transthiolation: Ubiquitin Carrier Proteins (E2s)

3.3.1. Mechanism

Although the E1–Ub thiol ester is analogous to aminoacyl tRNA with respect to the initial activation reaction, E1–Ub differs from aminoacyl tRNA in that E1–Ub is not the actual acyl donor in (iso)peptide bond synthesis. Purified E1 does not catalyze Ub transfer to protein substrates. Hershko *et al.*[33] used Ub affinity chromatography to generate two additional fractions, E2 and E3, which were required (along with E1) to reconstitute conjugate formation and Ub-dependent protein breakdown.

Like E1, E2 bound covalently to the column and could be eluted with thiols. E2 was sensitive to thiol alkylating reagents, but it was protected against inactivation by the combination of E1, Ub, and ATP. Unlike E1, E2 did not catalyze any ATP- and Ub-dependent isotope exchange reactions. Most of the E2 activity had a low molecular weight ($\sim$30 kDa). Since E2 did not bind to the column in the absence of E1 and ATP, it appeared that E2 might accept activated Ub from the E1–Ub thiol ester.

SDS–PAGE under nonreducing conditions and autoradiography showed that multiple low molecular weight E2–Ub adducts formed in the presence of E1 and radioiodinated Ub. In the presence of the E3 (ligase) fraction, Ub was transferred from one or more of these adducts to protein amino groups (Figure 3, ref. 33). Therefore, proteins in the E2 fraction acted as intermediate carriers of activated Ub as shown in Equation (4) (for simplicity, the Ub adenylate has been omitted). By analogy with the role of acyl carrier protein in fatty acid biosynthesis,[40] the proteins in the E2 fraction were called Ub carrier proteins.

$$\text{E1-S}\overset{\displaystyle O}{\overset{\|}{C}}\text{Ub} + \text{E2-SH} \rightleftharpoons \text{E1-SH} + \text{E2-S}\overset{\displaystyle O}{\overset{\|}{C}}\text{Ub} \qquad (4)$$

The proteins in the E2 fraction can be resolved by molecular weight fractionation,[41] and five distinct carrier proteins have been found to participate in the reaction shown in Equation (4). They range in native molecular weight from 25 to about 70 kDa, with most being homodimers. Although they are about equally abundant and exhibit similar activities as Ub acceptors [Equation (4)], only *one* of these five proteins—the smallest—is active in conjugating Ub to most protein substrates.[41] Conjugation usually requires a ligase (E3, below) in addition to this carrier protein.[33,41] The smallest carrier protein thus interacts uniquely.and specifically with the ligase.

However, at least two of the five carrier proteins (the smallest and one other) catalyze Ub transfer to a few basic protein substrates, including histones and yeast cytochrome c.[41] This reaction is ligase independent and produces conjugates generally having one molecule of Ub per molecule of protein. In the case of yeast cytochrome c, E2-catalyzed conjugation does not appear to convert the protein to a proteolytic substrate.[42] Whether these reactions have any biological relevance, for instance, in histone conjugation, is not known. All the carrier proteins catalyze Ub transfer to a variety of small primary amines, such as lysine and polyamines. All Ub transfer involving carrier proteins requires the Ub thiol ester form as an intermediate; continuous Ub transfer occurs only in the presence of E1, ATP, and Ub [Equations (2) and (4)].

3.3.2. Significance

Why do cells contain multiple Ub carrier proteins? Why is Ub not directly transferred from E1–Ub to proteins? There are at least two possible explanations for the existence of multiple E2s. The carrier protein might contribute specificity in conjugate formation. Ub transfer to small amines is a model for the biological reaction of Ub transfer to proteins. That all the carrier proteins catalyze small-amine ubiquitination is somewhat suggestive, as is the fact that more than one E2 catalyzes Ub transfer to proteins. So far, however, only the smallest carrier protein has been shown to interact with ligase in conjugating Ub to proteolytic substrates. Although this argues against a role for the larger E2s in protein turnover, relatively few substrates have been examined in this regard, and nearly all of them are nonphysiological. It is also possible that conjugation involving certain carrier proteins could be regulatory rather than degradative in function, or that it could require other, as yet unknown, ligases.

A second possibility is that some or all of the carrier proteins are precursors to the smallest E2. This is suggested by the observation that only the *smallest* carrier protein has so far been shown to be active in protein turnover. For so many intermediates to exist simultaneously and in equal abundance would be unusual, but it cannot be excluded. In this case, it would also appear to be undesirable. The rates of reaction of E2–Ub adducts with physiological concentrations of cellular nucleophiles are such that significant transfer of Ub to adventitious nucleophiles should occur.[43] If the larger E2s are really "pre-E2s," it would be more economical to have Ub-accepting activity, which leads to adventitious adduct formation, appear simultaneously with physiologically useful Ub-donating activity. Ub-regenerating enzymes can hydrolyze adventitious adducts, but the net result is ATP hydrolysis. It seems most reasonable that all the carrier proteins have a useful function to make this cost worthwhile to the cell.

These two hypotheses are not mutually exclusive. Further structural characterization of the carrier proteins will provide a test of the second hypothesis. The first will be more difficult to address in view of the possibility that specific substrates might be present only under certain conditions or in very low abundance.

While much remains to be discovered about specificity of Ub transfer to ultimate acceptor proteins, it is clear that binding of the protein substrate to ligase (E3) is important (Section 3.4). Such binding occurs in the absence of E1 and E2, so that when E3 is involved, the smallest carrier protein apparently does not play any role in determining conjugative specificity. Whether the carrier protein may direct Ub to certain sites on the substrate is not known. In most cases, there is a lack of evidence as to whether there are specific sites of ubiquitination. The carrier protein is, however, necessary for conjugation. Despite the fact that its subunit molecular weight is only 14 kDa, the smallest carrier protein must have binding sites for both E1 and E3. Affinity for both must be high, since submicromolar concentrations of carrier protein are effective in reconstituting conjugation in the presence of similar concentrations of E1 and E3.[33,41]

Synthesis of the antibiotic gramicidin S involves initial amino acid activation by a mechanism virtually identical to that of initial Ub activation. In gramicidin synthesis, the growing peptide chain undergoes a transthiolation, being passed from a thiol group of the activating enzyme to a thiol group of a phosphopantetheine prosthetic group.[39] Fatty acid synthesis also involves the use of pantetheine to carry an elongating molecule.[40] In both cases, the occurrence of pantetheine as carrier reflects the necessity for several enzymes contained within a multienzyme complex to have sequential access to a growing polymer. Whereas at first sight the occurrence of a "second" (pantetheine) thiol ester intermediate in these two processes may seem analogous to the occurrence of E2–Ub thiol esters in Ub conjugation, there are important differences. Although conjugate formation often involves transfer of multiple Ub molecules to a single protein substrate, there is no evidence at present that transfer occurs within a stable multienzyme complex or that there is any kind of regular occurrence of ubiquitination sites on the substrate protein. In fatty acid and gramicidin biosyntheses, the pantetheine moiety functions "hold on" to the substrate; in conjugate synthesis, this function appears to reside in the ligase.

3.4. Ubiquitin–Protein Ligase (E3): Mechanism and Significance

For most protein substrates, a Ub–protein ligase (E3) is required to observe transfer of Ub from E2–Ub to protein substrates. These results are consistent with E3 catalyzing the chemical step of (iso)peptide bond synthesis [Equation (5) and Figure 3]. Evidence that the conjugate product of this reaction contains Ub–protein (iso)peptide bonds has been summarized previously (see Section 2.3).

$$
\text{E2–S}\overset{\overset{\displaystyle O}{\displaystyle \|}}{\text{C}}\text{Ub} + \text{prot–NH}_2 \underset{\text{E3}}{\overset{\text{E3}}{\rightleftharpoons}} \text{E2–SH} + \text{prot–}\underset{\displaystyle H}{\overset{\overset{\displaystyle O}{\displaystyle \|}}{\text{N}}}\text{C}\text{Ub} \tag{5}
$$

Hershko *et al.*[33] partially purified E3 using Ub affinity chromatography. In contrast to E1 and the carrier proteins, E3 binds noncovalently to the matrix, although its affinity for Ub is low. E3 has a molecular weight of ~250 kDa and appears to contain a 200 kDa subunit.[28] Of the enzymes involved in conjugate formation, it has been the most resistant to complete purification. Multiple red cell Ub–protein ligases with different but overlapping substrate specificities have been reported.[44] The relation between these enzymes and the reticulocyte ligase is not clear. Possibly the reticulocyte preparation contains several different enzymes with similar molecular weights and similar affinities for Ub.

Very little is known about the mechanism of the chemical step catalyzed by the ligase. There are no known inhibitors of the enzyme, and there is no information about potential active-site groups. In addition, the complex and heterogeneous structures of the products usually formed by the conjugation reaction have made its study difficult. But considerable progress has been made in relating specificity in the E3-catalyzed reaction to specificity in conjugation and protein turnover.

The Ub-dependent proteolytic system exhibits a strong preference for substrates with unblocked amino termini. The only known exception

is an N-acetylated calmodulin.[27] This specificity has been demonstrated
in vitro using naturally and artificially N-terminally blocked proteins.[26]
Recent kinetic studies on ligase-dependent conjugation show that this
specificity reflects tight binding of substrates with unblocked amino termini to the ligase.[28]

Hershko *et al.*[28] demonstrated that good substrates dissociate very
slowly from the ligase. This was shown by "trapping" experiments with
the following design.[45] E3 and labeled lysozyme were preincubated in the
absence of other components necessary for Ub conjugation. Subsequent
addition of E1, E2, Ub, ATP, and a large excess of *unlabeled* lysozyme
resulted in preferential ubiquitination of the *labeled* lysozyme. The preferential conjugation to preincubated lysozyme, that is, "trapping" occurs,
because dissociation of E3 and labeled lysozyme is much slower ($t_{1/2} \approx$
3 min) than Ub transfer to bound lysozyme. Much less trapping occurred
if methylated lysozyme, which has a blocked N terminus and is not a
substrate for Ub-dependent proteolysis, was substituted for normal lysozyme. These results indicate that methylated lysozyme does not bind
tightly to E3. Similar results were obtained with other appropriate substrate pairs, for instance, yeast and horse heart cytochrome c (unblocked
versus acetylated N terminus). Additional experiments showed that oxidation of substrate methionine residues, an event that might occur during
oxidative damage to proteins, increases substrate binding to E3 and increases susceptibility to Ub-dependent proteolysis by five- to tenfold.
Therefore, substrate binding to E3 can lead to conjugate formation and
thus can determine specificity in Ub-dependent protein turnover. In addition, slow dissociation of substrate from E3 provides some rationale for
the frequent occurrence of very highly ubiquitinated conjugates, since
many Ub transfers may occur per substrate binding event.

What is the mechanistic basis of the ligase's preference for substrates
having free N termini? In particular, does the N terminus become conjugated to Ub? There is conflicting evidence on this point. Guanidination
of proteins preferentially blocks lysine amino groups, leaving amino termini intact.[46] Guanidinated lysozyme is degraded in a Ub-dependent fashion at a rate that can approach that of unmodified lysozyme, but very
little conjugation of Ub to guanidinated lysozyme is observed.[26,47] Yet,
conjugation does seem to be necessary for guanidinated lysozyme breakdown because the conjugating enzymes are required.[26] These results suggest that conjugation of Ub to substrate protein lysine residues may not
be necessary for substrate turnover and that conjugation to the substrate
N terminus may be particularly critical. However, N-terminal conjugates
have not been demonstrated directly. Ub conjugation to multiple lysine

residues, which seems to occur *in vivo*,[48] may provide a kinetic advantage in degradation of the conjugate.[26,47,48]

Bachmair *et al.*[48] attempted to test the hypothesis that N-terminal conjugation is important by producing Ub–β-galactosidase fusion proteins in yeast. They found that the Ub was removed almost instantaneously, but that subsequent turnover of β-galactosidase appeared to be Ub dependent. This result does not necessarily address the original hypothesis since the experiment may have created a nonphysiological situation. Because dissociation of substrates from E3 is slow,[28] a conjugate bearing a single N-terminal Ub may never exist free in solution under normal conditions. (In solution it will be subject to the action of isopeptidases.)

However, this experiment provided important information of another sort: that the half-life of the β-galactosidase produced upon Ub removal was a function of the identity of its N-terminal amino acid.[48] Whether this observation has a mechanistic basis in the reaction catalyzed by E3 is not known. That is, does the ligase discriminate among potential protein substrates based on the identities of their N-terminal amino acids? In addition to what has been discussed here, tRNA is required for conjugation of Ub to some, but not all, substrates.[49] (See Chapter 6 for a discussion of this issue.) It seems possible that amino acid transfer via aminoacyl tRNA to a target protein, perhaps at its N terminus, may create a "better" substrate for E3 and therefore for protein breakdown.[48] These ideas have exciting implications for regulation of intracellular protein turnover.

4. UNRESOLVED QUESTIONS

The three enzymes just described catalyze the conjugation of Ub to proteins, and the conjugate products are competent substrates for an ATP-dependent protease.[20,21] All three have been partially or completely purified, and an outline of the process of conjugate formation is now clear (Figure 3). The studies that led to this outline strongly support the original hypothesis that conjugation of a cellular protein to Ub is a necessary event in the Ub-dependent degradation of that protein. Many questions remain, however, and others have been raised by recent findings. Some of these may be addressed using the purified enzymes and relatively well-characterized substrates that are now available.

4.1. Mechanistic Questions

The ligase-catalyzed reaction is very important, as shown by results relating ligase–substrate binding to specificity in conjugation and breakdown.[28] But there is still very little understanding of how Ub transfer between E2–Ub and the protein substrate actually occurs. Slow substrate dissociation does not adequately explain why multiple Ub molecules are usually transferred. Does the ligase have a single active site that binds E2–Ub and the protein? Does the protein move relative to this site, successively exposing sites for Ub transfer? Or does the protein have a single, static binding mode, and the ligase somehow mediate Ub transfer from E2–Ub at many different "active" sites? Is there any specificity in Ub transfer with regard to sites on a given substrate molecule? What role does the substrate amino terminus play? It is clearly important in substrate binding. Is it also a site for initial, critical Ub transfer?

There is also little understanding of the molecular basis of the various interactions that occur during conjugate formation. Two of the enzymes involved in conjugate formation must interact with two other macromolecules: the carrier proteins must interact successively (perhaps even simultaneously) with E1–Ub and with ligase; the ligase must interact simultaneously with E2–Ub and with the protein substrate. What sorts of amino acid contacts are important in these various interactions?

4.2. Biological Questions

What is the biological rationale for the existence of multiple forms of certain enzymes in the conjugation pathway? There are five carrier proteins, only one of which has so far been shown to function in protein turnover. What are the others doing? If there are multiple ligases, what are their roles? It is attractive to use "specificity" as a rationale for multiple enzyme forms, but more evidence is needed.

The enzymes discussed here all appear to be cytoplasmic. But in addition to cytoplasmic proteins, there are Ub conjugates in the nucleus[5] and at the cell surface.[6,7] In addition, reticulocyte mitochondrial proteins are conjugated to Ub.[50] Whereas conjugation of Ub to mitochondrial proteins appears to be involved in their degradation,[18] the functional significance of conjugation to histones and cell surface proteins is not yet clear. Cytoplasmic enzymes could conjugate cytoplasm-exposed portions of membrane proteins. It is harder, however, to see how these enzymes could access proteins within the nucleus or external domains on proteins destined for the cell surface, since the latter are usually cotranslationally inserted into the rough endoplasmic reticulum membrane. The same E1 appears to function in all conjugative processes.[16] Are there other carrier proteins and/or ligases yet to be discovered?

Finally, if some cellular conjugates are not degradative intermediates, then what is the basis for discrimination among conjugates with different fates? Will all specificity in processes involving Ub be accounted for by specificity in conjugate formation? Much more information, and possibly new approaches, will be required to answer these biological questions.

Note added in proof: Sequencing of the DNA encoding a 20 kDa Ub carrier protein from yeast has shown it to be identical to the yeast DNA repair enzyme *rad6* (Jentsch, S., McGrath, J. P., and Varshavsky, A., 1987, The yeast DNA repair gene *RAD6* encodes a ubiquitin-conjugating enzyme, *Nature* **329:** 131–134).

REFERENCES

1. Simpson, M. V., 1953, The release of labeled amino acids from the proteins of rat liver slices, *J. Biol. Chem.* **201:** 143–145.
2. Schweiger, H. G., Rapoport, S., and Scholzel, E., 1956, Role of nonprotein nitrogen in the synthesis of haemoglobin in the reticulocyte *in vitro, Nature* **178:** 141–142.
3. Tanaka, K., Waxman, L., and Goldberg, A. L., 1983, ATP serves two distinct roles in protein degradation in reticulocytes, one requiring and one independent of ubiquitin, *J. Cell. Biol.* **96:** 1580–1585.
4. Ciechanover, A., Finley, D., and Varshavsky, A., 1984, Ubiquitin dependence of selective protein degradation demonstrated in the mammalian cell cycle mutant ts85, *Cell* **37:** 57–66.
5. Busch, H., and Goldknopf, I. L., 1981, Ubiquitin–protein conjugates, *Mol. Cell. Biochem.* **840:** 173–187.
6. Siegelman, M., Bond, M. W., Gallatin, W. M., St. John, T., Smith, H. T., Fried, V. A., and Weissman, I. L., 1986, Cell surface molecule associated with lymphocyte homing is a ubiquitinated branched-chain glycoprotein, *Science* **231:** 823–829.
7. Yarden, Y., Escobedo, J. A., Kuang, W.-J., Yang-Feng, T. L., Daniel, T. O., Tremble, P. M., Chen, E. Y., Ando, M. E., Harkins, R. N., Francke, U., Fried, V. A., Ullrich, A., and Williams, L. T., 1986, Structure of the receptor for platelet-derived growth factor helps define a family of closely related growth factor receptors, *Nature* **323:** 226–232.
8. Hershko, A., Heller, H., Ganoth, D., and Ciechanover, A., 1978, in *Protein Turnover and Lysosomal Function* (H. L. Segal and D. Doyle, eds.), Academic Press, New York, pp. 149–169.
9. Etlinger, J. D., and Goldberg, A. L., 1977, A soluble ATP-dependent proteolytic system responsible for the degradation of abnormal proteins in reticulocytes, *Proc. Natl. Acad. Sci. U.S.A.* **74:** 54–58.
10. Ciechanover, A., Hod, Y., and Hershko, A., 1978, A heat-stable polypeptide component of an ATP-dependent proteolytic system from reticulocytes, *Biochem. Biophys. Res. Commun.* **81:** 1100–1105.
11. Ciechanover, A., Elias, S., Heller, H., Ferber, S., and Hershko, A., 1980, Characterization of the heat-stable polypeptide of the ATP-dependent proteolytic system from reticulocytes, *J. Biol. Chem.* **255:** 7525–7528.
12. Wilkinson, K. D., Urban, M. K., and Haas, A. L., 1980, Ubiquitin is the ATP-dependent factor I of rabbit reticulocytes, *J. Biol. Chem.* **255:** 7529–7532.
13. Ciechanover, A., Heller, H., Elias, S., Haas, A. L., and Hershko, A., 1980, ATP-

dependent conjugation of reticulocyte proteins with the polypeptide required for protein degradation, *Proc. Natl. Acad. Sci. U.S.A.* **77:** 1365–1368.

14. Hershko, A., Ciechanover, A., Heller, H., Haas, A. L., and Rose, I. A., 1980, Proposed role of ATP in protein breakdown: Conjugation of proteins with multiple chains of the polypeptide of ATP-dependent proteolysis, *Proc. Natl. Acad. Sci. U.S.A.* **77:** 1783–1786.

15. Hershko, A., Eytan, E., Ciechanover, A., and Haas, A. L., 1982, Immunochemical analysis of the turnover of ubiquitin–protein conjugates in intact cells, *J. Biol. Chem.* **257:** 13964–13970.

16. Finley, D., Ciechanover, A., and Varshavsky, A., 1984, Thermolability of ubiquitin-activating enzyme from the mammalian cell cycle mutant ts85, *Cell* **37:** 43–55.

17. Ozkaynak, E., Finley, D., and Varshavsky, A., 1984, The yeast ubiquitin gene: Head-to-tail repeats encoding a polyubiquitin precursor protein, *Nature* **312:** 663–666.

18. Rapoport, S., Dubiel, W., and Muller, M., 1985, Proteolysis of mitochondria in reticulocytes during maturation is ubiquitin-dependent and is accompanied by a high rate of ATP hydrolysis, *FEBS Lett.* **180:** 249–252.

19. Chin, D. T., Kuehl, L., and Rechsteiner, M., 1982, Conjugation of ubiquitin to denatured hemoglobin is proportional to the rate of hemoglobin degradation in HeLa cells, *Proc. Natl. Acad. Sci. U.S.A.* **79:** 5857–5861.

20. Hershko, A., Leshinsky, E., Ganoth, D., and Heller, H., 1984, ATP-dependent degradation of ubiquitin–protein conjugates, *Proc. Natl. Acad. Sci. U.S.A.* **81:** 1619–1623.

21. Hough, R., and Rechsteiner, M., 1986, Ubiquitin–lysozyme conjugates: Purification and susceptibility to proteolysis, *J. Biol. Chem.* **261:** 2391–2399.

22. Hough, R., Pratt, G., and Rechsteiner, M., 1986, Ubiquitin–lysozyme conjugates: Identification and characterization of an ATP-dependent protease from rabbit reticulocyte lysates, *J. Biol. Chem.* **261:** 2400–2408.

23. Wilkinson, K. D., and Audhya, T. K., 1981, Stimulation of ATP-dependent proteolysis requires ubiquitin with the COOH-terminal sequence Arg-Gly-Gly, *J. Biol. Chem.* **256:** 9235–9241.

24. Hershko, A., Ciechanover, A., and Rose, I. A., 1981, Identification of the active amino acid residue of the polypeptide of ATP-dependent protein breakdown, *J. Biol. Chem.* **256:** 1525–1528.

25. Hershko, A., and Heller, H., 1985, Occurrence of a polyubiquitin structure in ubiquitin–protein conjugates, *Biochem. Biophys. Res. Commun.* **128:** 1079–1086.

26. Hershko, A., Heller, H., Eytan, E., Kaklij, G., and Rose, I. A., 1984, Role of the alpha-amino group of protein in ubiquitin-mediated protein breakdown, *Proc. Natl. Acad. Sci. U.S.A.* **81:** 7021–7025.

27. Gregory, L., Marriott, D., West, C. M., and Chau, V., 1985, Specific recognition of calmodulin from *D. discoideum* by the ATP, ubiquitin-dependent degradative pathway, *J. Biol. Chem.* **260:** 5232–5235.

28. Hershko, H., Heller, H., Eytan, E., and Reiss, Y., 1986, The protein substrate binding site of the ubiquitin–protein ligase system, *J. Biol. Chem.* **261:** 11992–11999.

29. Haas, A. L., Murphy, K. E., and Bright, P. M., 1985, The inactivation of ubiquitin accounts for the inability to demonstrate ATP, ubiquitin-dependent proteolysis in liver extracts, *J. Biol. Chem.* **260:** 4694–4703.

30. Waxman, L., Fagan, J. M., Tanaka, K., and Goldberg, A. L., 1985, A soluble ATP-dependent system for protein degradation from murine erythroleukemia cells: Evidence for a protease which requires ATP hydrolysis but not ubiquitin, *J. Biol. Chem.* **261:** 11994–12000.

31. Vierstra, R. D., 1986, Demonstration of ATP-dependent, ubiquitin-conjugating activities in higher plants (abstr. 696), *Fed. Proc.* **45:** 1599.

32. Ciechanover, A., Elias, S., Heller, H., and Hershko, A., 1982, "Covalent affinity" purification of ubiquitin-activating enzyme, *J. Biol. Chem.* **257:** 2537–2542.

33. Hershko, A., Heller, H., Elias, S., and Ciechanover, A., 1983, Components of ubiquitin-protein ligase system: Resolution, affinity purification, and role in protein breakdown, *J. Biol. Chem.* **258:** 8206–8214.

34. Ciechanover, A., Heller, H., Katz-Etzion, R., and Hershko, A., 1981, Activation of the heat-stable polypeptide of the ATP-dependent proteolytic system, *Proc. Natl. Acad. Sci. U.S.A.* **78:** 761–765.

35. Haas, A. L., and Rose, I. A., 1982, The mechanism of ubiquitin activating enzyme: A kinetic and equilibrium analysis, *J. Biol. Chem.* **257:** 10329–10337.

36. Haas, A. L., Warms, J. B. V., and Rose, I. A., 1983, Ubiquitin adenylate: Structure and role in ubiquitin activation, *Biochemistry* **22:** 4388–4394.

37. Haas, A. L., Warms, J. V. B., Hershko, A., and Rose, I. A., 1982, Ubiquitin-activating enzyme: Mechanism and role in protein–ubiquitin conjugation, *J. Biol. Chem.* **257:** 2543–2548.

38. Jakubowski, H., and Fersht, A. R., 1981, Alternative pathways for editing non-cognate amino acids by aminoacyl–tRNA synthetases, *Nucleic Acids Res.* **9:** 3105–3117.

39. Lipmann, F., 1971, Attempts to map a process evolution of peptide biosynthesis, *Science* **173:** 875–884.

40. Alberts, A. W., Goldman, P., and Vagelos, P. R., 1963, The condensation reaction of fatty acid synthesis, *J. Biol. Chem.* **238:** 557–565.

41. Pickart, C. M., and Rose, I. A., 1985, Functional heterogeneity of ubiquitin carrier proteins, *J. Biol. Chem.* **260:** 1573–1581.

42. Vella, A. T., and Pickart, C. M., unpublished experiments.

43. Pickart, C. M., and Rose, I. A., 1985, Ubiquitin carboxyl-terminal hydrolase acts on ubiquitin carboxyl-terminal amides, *J. Biol. Chem.* **260:** 7903–7910.

44. Lee, P. L., and Midelfort, C. F., Murakami, K., and Hatcher, V. B., 1986, Multiple forms of ubiquitin–protein ligase: Binding of activated ubiquitin to protein substrates, *Biochemistry* **25:** 3134–3138.

45. Rose, I. A., O'Connell, E. L., Litwin, S., and Bar-Tana, J., 1974, Determination of the rate of hexokinase–glucose dissociation by the isotope-trapping method, *J. Biol. Chem.* **249:** 5163–5168.

46. Kimmel, J. R., 1967, Guanidination of proteins, in *Methods in Enzymology*, Vol. 11 (C. H. W. Hirs, ed.), Wiley, New York, pp. 584–589.

47. Chin, D. T., Carlson, N., Kuehl, L., and Rechsteiner, M., 1986, The degradation of guanidinated lysozyme in reticulocyte lysate, *J. Biol. Chem.* **261:** 3883–3890.

48. Bachmair, A., Finley, D., and Varshavsky, A., 1986, In vivo half-life of a protein is a function of its amino-terminal residue, *Science* **234:** 179–186.

49. Ferber, S., and Ciechanover, A., 1986, Transfer RNA is required for conjugation of ubiquitin to selective substrates of the ubiquitin- and ATP-dependent proteolytic system, *J. Biol. Chem.* **261:** 3128–3134.

50. Haas, A. L., and Bright, P. M., 1985, The immunochemical detection and quantitation of ubiquitin–protein conjugates, *J. Biol. Chem.* **260:** 12464–12473.

Chapter 4

Ubiquitin/ATP-Dependent Protease

Ronald F. Hough, Gregory W. Pratt, and Martin Rechsteiner

1. INTRODUCTION

In 1980, Hershko et al.[1] proposed that attachment of ubiquitin (Ub) to substrate proteins marks those proteins for destruction. The marking hypothesis, which is shown schematically in Figure 1, has received substantial support during the past seven years. Five studies have demonstrated a good correlation between ubiquitination of a protein and its rapid proteolysis. First, when hemoglobin is injected into cultured mammalian cells, which are then treated with phenylhydrazine, globin is rapidly degraded. The concentration of globin–ubiquitin conjugates that form upon denaturation of hemoglobin is proportional to the rate of globin degradation.[2] Second, proteins that incorporate amino acid analogs are generally degraded rapidly, and the concentration of Ub conjugates increases in Ehrlich ascites cells exposed to analogs.[3] Third, Dictyostelium calmodulin is ubiquitinated at lysine 115 and subsequently degraded after being added to reticulocyte lysate; bovine calmodulin, which contains a methylated lysine 115, is not conjugated to Ub and is more stable.[4] Fourth, phytochrome, a cytoplasmic light receptor in plants, exists in two inter-

RONALD F. HOUGH, GREGORY W. PRATT, and MARTIN RECHSTEINER • Department of Biochemistry, School of Medicine, University of Utah, Salt Lake City, Utah 84132.

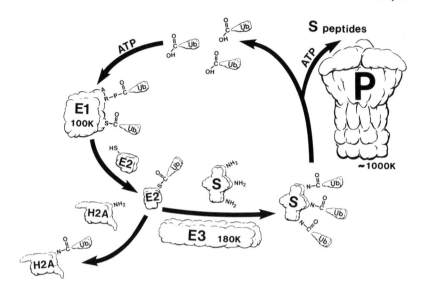

Figure 1. Schematic representation of Ub activation and ATP-dependent proteolysis of conjugated substrates. According to our present understanding, the carboxyl terminus of Ub is activated by E1 and transferred to one of several small carrier proteins (E2s) in the form of a reactive thiol ester. Ub can then be transferred directly to lysine amino groups or histones (H2A) or in the presence of a third factor (E3) Ub can be transferred to lysine amino groups on the proteolytic substrate (S). The conjugated substrate is subsequently hydrolyzed by a large ATP-dependent protease (P), and Ub is recycled.

convertible forms that differ in half-life by more than 100-fold. When dark-grown oat seedlings receive a light-flash, rapid degradation of phytochrome follows. At the same time, a portion of the phytochrome molecules become multiply ubiquitinated.[5] Fifth, it has been found that upon expression of ubiquitin–β-galactosidase fusion proteins in yeast, ubiquitin is rapidly removed from the N terminus of β-galactosidase.[6] When site-directed mutagenesis is used to produce enzymes with different residues at the N terminus, the resulting proteins vary considerably in stability. Those with Met, Ser, Ala, Thr, Val, or Gly are stable; those with N-terminal Phe, Leu, Asp, Lys, or Arg were degraded with half-lives less than 3 min. Western blots show that the latter set of proteins are ubiquitinated, some with as many as 15 attached ubiquitins.[6]

Supporting evidence of a different kind has been obtained in studies on a mutant mouse lymphoma line, ts85. Originally isolated as temperature sensitive in G_2 traverse, the ts lesion was subsequently shown to be a thermolabile E1, the Ub activating enzyme.[7] Incubation of mutant cells or extracts at 39°C, the nonpermissive temperature, led to inactivation of ubiquitination reactions *in vitro*. The intracellular turnover of abnormal or short-lived proteins was also inhibited at nonpermissive temperatures.[8] Although often cited as evidence for the marking hypothesis, these studies do not address the model per se. Strictly speaking, they only show that Ub conjugation is necessary for intracellular proteolysis. The ubiquitinated target could be a protease, kinase, inhibitor, or activator. In fact, ts85 cells synthesize heat-shock proteins at 39°C, and two studies report that decreased proteolysis is a characteristic of the heat-shock response.[9,10] The inability of ts85 cells to degrade short-lived proteins at 39°C could therefore reflect activation of the heat-shock response rather than the inability to conjugate Ub to protein substrates.

Recognition of ubiquitinated substrates by a specific protease(s) is implicit in the marking hypothesis. Recently, this central feature of the model has been confirmed. Haas and Rose[11] showed that addition of hemin to reticulocyte lysates inhibits proteolysis, but not ubiquitin conjugation. We exploited this observation to generate Ub–lysozyme conjugates in hemin-inhibited lysates.[12] Upon return to fresh lysate, the partially purified conjugates were degraded ten times faster than free lysozyme. More important, these unique substrates allowed us to identify a large, Ub/ATP-dependent protease that degrades Ub–lysozyme conjugates but not unmodified lysozyme.[13] In this chapter, we describe the properties of the Ub/ATP-dependent protease and a somewhat smaller but similar enzyme not stimulated by ATP or ubiquitinated substrates. We begin, however, with a brief review of some unusual properties of Ub–lysozyme conjugates and some observations not fully explained by the marking hypothesis.

2. UBIQUITIN–LYSOZYME CONJUGATES

2.1. Unexpected Properties

Lysozyme, RNase, and bovine serum albumin (BSA) are among the best substrates for Ub-mediated proteolysis in reticulocyte lysates. BSA, which has a molecular weight of 68,000 and an isoelectric point of about 5, would seem to have little in common with the two smaller, positively charged enzymes. All three proteins do, however, contain disulfide bonds and unblocked α-amino termini, and the latter are thought to be an important signal for E3-mediated ubiquitination (see Chapters 3, 11, and 12). BSA is an excellent substrate, but conjugates between the unmodified serum protein and Ub have never been convincingly demonstrated. Thus, most investigators have chosen lysozyme for preparation of Ub substrates.

Although Ub forms thiol ester linkages to cysteine residues during activation, in conjugates it has only been attached to lysine ε-amino groups. Lysozyme contains six lysine residues, so according to the simplest form of the marking hypothesis, Ub–lysozyme conjugates should not have molecular weights greater than 65,000 (8500 from each ubiquitin and 14,000 from lysozyme). Yet, when Hershko et al.[14] produced Ub–lysozyme conjugates in recombined fractions of E1, E2, and E3, almost 20% of the conjugates migrated on SDS–PAGE with apparent molecular weights greater than 70,000. Hough and Rechsteiner[12] trapped Ub–lysozyme conjugates in hemin-inhibited whole reticulocyte lysate, and using sodium dodecyl sulfate–polyacrylamide gel electrophoresis (SDS–PAGE), gel filtration, and sedimentation, they demonstrated that such conjugates have molecular weights greater than 200,000.

The basis for the anomalous size of Ub–lysozyme conjugates is not known. Rechsteiner et al.[15] suggested that Ub might conjugate to itself, thereby producing polyubiquitin chains. This explanation is favored by Hershko and Heller,[16] who showed that large conjugates were not generated from Ub molecules whose lysine residues were converted to homoarginines by guanidination. However, disassembly of high molecular weight conjugates in ATP-depleted lysates produces intermediates varying in molecular mass by 34, 22, and 12 kDa, values unexpected from a simple hyperubiquitination model.[12] Moreover, the ratio of [^{125}I]ubiquitin to [^{131}I]lysozyme in double-labeled conjugates is not consistent with hyperubiquitination. It is possible that transglutaminase cross-links ubiquitinated substrates to other proteins. Alternatively, Ub may introduce reactive thiol esters into other proteins[17] and these proteins could, like α-2-macroglobulin or complement components C-3 and C-4, form cross-links with ubiquitinated substrates. As another possibility, hemin may introduce or trap covalent linkages between E3 and lysozyme. Recent studies[18] show E3 to be a 180 kDa protein so a 1:1 covalent adduct of E3 and lysozyme plus several Ub molecules would approximate the highest molecular weight observed for Ub–lysozyme conjugates. Whatever the eventual explanation may be for the unexpected molecular weights of Ub–lysozyme conjugates, their large size is not consistent with a simple model for Ub marking.

Several groups have reported that substrates with blocked amino groups are degraded in an ATP-dependent process.[2,19–22] By itself, this does not affect the model; Ub-mediated proteolysis need not be the only pathway for degrading cytosolic proteins. In fact, the turnover of ornithine decarboxylase[23] or HMG CoA reductase[24] is unimpaired in ts85 cells at the nonpermissive temperature. If intracellular Ub conjugation is indeed abolished in ts85 cells at 39°C, there must be alternate proteolytic pathways.

One result difficult to reconcile with the substrate marking model was obtained by Chin et al.[21] They found that normal lysozyme and guanidinated (G) lysozyme were degraded at similar rates in lysate. Whereas Ub conjugates of normal lysozyme were readily apparent, covalent adducts between G-lysozyme and Ub could not be detected even when care was taken to preserve labile bonds, such as thiol esters. Nevertheless, degradation of G-lysozyme was Ub-dependent. This unexpected result might be explained by assuming that E3 can present substrates lacking ε-NH_2 groups, such as guanidinated lysozyme, to the ATP-dependent protease. The requirement for Ub could then be explained by assuming that its conjugation to the α-amino terminus of G-lysozyme is required to initiate E3 binding.[22] Or one could assume other functions for Ub in the pathway. Several possibilities come to mind. Ub may play an essential role in the proteolytic step. Alternatively, Ub may be necessary to regenerate active E3 molecules. If E3 occasionally becomes "entangled" with a substrate protein, conjugation of Ub to the substrate could promote dissociation. In this view, ubiquitination would not be mandatory for proteolysis, rather it would serve as an "escape" reaction.

Yet another possibility has been proposed by Speiser and Etlinger.[25] They treated reticulocyte lysate with ammonium sulfate and obtained a fraction after DEAE chromatography that rapidly degraded casein in the absence of ATP. Adding back an inhibitor fraction obtained by ammonium sulfate precipitation resulted in a marked decrease in casein degradation, both in the presence and absence of ATP. However, when the Ub fraction was added to the inhibited protease and ATP, casein was degraded in an ATP-dependent reaction. These investigators proposed that the Ub fraction represses an endogenous protease inhibitor. Recently, Murakami and Etlinger[26] reported purification of the inhibitor, which consists of six subunits of M_r 40,000 each. Interestingly, the hexamer inhibited two proteases, a high-molecular-weight reticulocyte enzyme and calpain, both noncompetitively.

2.2. Purification and Substrate Characteristics

Lysozymes' small size and positive charge facilitate separation of Ub–lysozyme conjugates from the unmodified starting protein. Since each added Ub produces a significant increase in molecular weight and removes a positive charge, one expects larger, less basic derivatives of radiolabeled lysozyme. Hershko et al.[14] exploited the size increase and obtained conjugates by chromatography on Sephadex G150 in 5% formic acid. Hough and Rechsteiner[12] based their purification on both size and charge. Radioiodinated lysozyme was incubated for 2 hr in whole reticulocyte lysate containing 100 µM hemin, and the reaction mixture was then chromatographed on carboxymethyl cellulose to remove lysozyme and on a Fractogel molecular sieve column to obtain high-molecular-weight species. Recently, Waxman et al.[27] used a crude conjugating fraction to produce Ub–lysozyme conjugates in individual column fractions.

As noted earlier, the size of the resulting Ub–lysozyme conjugates on SDS–PAGE varies considerably depending on the method of preparation. Conjugates generated from reconstituted E1, E2, and E3 range from 30,000 to more than 150,000 on SDS–PAGE[14]; those prepared from a crude conjugating fraction from reticulocytes are between 20,000 and 40,000,[27] whereas those purified from whole reticulocyte lysate are almost all greater than 100,000.[12] Large conjugates are also obtained in lysates of rabbit liver and rat skeletal muscle.[28] Although it is clear that smaller conjugates can be substrates for the Ub/ATP-dependent protease,[14,27] only the larger conjugates (>100,000) have been shown to be rapidly degraded.[12] Hence, bigger conjugates may be preferential substrates for the ATP-dependent protease.

Figure 1 presents a hypothetical substrate with three attached Ub molecules. However, it should be clear by now that the structure shown there is an oversimplification. Conjugates produced in whole lysate are much larger. Although we do not understand their large size, hemin-trapped conjugates have contributed strong support for the marking hypothesis. Upon return to fresh reticulocyte lysate, Ub–lysozyme conjugates are degraded ten times faster than lysozyme (see Figure 2 taken from ref. 12). The observation that conjugates are *preferred* substrates would appear to confirm the central feature of the model. That is, conjugation of Ub to a protein enhances its susceptibility to proteolysis. However, other cellular proteins may be linked to the substrate as well, so one cannot unequivocally attribute the *enhanced* rates of proteolysis to Ub. The same studies[12] confirm another step in the proposed reaction sequence. Whereas double-label conjugates formed from $[^{125}I]Ub$ and $[^{131}I]$lysozyme gave rise to $[^{131}I]$peptides, all ^{125}I emerged from the conjugates as intact Ub. The direct demonstration that Ub recycles intact is an important point since it eliminates the possibility that many of the radiolabeled species observed when $[^{125}I]Ub$ is added to lysate actually represent conjugates in which Ub is a substrate.

The hydrolysis curve in Figure 2 and similar results from Hershko et al.[14] demonstrate that only a portion of the conjugated lysozyme is converted to soluble products. SDS–PAGE analyses of samples taken at various times during the reaction revealed that intact lysozyme is also

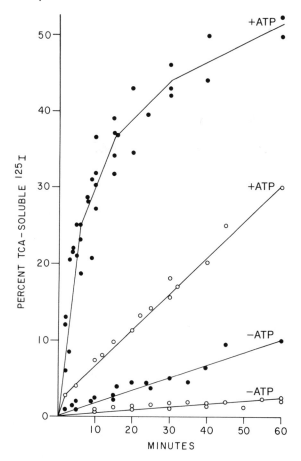

Figure 2. Degradation of Ub–lysozyme conjugates and free lysozyme in reticulocyte lysates. Lysozyme conjugates (●) and free lysozyme (○) were incubated in reticulocyte lysates, and their degradation was measured by acid precipitation. The curves denoted by $+$ATP show rates of proteolysis in the presence of ATP, whereas the curves denoted $-$ATP show ATP-independent proteolysis. Each curve is a composite from four separate experiments.

regenerated from the conjugates. Apparently, this results from the action of enzymes, presumably isopeptidases (see Chapter 5), that disassemble the conjugates before proteolysis takes place. Both proteolysis and disassembly are observed in the presence of ATP; only disassembly occurs in the absence of nucleoside triphosphates (see Figure 3). Although disassembly reactions might have complicated the use of lysozyme–Ub conjugates as proteolytic substrates, they did not, in practice, prevent the identification of Ub-dependent protease in reticulocyte lysate.

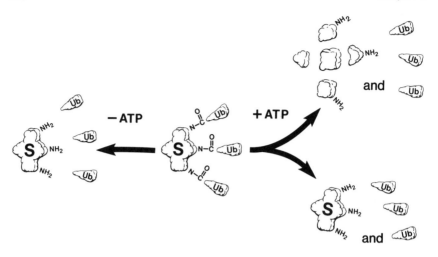

Figure 3. Schematic representation of the fate of ubiquitinated substrates in the presence or absence of ATP.

3. IDENTIFICATION OF AN ATP-DEPENDENT PROTEASE FROM RETICULOCYTES

In 1953, Simpson showed that protein breakdown in rat liver slices required ATP.[29] When almost 30 years later it was found that Ub activation consumed ATP,[30] the energy requirement for intracellular proteolysis appeared to have been explained. However, in 1984, Hough and Rechsteiner[31] and Hershko *et al.*[14] showed that degradation of Ub–lysozyme conjugates also required nucleoside triphosphates. In addition, studies using proteins with blocked amino groups as proteolytic substrates provided evidence for an ATP requirement other than Ub activation.[19] This "second" ATP-utilizing reaction was likely to be an energy-dependent protease.

A Ub/ATP-dependent protease was identified by Hough *et al.*[13] using Ub–lysozyme conjugates as substrate. They adsorbed glycerol-stabilized rabbit reticulocyte lysate onto DEAE-Fractogel and developed the column with a linear KC1 gradient; two peaks of conjugate-degrading activity eluted between 0.18 and 0.28 M KC1. Cleavage of conjugated lysozyme molecules occurred only in the presence of ATP, and free lysozyme molecules were not degraded under any condition tested. Centrifugation of each peak of activity on glycerol gradients revealed that both proteases sedimented with an apparent S value of 26. They concluded that they had identified isoforms of a very large ATP-dependent proteolytic complex that degrades ubiquitinated substrates—that is, the protease implicit in the Ub-marking hypothesis.

Additional studies using the partially purified enzymes strengthened this conclusion. Proteolysis of Ub–lysozyme conjugates was maximal at pH 7.8, was inhibited by hemin, N-ethylmaleimide, or aurintricarboxylic acid, and proceeded with an apparent Arrhenius activation energy of 27 ± 5 kcal per molecule. These properties are similar to those observed in whole lysate, thereby supporting the idea that the 26 S protease catalyzes the "second" ATP hydrolysis reaction in the Ub-dependent proteolytic pathway.

In their initial characterization of the Ub/ATP-dependent protease, Hough et al.[13] used Ub–lysozyme conjugates as the only substrate. Since the conjugate fraction contained a number of proteins, they could not eliminate the possibility that the protease, in fact, activated a latent protease present in the conjugate fraction. For this reason, they surveyed a spectrum of fluorogenic peptides as potential substrates and discovered that peptides containing phenylalanine, arginine, or tyrosine adjacent to the leaving group were hydrolyzed by the Ub/ATP-dependent enzyme; Suc-Leu-Leu-Val-Tyr-MCA (MCA = 4-methylcoumaryl-7-amide) is a particularly good substrate. Using both conjugates and fluorogenic peptides as substrates, they recently purified two large proteases from reticulocyte lysate.[32] One is a 26 S, Ub/ATP-dependent protease identified earlier; the other is a 20 S multisubunit protease previously identified by a number of investigators.

4. PURIFICATION OF THE 26 S ATP-DEPENDENT PROTEASE AND A 20 S MULTISUBUNIT PROTEASE FROM RETICULOCYTES

The two proteases were purified from rabbit reticulocytes in five steps that specifically avoided ammonium sulfate precipitation and maintained 20% glycerol (v/v) in all buffers. DEAE chromatography at pH 7.0, gel filtration, DEAE chromatography at pH 8.0, chromatography on hydroxylapatite, and sedimentation on glycerol gradients resulted in a 400-fold purification of the two high molecular weight proteases.[32] Purity of the enzymes at the last step (glycerol gradients) was assessed by nondenaturing PAGE, agarose gel electrophoresis, and isoelectric focusing. Nondenaturing PAGE revealed two major protein complexes in the 26 S region on glycerol gradients; a single major band was present in the 20 S region. Both species of the larger enzyme and the smaller enzyme hydrolyze Suc-Leu-Leu-Val-Tyr-MCA as measured by direct overlay of nondenaturing PAGE gels with substrate (see Figure 4). Densitometry

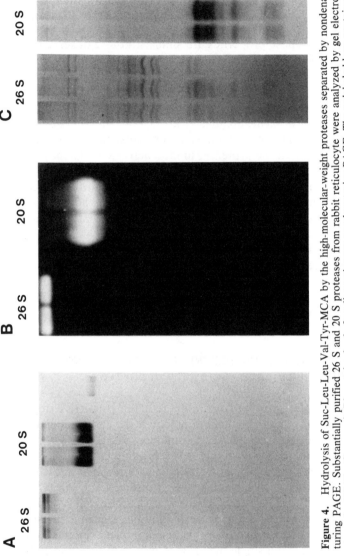

Figure 4. Hydrolysis of Suc-Leu-Leu-Val-Tyr-MCA by the high-molecular-weight proteases separated by nondenaturing PAGE. Substantially purified 26 S and 20 S proteases from rabbit reticulocyte were analyzed by gel electrophoresis. (A) The Coomassie Blue-stained gel after discontinuous nondenaturing PAGE. The unlabeled lane contained thyroglobulin (670,000). (B) The same gel overlaid with a solution of 0.1 mM Suc-Leu-Leu-Val-Tyr-MCA containing 2 mM ATP and incubated at 37°C. The hydrolysis product (free MCA) was visualized by UV irradiation of the gel. Although not clearly shown in the above photograph, the 26 S lanes contained two active bands that corresponded to the major protein bands in (A), and the 20 S lanes contained a single band of activity. The active regions were cut from the gel, incubated in SDS sample buffer, and subjected to SDS–PAGE. (C) The silver-stained profiles of these regions after electrophoresis in 10% polyacrylamide. The three lanes shown for the 26 S protease are from three equal 2 mm slices that correspond, left to right, to the top band and two sections of the lower band. The two lanes of the 20 S protease are two equal 4 mm slices of the active region from top to bottom, shown left to right. Reprinted from ref. 32, with permission.

scans of other nondenaturing PAGE gels showed that greater than 90% of the protein migrated with the 26 S species. Similarly, about 80% of the protein in the 20 S fractions migrated with peptide hydrolyzing activity. Only the 26 S component shows a preference for Ub–lysozyme conjugates as substrate. However, the two proteolytic complexes may share common subunits. For this reason, we compare the two proteases once the general properties of high molecular weight proteases have been discussed.

5. HIGH-MOLECULAR-WEIGHT CYTOSOLIC PROTEASES

High-molecular-weight cytosolic proteinases have been isolated from yeast,[33–35] fish skeletal muscle,[36] mammalian reticulocytes,[13,27, 32,37,38] erythrocytes,[26,39,40] pituitary,[41–43] lens,[44–46] skeletal muscle,[47–53] liver,[54–62] placenta,[63,64] heart,[65] lung,[66] and uterus.[67] The reported molecular weights for these enzymes range from >400,000 to 1,500,000. Although sizing errors may account for some of the variability in the literature (see refs. 58 and 60 for discussion), it is now evident that there are at least three distinct high-molecular-weight proteases in the cytosol of eukaryotic cells. Still, many of the activities described so far are probably identical or closely related to the *multicatalytic protease complex*. This term, coined by Wilk and Orlowski,[41] describes a high M_r neutral endopeptidase purified from bovine pituitary. Other proteases have been considered equivalent to this protease on the basis of their size, stability, preferred substrates, and similar subunit structure (i.e., bands between 18,000 and 36,000 on SDS–PAGE). For simplicity, we refer to these enzymes as 20 S proteases and later as macropain (see Section 5.2).

The rabbit reticulocyte 26 S, Ub/ATP-dependent protease is distinct from the 20 S protease. Waxman *et al.*[27] and Fagan *et al.*[28] also demonstrated the existence of these two, or similar, enzyme activities in reticulocytes and extracts from rabbit liver and muscle, respectively. Thus, the 26 S protease represents a unique enzyme that degrades ubiquitinated protein substrates in an ATP-dependent reaction. A third, arbitrary class of high M_r proteases includes enzymes stimulated or stabilized by ATP. It must be emphasized, however, that all the proteases are remarkably similar, and they may turn out to be different forms of a core proteolytic complex.

5.1. Subunit Structure of the 26 S Protease from Reticulocytes

SDS–PAGE analyses of individual fractions obtained at later stages in purification revealed a characteristic set of polypeptides whenever

ATP-dependent proteolysis was present.[32] The M_r values for the subunits of the 26 S protease ranged from 34,000 to 110,000 (see Figure 4C). Since nondenaturing PAGE indicated that the 26 S protease was sufficiently pure to gain useful information on the relative amounts of individual subunits, scanning densitometry was performed on SDS–PAGE samples. This analysis showed that the 26 S protease contains two large subunits with apparent molecular weights of 110,000 and 100,000, six or more subunits between 46,000 and 62,000, and may contain several subunits between 21,000 and 34,000 (see section 5.6). Of course, we presume that individual bands on SDS–PAGE do not arise by modification of prevalent subunits or by proteolysis during purification. Support for the latter assumption is provided by constant ratios of the various subunits from one preparation to another and by the apparent absence of autodigestion after purification. If one assumes single copies of the 110,000 and 100,000 subunits in each 26 S holoenzyme, an apparent molecular weight of 1,000,000 can be calculated from densitometric scans.[32] This value is in reasonable agreement with size estimates by gel filtration, sedimentation, and electrophoresis on nondenaturing gels.

5.2. General Features of the 20 S Protease

5.2.1. Subunit Composition

The 20 S protease was first identified in extracts from fish muscle,[68] and by 1980, Hase *et al.*[36] had substantially characterized the enzyme. The carp protease has a molecular weight of 477,000–680,000, depending on the method used to estimate its size and the number of subunits retained during purification. It contains at least four distinct subunits in the range 18,000–36,000, as shown by SDS–PAGE. The enzyme degrades casein with a pH optimum of 8.0 and a temperature optimum of 65°C, is stimulated by 2-mercaptoethanol and EDTA, and is reversibly inhibited by KCl, LiBr, and calcium acetate. The pI of the fish enzyme is 5.2, whereas the dissociated core catalytic complex is slightly more acidic (pI = 5.0). Dissociation of the complex into four major subunits (α = 36,000, β = 32,000, γ = 24,000, and δ = 18,000) was achieved by treatment with 2-mercaptoethanol followed by hydroxylapatite chromatography to remove the α subunit, acidification to pH 4.5 in the presence of 0.4 M LiCl followed by centrifugation to remove the B subunit, and treatment of the mercaptoethanol-purified complex with 4.5 M urea followed by gel filtration to separate the γ and δ subunits. Hase *et al.*[36] concluded from these studies that the native structure of the protease is $[\alpha\beta\gamma_2\delta_2]_4$, but that the amount of the α subunit can vary such that structures with a molecular weight of 680,000 containing 26 subunits arranged $\alpha_6[\beta\gamma_2\delta_2]_4$ are also found. After dissociation of the α and β subunits, the core particle $[\gamma_2\delta_2]$ is thought to form a complex (M_r 650,000) composed of 32 subunits arranged $[\gamma\delta]_{16}$. Hase et al.[36] assigned the following tentative functions to the subunits: α is important for subunit assembly, but it contains no catalytic activity; β is an inhibitor in the quaternary structure of the enzyme; γ coordinates with the δ subunit and is necessary for δ to retain an active conformation; δ contains the catalytic site.

Not all investigators report four subunits; many obtain eight to ten bands by SDS–PAGE. Variability in the size and subunit structure reported for the 20 S complex is conceivably due to structural differences in the enzyme from various tissues. Alternatively, differences in purification protocols, methods of analysis, or assay procedures may be primary causes of the discrepancy. Ray and Harris[45] have shown that bovine lens neutral endopeptidase is very similar to the bovine pituitary enzyme. Both contain eight or more subunits in the range 24,000–32,000. It is noteworthy that early reports on the pituitary enzyme demonstrated only three to five bands by SDS–PAGE.[41,42] The difference is probably attributable to the electrophoretic procedures employed. Generally, the SDS–PAGE bands are closely spaced in four or five regions, but they can be separated on low porosity or gradient gels into eight to ten bands, often as a series of doublets. Similar subunit compositions have been demonstrated in various tissues. Hough et al.[32] found eight to ten bands, M_r 21,000–32,000 from the 20 S protease of rabbit reticulocytes. This subunit composition appears identical to that obtained by McGuire and DeMartino[40] for the 20 S enzyme from human erythrocytes. Hence, the lower two bands 21,000–22,000 may be unique to red cells, although Zolfaghari et al.[66] have reported their presence in the human lung 20 S protease.

Purified enzyme preparations may contain higher-molecular-weight bands with apparent M_r of 65,000–70,000,[34,63,66] 105,000,[63] and 120,000.[61] It is not clear whether these higher M_r bands are contaminants or whether their association with the complex is significant. In summary, although certain differences can be expected depending on the source of the enzyme, subunit compositions of the 20 S particles are remarkably similar regardless of the source. Most of the apparent differences may be due to alternate methods of analysis.

5.2.2. General Structure

Electron microscopic studies by Hase et al.[36] revealed that the 20 S carp protease is a 10 × 15 nm tube-like object composed of four rings;

the inside diameter was estimated to be 2.5 nm. The LiCl-treated complex also appeared tube-like with four rings, similar to the native enzyme, except that the rings appeared to be smoother, suggesting a structure composed of four octagonal layers $[\gamma_4\delta_4]_4$. Recent studies on the rat skeletal muscle protease confirm the cylindrical structure consisting of four rings.[69] Data for the rat muscle enzyme reveal a 650,000 Da particle, 9.6–11 nm in diameter, with a length of 14.3–16 nm and an internal diameter of 3.9–4.3 nm. The rat skeletal muscle enzyme has a pI of[48] 5.1–5.2 and was first identified as high molecular mass[47] cysteine proteinase II/III based on comigrating activities for different synthetic peptide substrates. The liver protease, studied by Tanaka et $al.,$[60] was reported to have a molecular weight of 720,000–760,000, sedimentation coefficient of 19.8 S, pI of 5.0, and an extinction coefficient of 9.6, $E_{280\,nm}$ for a 1% solution. Tanaka et $al.$[60] concluded that the liver enzyme does not contain nucleic acid because of a negative orcinol reaction and the lack of a discernable A_{260} peak or shoulder.

5.2.3. Substrate Specificity

Hase et $al.$[36] used casein as the substrate for assaying the carp 20 S protease, and Wilk and Orlowski[41–43] assayed the pituitary enzyme using bradykinin, angiotensin II, LHRH, neurotensin, substance P, and the synthetic peptides Bz-Phe-Val-Arg-pNA, Cbz-Gly-Gly-Leu-pNA, Cbz-Gly-Gly-Tyr-Leu-pNA, Cbz-Gly-Gly-Tyr-Ala-pNA, and Cbz-Leu-Leu-Glu-2NA. Analysis of the primary and secondary cleavage sites on the synthetic and naturally occurring peptides demonstrated preferential cleavage at the carboxyl side of hydrophobic residues. These data foreshadowed subsequent findings that N-blocked peptides containing tyrosine or phenylalanine at the P_1 site are rapidly hydrolyzed by the protease.[32,34,38–40,48–51,59]

Ray and Harris[46] demonstrated that the immunologically indistinguishable lens and pituitary 20 S enzymes also degrade α_2-crystallin with a temperature optimum of 60°C. Although subtle differences were observed in regard to the effects of SDS on hydrolysis of Z-Leu-Leu-Glu-2NA and Z-Gly-Gly-Leu-pNA at various temperatures, the enzymes appear to be the same. Furthermore, they reported that the lens protease is identical to the Mg^{2+}- or Ca^{2+}-dependent lens neutral endopeptidase identified by Blow et $al.,$[44] but upon complete purification, the Ca^{2+} or Mg^{2+} dependence was lost.[45] While the significance of this finding is presently unclear, it may prove important since considerable variation has been reported regarding the effects of monovalent and divalent cations on enzyme activity. In fact, it is likely that the protease is regulated by a number of compounds, and depending on the purification protocol, different results may be obtained.

Rivett identified a third protein substrate when she showed that rat liver 20 S enzyme preferentially degrades, oxidized *E. coli* glutamine synthetase.[57,58] A detailed study of the degradation products arising from the insulin B chain revealed major cleavage sites at the carboxyl group of Gln_{-4}, Glu_{-13}, Leu_{-15}, Leu_{-17}, and Cys_{-19}. Thus, the purified 20 S enzyme has been shown to degrade three protein substrates (casein, crystallin, and oxidized glutamine synthetase), a variety of natural peptides, including the B chain of insulin and several neuropeptides, and a number of synthetic peptides.

5.2.4. Inhibitors

Based on their inhibitor sensitivity,[70] endopeptidases are generally assigned to one of four catalytic classes: serine, cysteine, acid, or metalloproteases. These categories are not absolute, and final judgment on the reaction mechanism should be withheld until appropriate kinetic and sequencing studies have been performed with highly purified enzymes. Still, when describing a new protease most investigators provide inhibitor data for preliminary classification. In regard to the 20 S protease, inhibitor studies are not straightforward owing to the complexity of the enzyme and the distinct possibility that it is a "multicatalytic protease." Despite these complications, most experiments indicate that the 20 S complexes are cysteine proteinases. These enzymes are inhibited by several organomercurials, thiol reagents such as N-ethylmaleimide (NEM), 5,5'-dithiobis-2-nitrobenzoic acid (DTNB), and iodoacetic acid, divalent heavy metals introduced as the chloride salt, $FeCl_2$, $FeCl_3$, $ZnCl_2$, $CoCl_2$, $NiCl_2$, and $SnCl_2$, the microbial inhibitors chymostatin, leupeptin and antipain, certain chloromethyl ketones and synthetic aldehydes, and hemin. In addition, there is general agreement that reducing agents such as cysteine, dithiothreitol (DTT), 2-mercaptoethanol, detergents (SDS, Triton-X-100), fatty acids (oleic, linoleic), and various other compounds may "activate" the enzyme. On the other hand, Wilk and Orlowski[41] showed that only the "trypsin-like" active site of the proposed multicatalytic enzyme is inhibited by iodoacetamide and stimulated by dithiothreitol. In contrast, the "chymotrypsin-like" and "peptidyl–glutamyl" sites are essentially unaffected by either compound.

There is a minority opinion that the 20 S enzyme is a serine proteinase.[38,59,62] Although Ishiura *et al.*[38,53,62,64] invariably report that "ingensin" is a serine proteinase based on inhibition by diisopropylfluorophosphate (DFP), their data are unconvincing. For example, variable amounts of high molecular weight bands greater than 35,000 are present, and in one case, ingensin A purified from rat liver[61] consisted almost entirely of a 120,000 subunit on SDS–PAGE although by nondenaturing PAGE it was indistinguishable from ingensin B (SDS–PAGE subunits of M_r = 20,000–30,000). In another report,[63] the major subunit present in what is presumably purified human placenta ingensin B migrates on SDS–PAGE at about 70,000. To complicate matters even further, there may have been reporting errors since a comparison of inhibitor data for porcine skeletal muscle ingensin B (Table 2, ref. 53) and human placenta ingensin B (Table 3, ref. 63) shows that six rows and two columns of figures are identical. It is also curious that Ishiura and colleagues find hydrolysis of Suc-Leu-Leu-Val-Tyr-MCA by reticulocyte ingensin to be unaffected by 0.1 mM p-chloromercuribenzoic acid (pCMB)[38] because every other description of the 20 S protease (including other studies by these authors) reports sensitivity to micromolar concentrations of organomercurials. Thus, several controversies remain: (1) Do monovalent and divalent cations inhibit, activate, or have no effect on proteolytic activity? (2) Is the enzyme a cysteine or serine proteinase? (3) Do nucleotides and nucleotide analogs stimulate, inhibit, or have no effect on proteolytic activity? Although there are insufficient data to consider the role of cations, the latter two questions are discussed in Sections 5.2.5 and 5.3, respectively.

5.2.5. Cysteine, Serine, or Multicatalytic Protease?

Wilk and Orlowski[41] propose three catalytic sites based on the hydrolysis of synthetic fluorogenic peptides: a peptidyl–glutamyl site (Z-Leu-Leu-Glu-2NA), a chymotrypsin-like site (Z-Gly-Gly-Leu-pNA), and a trypsin-like site (Z-D-Ala-Leu-Arg-2NA). Their model is based on (1) comigration of peptide hydrolyzing activities throughout purification, (2) concentration-dependent Na^+/K^+ inhibition of all three activities, (3) the failure to separate the activities, and (4) differential response of peptide hydrolysis to inhibitors and activators. For example, SDS (0.02%) stimulated the Glu site while inhibiting the Leu and Arg sites. Leupeptin specifically inhibited the Arg site; Z-Gly-Gly-Leucinal inhibited the Leu site and stimulated the Arg site with no affect on the Glu site. Additional evidence for a multicatalytic protease was provided by Dahlmann et al.,[48] who showed different pH optima for substrate hydrolysis and differential effects of inhibitors. In addition, they reported that the activity of the rat skeletal muscle enzyme against [^{14}C]methylcasein and several peptides could be substantially increased with low levels of SDS and physiological concentrations of fatty acids, particularly oleic acid and linoleic acid.[49]

It is tempting to conclude that the complex is indeed a multicatalytic enzyme that does not fit neatly into the fourfold classification scheme for proteases. However, the broad substrate specificity of the enzyme is also typical of papain, which clearly has a single catalytic site. Simultaneous inhibition or activation is consistent with a single active site, so the best evidence for multiple sites is the selective inhibition of hydrolysis of one peptide with no effect or stimulation of the hydrolysis of other peptides. This was shown by Wilk and Orlowski[41] in their pioneering work. Nevertheless, thorough kinetic studies with various substrates and inhibitors like those recently reported for calpains I and II by Sasaki $et\ al.$[71,72] will be required to prove the existence of distinct catalytic sites. Moreover, unambiguous identification of the separate sites within the complex would be ideal. To date, only Hase $et\ al.$[36] have had limited success in identifying the subunit responsible for casein hydrolysis by the carp muscle enzyme. Perhaps the wisest position is that taken by McGuire and DeMartino[40] who state: "The actual relationship of these substrates to the putative catalytic sites is unknown. Until these sites are identified, isolated and characterized, we believe that the issue of multiple catalytic sites remains unresolved."

With the caveat that multiple sites of different catalytic mechanisms may exist in the 20 S complex, most available data suggest that the enzyme is a cysteine proteinase. As already stated, broad substrate specificity is characteristic of several proteases, particularly the plant sulfhydryl enzymes, papain and ficin. Different pH optima for various substrates may be determined by the leaving group in synthetic peptides[66] or residues adjacent to the cleavage site.[32–35,39–51,58,59,66] Since the 20 S enzyme has a high temperature optimum[36,46] and most studies have thus far been performed at physiological temperatures, the putative inhibitors and activators, including changes in pH, may affect catalysis by driving conformational changes or by removing protein inhibitors.

The evidence for classifying the 20 S enzyme as a cysteine proteinase is incomplete, but the alternative view that it is a serine protease is even less convincing. Although trypsin and chymotrypsin are inhibited at micromolar concentrations of DFP, the reported concentrations of DFP needed to inhibit the 20 S enzyme range from 0.5 mM[38] to 50 mM.[61] The significance of DFP inhibition at such high levels is questionable especially because commercial preparations of DFP may be contaminated by inhibitors of cysteine proteinases.[73] More important, the known inhibition of esterase activity by DFP[74] and the clear demonstration that DFP phosphorylates the cysteine proteinases, papain,[75] chymopapain,[76] ficin,[73] and bromelain,[77,78] indicate that DFP may inhibit the 20 S protease by mechanisms unrelated to titration of the active site(s). Therefore, evidence of labeling by radioactive DFP[59] should be viewed with caution. A further argument against the 20 S enzyme being a serine protease is the fact that 100 μM 3,4-dichloroisocoumarin, an inhibitor of serine proteinases,[79] actually stimulated the reticulocyte enzyme more than twofold.[27]

Thus, the weight of evidence favors the view that the 20 S enzyme is a cysteine protease. With the exception of reports by Waxman et al.[27] and Tanaka et al.[59] that N-ethylmaleimide (NEM) stimulates proteolysis by the rabbit reticulocyte or rat liver 20 S proteases, all other reports show substantial inhibition by thiol reagents. Since this observed NEM stimulation was only demonstrated in crude protease fractions,[59] the suggestion that activation by NEM in crude rat liver cytosol is due to the 20 S protease seems unwarranted, particularly since the "activated" enzyme is sensitive to NEM. Other caseinolytic activities most certainly exist, and they may account for the NEM activation seen in crude cytosol. These NEM data mimic, to some extent, the results obtained with partially purified, destabilized high molecular weight protease fractions from reticulocytes.[32] This resultant activity degrades [^{125}I]lysozyme, can be stimulated by low levels of NEM and iodoacetamide, and is inhibited by aprotinin (R. Hough, unpublished results). The source of this activity is presently unknown, but it may not be related to the purified 20 S complex.

5.2.6. One Enzyme or Several?

Is the 20 S protease a protein complex of defined composition? Or are there several classes of 20 S protease? Perhaps, as suggested by Achstetter et al.,[34] more sophisticated fractionation procedures will finally separate comigrating enzymes. Their work on the 20 S yeast protease, yscE, provides a good example of the problems encountered in trying to separate several enzymatic activities. The yeast enzyme does not normally bind DEAE–Sepharose CL-6B in low ionic strength buffer, but it will bind in 0.1 M Tris–HCl (pH 7.2), 0.15 M NaCl, and it can subsequently be eluted at higher salt concentrations as two apparent isomers of a single proteolytic complex. On SDS–PAGE the enzyme shows bands at 70,000 and eight to ten subunits less that 35,000, but it migrates as a single band on nondenaturing gels at pH 7.9 or 8.9. The 95% pure enzyme degrades Cbz-Gly-Gly-Leu-NA, Cbz-Ala-Ala-Leu-NA, and Suc-Phe-Leu-Phe-NA. However, by electrophoretic separation of the purified enzyme and analysis of proteolysis in very thin gel slices, Achstetter et al.[34] showed that the activities that hydrolyzed Bz-Phe-Val-Arg-NA and Suc-Tyr-Leu-Val-NA, previously thought to comigrate with the proteolytic complex, were now resolved. Low levels of contaminants undetected or barely visible by Coomassie Blue staining may have been responsible. These activities were not always present, and the predominant enzyme displayed a primary specificity for hydrophobic amino acids at the P_1 site.

The low solubility of the substrates did not permit the determination of the Michaelis–Menten constant, but Lineweaver–Burk plots were nonlinear, suggesting that the yeast enzyme is activated by substrate binding. In addition, the major caseinolytic activity, which was stimulated 1.6–2-fold in the presence of ATP and a regenerating system, was shown by gel filtration to be of similar size as yscE.[33] However, it appeared to be greater than 670,000 (radius > 85 Å), whereas yscE was estimated to be 600,000 (radius of 75 Å), and hydrolyzed casein only slightly. Furthermore, the "ATP stimulation" of the high M_r protease was due primarily to phosphoenolpyruvate in the regenerating system. Similar findings were reported by DeMartino and Goldberg[54] and Rose et al.[55] for proteases partially purified from rat and mouse liver, respectively.

5.3. ATP-Dependent, Stimulated, or Stabilized High M_r Proteases

In contrast to the 26 S ATP-dependent protease, which appears to require hydrolyzable nucleotides for the degradation of Ub–lysozyme conjugates,[13,14] other large proteases may only be stabilized against thermal inactivation by nucleotides. This was evident in 1979, when Rose et al.[55] discovered that the apparent ATP stimulation of globin degradation by the mouse liver enzyme ($M_r > 400,000$) was not due to ATP but rather to creatine phosphate added to the incubation mixture as part of the ATP regenerating system. It was found that the sulfhydryl protease, normally inactivated above 20°C, was stabilized at 37°C in the presence of the metal chelators, glycerate-3-P, creatinine-P, and citrate. At about the same time, DeMartino and Goldberg[54] reported that proteolysis of [^{14}C]methylglobin and [^{125}I]hemoglobin by a rat liver protease (M_r 550,000) was stimulated by ATP, other nucleoside triphosphates, ADP, pyrophosphate, and nonionic detergents (i.e., Triton-X-100). However, a requirement for ATP hydrolysis was not demonstrated. The role(s) for nucleotides and other metabolically important cofactors could multiply rapidly with increased understanding of the complexity of these high-molecular-weight proteases. For example, the rat liver enzyme showed a greater "stimulation" by ATP (5 mM) at higher concentrations of Triton-X-100.[54] This was due in part, however, to inhibition by the detergent at concentrations greater than the optimum, as pointed out at the time.[55]

Another ATP-stimulated protease has been partially purified from rat muscle.[50] At low levels of ATP, stimulation was additive with that produced by Triton-X-100; at higher ATP levels (5 mM) and optimal detergent levels (0.1%), the enzyme was inhibited twofold by ATP. The rat muscle protease is seemingly identical to enzymes previously purified by these authors from human skeletal muscle[51] and erythrocytes.[39] It is noteworthy that this enzyme was stimulated or inhibited by all the compounds that affect the 20 S protease from reticulocytes,[32] including enhanced activity in the presence of propanol (R. Hough, unpublished results). Thus, the rat muscle enzyme is probably active in the absence of ATP.

Dahlmann et al.[47,80] have identified an enzyme that they call proteinase I (M_r 750,000) based on its elution position from DEAE–cellulose and its specificity for the synthetic peptide, Z-Phe-Arg-NMec. The protease is stabilized by ATP against thermal inactivation, is dramatically stimulated by dithiothreitol or cysteine, is inhibited by thiol reagents, antipain, leupeptin, chymostatin, and zinc, and has a pH optimum of 8–9. It has been resolved from two other high molecular weight proteases by fast protein liquid chromatography on Mono Q.[80] Proteinase I is actually the second high M_r proteinase to elute from Mono Q; the first hydrolyzes casein, Bz-Val-Gly-Arg-NMec, Z-Phe-Arg-NMec, contains no Phe-X splitting activity, and has a molecular weight of approximately 650,000. The second enzyme, proteinase I, constitutes the main Z-Phe-Arg-NMec hydrolyzing activity. The third proteinase, which exhibited the major caseinolytic and peptidase activities, was identical to proteinase(s) II/III described previously.[47] The latter enzyme is identical to the 20 S multisubunit complex.[48]

An enzyme purified from rat heart (M_r 500,000) displays similar properties as proteinase I.[65] However, DeMartino suggests that ATP does more than merely stabilize the heart enzyme against thermal inactivation, although phosphate bond hydrolysis does not appear to be required for degradation of [^{14}C]methylglobin or [^{14}C]methylcasein. The proteinase purified by Ismail and Gevers[52] may be related to the Dahlmann et al. proteinase I and the DeMartino heart enzyme. It also appears to be a cysteine proteinase (M_r 500,000) stabilized by low concentrations of ATP (0.1 mM) against thermal inactivation. The proteinase is markedly inhibited by zinc but not by the microbial inhibitors leupeptin, antipain, or chymostatin. It is important to reiterate that the nature of the substrate may, to a large degree, determine the inhibitor results. Proteinase I was assayed with a synthetic N-blocked dipeptidyl-NMec, whereas Ishmail and Gevers used native and denatured proteins as the substrates (inter alia, casein, globin, and hexokinase).

Immunological cross-reactivity between the skeletal muscle proteinase I and the cardiac muscle proteinase has been reported.[80] This result is surprising since the two enzymes differ substantially by substrate specificity and molecular weight. On the other hand, McGuire and De-Martino[40] claim that "macropain" purified from human erythrocytes was previously identified as the smaller ATP-stimulated proteases from rat liver[54] and rat heart.[65] Unfortunately, there are no reports on the subunit structure of ATP-stimulated proteases from eukaryotic cells that would help to explain these discrepancies. One can infer that, under certain conditions,[50] the 20 S multisubunit complex is stimulated by ATP, but most of the evidence suggests the complex is unaffected, or slightly inhibited by nucleotides.[27,32,35,38,40,47,51,58] Therefore, either multiple high molecular weight proteases of similar size and with similar characteristics exist in eukaryotic cells, or multiple forms of a core proteolytic complex abound. Perhaps, in association with other proteins, this hypothetical core complex performs different functions (see Section 5.6).

5.4. Stability of the 26 S and 20 S Proteases

The 20 S protease is quite stable at 37°C; in fact, it appears to be maximally active against casein as the substrate between 60 and 65°C.[36,46] Studies exploiting the high-temperature optimum might prove useful in sorting out the effects produced by compounds, such as SDS, oleic acid, or various cations. If these agents act by changing the conformation of the 20 S particle or by removing inhibitors, their effects may be much reduced at high temperature. A distinguishing feature of the 26 S ATP-dependent protease and other ATP-stimulated proteases is thermal lability in the absence of nucleotides and other stabilizing factors.[32,47,52,55] This key difference may permit one to quantify each enzyme in mixtures and may provide clues about the regulation of proteolysis.

5.5. Subcellular Location of 26 S and 20 S Proteases

Erythrocytes and reticulocytes contain small amounts of cytoskeleton and virtually no internal membrane organelles. Hence, it is not surprising that by all criteria the 26 S and 20 S proteases appear to be soluble enzymes. Likewise, there is little evidence that the 20 S protease is bound to specific organelles in other tissues. It is almost always found in the postmicrosomal supernate. Moreover, antibody localization studies suggest a uniform distribution of 20 S protease in rat liver cells.[60]

Although there is no definitive information on the distribution of 26 S protease in cells other than RBCs or reticulocytes, there is a hint that the enzyme may be particulate in nucleated erythroid precursors. Rieder *et al.*[81] have reported a particulate ATP-dependent protease in erythroleukemia cells. The enzyme, which was stimulated as much as fourfold by ATP, was isolated on a 30% sucrose cushion. It did not appear to be associated with membranes, cytoskeleton, or polysomes. To complicate

Table I. Distinguishing Characteristics of the Two Purified High-Molecular-Weight
Proteases from Rabbit Reticulocyte Lysate

	ATP dependent	ATP independent
Native molecular weight	1,000,000	700,000
Sedimentation coefficients	26 S	20 S
Subunit composition (M_r)	34,000–110,000	21,000–32,000
Isoelectric point	4.6–5.1	5.2–5.4
Preference for Ub–lysozyme conjugates as substrate	Yes	No
Nucleotide stimulation	Yes	No
Detergent activation	No	Yes
Stability at 37°C	No	Yes

matters even further, Waxman et al.[82] have described a similar ATP-dependent enzyme in erythroleukemia cells. In contrast to the results of Rieder et al.,[81] the latter investigators report that the enzyme is soluble. Given these contradictory results and the tenuous connection between the erythroleukemia proteases and the 26 S complex, it is safe to say that the location of the 26 S protease in nucleated eukaryotic cells is not known.

5.6. Structural Models for the 26 S and 20 S Proteases

The reticulocyte 26 S and 20 S enzymes are so similar in their chromatographic behavior that each is the only significant contaminant of the other during the final stages of purification.[32] The enzymatic properties of the proteases are also very similar. Both hydrolyze [^{125}I]α-casein but do not degrade lysozyme or bovine serum albumin. Both prefer fluorogenic peptides containing aromatic residues adjacent to the MCA leaving group. The pH optima for hydrolysis of fluorogenic peptides are also similar for both proteases.[32]

The two enzymes differ, however, with respect to ATP-dependent proteolysis of [^{125}I]lysozyme–Ub conjugates and nucleotide stimulation of peptide hydrolysis.[32] The 26 S enzyme preferentially degrades ubiquitinated lysozyme whereas the 20 S enzyme does not. There are also key differences between the proteases with regard to inhibitors and activators. The smaller enzyme is stimulated by levels of SDS and oleic acid that completely inhibit the larger enzyme.[32] The larger enzyme is more sensitive to PMSF suggesting that a serine active site may also be present. So, despite a number of parallels in the behavior of the 26 S and 20 S enzyme, we consider them to be distinct enzymes (see Table I).

In addition to the 26 S particle capable of degrading Ub–lysozyme

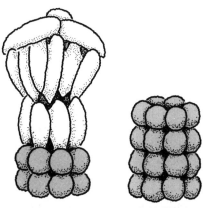

Figure 5. Structural models for the 26 S and 20 S reticulocyte proteases. The 20 S protease is shown on the right as a cylinder containing 24 subunits of M_r 20,000–35,000. This structure is consistent with the subunit composition of the protease and the electron microscopic studies of Hase *et al.*[36] and Kopp *et al.*[69] The 26 S protease is shown to the left as a mushroom-shaped particle containing 12 subunits (shaded spheres) shared with the 20 S protease and 15 additional subunits with M_r values between 45,000 and 110,000. This structure is based on electron microscopic images observed by Shelton *et al.*[103] and the subunit compositions determined by Hough *et al.*[32] (see Figure 4C). In Table III and Figure 11 of ref. 32, SDS–PAGE bands between 21,000 and 32,000 were specifically excluded from the subunit composition estimates for the 26 S protease since they were assumed to be contaminants. In the structures shown above we consider the possibility that the two protease complexes share a set of subunits. Thus, the actual molecular weight of the 26 S protease may be greater than 1,000,000.

conjugates, Hough *et al.*[32] isolated a slower sedimenting protein complex that contains the 46,000–62,000 M_r subunits and the two heavy chains but apparently lacks subunits between 21,000 and 32,000. The presence of a slower sedimenting form of the 26 S protease and the similarities of the 26 S protease to the 20 S protease suggest a possible arrangement for the various subunits in the 26 S particle. According to the models shown in Figure 5, the 26 S ATP-dependent protease would contain subunits (M_r 21,000–32,000) also present in the 20 S protease. Loss of these subunits would result in a complex unable to hydrolyze the fluorogenic peptide, Suc-Leu-Leu-Val-Tyr-MCA. Whether the subcomponent that lacks the 21,000–32,000 subunits can produce limited cleavage of Ub–lysozyme conjugates has not been determined. It must be emphasized that the proposed structure of the 26 S enzyme shown in Figure 5 is merely our current working hypothesis. We have not rigorously excluded the possibility that the 21,000–32,000 subunits present in the 26 S region on nondenaturing gels arise from a dimer of the 20 S. If this is the case, then the Ub/ATP-dependent protease may only contain subunits with M_r values between 34,000 and 110,000.

6. OTHER PROTEASES STIMULATED BY ATP

6.1. Adrenal Mitochondrial Enzyme

Several ATP-stimulated proteases have been characterized from sources other than erythroid cells. Watabe and Kimura[83] identified an

ATP-stimulated protease from bovine adrenal cortex mitochondria using [^{14}C]methylcasein as a substrate. The requirement for nucleoside triphosphate was later shown to depend on the substrate.[84] Casein hydrolysis was stimulated 25-fold by Mg–ATP and to lesser extent by other nucleotides; ADP, AMP, and inorganic phosphates would not substitute for nucleotides. By contrast, when insulin or angiotensinogen was used as substrate, inorganic triphosphate and pyrophosphate stimulated the enzyme as much as ATP.

Watabe and Kimura have purified the mitochondrial enzyme to near homogeneity.[84] The protease elutes from gel filtration columns with an apparent molecular weight of 650,000, and SDS–PAGE analysis reveals a prominent protein subunit at 108,000. It thus appears to be a hexamer. In this regard, the enzyme is similar to *E. coli* protease La, which is a tetramer of subunits of about 100,000 in molecular weight (see Section 6.3). The mitochondrial enzyme also resembles protease La in three other respects. Small peptides enhance ATPase activity, the K_m for ATP is about 5 μM, and the enzyme is inhibited by vanadate although at much higher levels than required to inhibit protease La.

6.2. *Escherichia coli* Multicomponent Pathway

E. coli contains two ATP-dependent proteolytic systems, protease La and a multicomponent pathway recently described by Maurizi and his collaborators.[85] Casein hydrolysis by the latter system requires at least two macromolecular species and ATP, which stimulates proteolysis almost 20-fold. Nucleotides other than ATP such as GTP, CTP, β,γ-methylene ATP, or ADP–SO$_4$ are completely ineffective. The two required macromolecules, termed components I and II, can be separated by phosphocellulose chromatography. Component I elutes from the gel filtration resin, TSK 3000, with an apparent M_r of 150,000–200,000 and the molecular weight of component II is about 100,000. Component II appears to contain the ATP-binding site since it is retained on an ATP-affinity column, whereas component I does not bind. Neither component alone is capable of degrading casein, and mixing experiments indicate that both components must be present simultaneously for casein hydrolysis to occur. The continuous presence of ATP is also required for proteolysis.

Somewhat surprisingly, casein is the only protein substrate so far identified. Several native λ proteins (O, N, cI, X$_{is}$), the galactose repressor, and fragments of O protein and cI are not cleaved. Several fluorogenic peptides that are good substrates for protease La are not hydrolyzed by the I/II/ATP pathway. This adds to the evidence that the newly discovered proteolytic pathway is unrelated to protease La. Casein hydrolysis is not inhibited by a variety of serine protease inhibitors including PMSF or the trypsin inhibitors from soybean, pancreas, or egg white. Proteolysis is sensitive, however, to N-ethylmaleimide and the organomercurial Neohydrin. Hence, the enzyme may use an active sulfhydryl for bond cleavage.[85]

6.3. *Escherichia coli* Protease La

Protease La is an extensively studied, ATP-stimulated protease in *E. coli*. Four identical subunits, encoded by the *lon* gene,[86,87] associate to form the holoenzyme. Protease La is a heat-shock protein[88] that degrades denatured proteins as well as certain natural substrates including Sul A and λN protein.[89] Although the enzyme will cleave fluorogenic peptides in the presence of nonhydrolyzable ATP analogs, it was reported that ATP hydrolysis is required for the degradation of proteins[90] and that protein substrates enhance ATPase activity[91]; in contrast, proteins resistant to protease La do not enhance ATP hydrolysis. Apparently, 2 ATP molecules are consumed for each peptide bond cleaved.[92] The presence of protein substrates can stimulate hydrolysis of smaller peptides by two- to ten-fold. Although protease La clearly plays an important physiological role, it is not an essential *E. coli* enzyme since mutants carrying a deletion in *lon* are viable.

7. THE ROLE OF ATP IN PROTEOLYSIS

ATP can drive chemical reactions by directly participating in bond rearrangements, for example, adenylation of the carboxyl group of glycine 76 during Ub activation (see Figure 1). The energy in ATP can be coupled to the transport of small ions by the repetitive phosphorylation of specific residues on membrane embedded transport proteins[93]; cyclic phosphorylation events produce conformational changes that lead to the net movement of ions across lipid bilayers. Hydrolysis of high-energy phosphate bonds can also produce directional assembly and disassembly in polymers formed from actin,[94] tubulin,[95] and rec A.[96] Finally, ATP (or GTP) hydrolysis can produce the directed movement of one macromolecule relative to another. For example, myosin moves on actin filaments, dynein moves on microtubules, and mRNAs move relative to the ribosome.

Which of the above mechanisms might explain the energy requirement for ATP-dependent proteases? For the 26 S protease, we simply do not know; it is not even firmly established that bond hydrolysis is required. For protease La, the answer is none of them. Despite earlier claims to the contrary,[90] recent data show that La can degrade casein[97] or N protein[89] in the presence of nonhydrolyzable ATP analogs. This does not rule out the possibility that ATP hydrolysis promotes release of cleaved polypeptide products, but it does eliminate phosphate bond breakage as a *mandatory* step in the reaction sequence. Simple binding of ATP or ATP analogs can sustain repeated peptide bond cleavages by protease La. Thus, accounts[98] that "this remarkable enzyme degrades a polypeptide chain into small, acid-soluble peptides, starting from one end and finishing off the chain it is degrading before it starts another," and "this processive degradation requires sustained ATP hydrolysis which allows rapid translocation of the enzyme along its polypeptide substrate" are probably incorrect. In light of the fact that the mechanism by which myosin converts ATP hydrolysis to force production remains obscure despite 25 years of intense study,[99] it may be some time before we understand the function of ATP in the degradation of Ub–protein conjugates.

8. NOMENCLATURE: MACROPAIN AND MEGAPAIN

Naming enzymes can be a tricky business. One wants a short word that embodies crucial features of the protein's activity. In this regard, calpain is an excellent name for the calcium-activated protease widespread among eukaryotes. The term calpain invokes a connection between *cal*cium and the sulfhydryl protease pa*pain*. In fact, recent sequence analysis shows that the 80 kDa subunit of calpain contains four domains apparently joined during the course of evolution by gene fusion.[100] One domain is related to papain and another is a member of the calmodulin family of proteins. Hence, calpain is a perfect choice.

The 26 S and 20 S proteases described above have not been officially named. Several names have been proposed for the 20 S protease: cation-sensitive neutral endopeptidase,[42] multicatalytic proteolytic complex,[41] high molecular weight cysteine proteinase,[47] ingensin,[38] and macropain.[40] We recently suggested[32] that the 20 S protease be called "700 kDa multisubunit proteinase." Upon further reflection, we believe that macropain is a better choice. The name implies a large, sulfhydryl protease, and most evidence suggests the involvement of cysteine in the catalytic mechanism. Calling the 700 kDa protease macropain leaves megapain as a possible name for the larger, 26 S ATP-dependent enzyme. This proteolytic complex also appears to use sulfhydryls for peptide bond cleavage, and its larger size, on the order of a million daltons, certainly justifies the prefix mega. Other features of enzyme function can be related to the proposed names. There is evidence that macropain preferentially degrades oxidized proteins,[57,58] and megapain is clearly stimulated by ATP.[32] The convenient correspondence between macro and oxidized on one hand, and mega and ATP on the other, should facilitate remembering which enzyme is which.

A major reservation about macropain and megapain is the possibility that the 20 S and 26 S enzymes will prove to be serine[101] or multicatalytic proteases after all. Should this happen, it may not be too difficult to adopt macrosin or macrobites and megasin or megabites as alternative titles. In any event, the short names proposed here should prove far more useful than such cumbersome phrases as Ub/ATP-dependent protease or high molecular weight cysteine protease and should prove far more sonorous than recently proposed acronyms.

9. MACROPAIN AND PROSOMES

The properties listed above for the protease macropain are remarkably similar to properties ascribed to another cytoplasmic particle, the prosome. In 1984, Schmid et al.[102] coined the term prosome to identify a novel ribonucleoprotein that sediments at 19 S. These particles, seen by a number of investigators over the past two decades,[103,104] are characterized by their extraordinary stability since they can be purified on gradients containing 1% Sarkosyl, 0.5 M KCl, or 1 M urea. Like macropain, they are composed of at least 10 proteins with M_r values between 19,000 and 35,000; by two-dimensional PAGE prosomes contain 25 proteins of which 16 are highly conserved from one species to another.[105]

Prosomes are known to inhibit mRNA translation.[102] Although this activity is usually attributed to the small RNAs associated with them, it could also be due to proteolytic activity. These particles have also recently been implicated in tRNA processing.[106] Thus, all available evidence strongly suggests either that prosomes and macropain are identical or that macropain is a member of a class of particles that process or degrade nucleic acids and proteins.

10. ALTERNATE PROTEOLYTIC PATHWAYS

Macropain and megapain, the two enzymes discussed most extensively in this chapter, plus calpain constitute the three proteases currently thought to degrade intact cytosolic proteins. Obvious questions are raised by the presence of multiple cytoplasmic enzymes: How many proteolytic pathways are there? Do the three enzymes act in sequence so that each participates in the degradation of every cytosolic protein? Are some cellular proteins degraded exclusively by just one of these enzymes? Do different marking reactions funnel a specific cytosolic protein to one or another of these proteases? Macropain has been implicated in the processing of neuropeptides in the pituitary,[41] turnover of lens proteins,[45] nonlysosomal and diabetes-induced protein degradation in skeletal muscle,[80] degradation of oxidized glutamine synthetase in liver,[57,58] and possibly cleavage of steroid receptors.[56,67] There is evidence that calpains are involved in the degradation of cytoskeletal proteins.[107] Similarly, megapain appears to be responsible for the degradation of newly synthesized abnormal proteins.[8] However, present data do not preclude an involvement of Ub and megapain in the degradation of cytoskeletal proteins after initial cleavages by calpain. Clearly, a major challenge for future studies will be to determine the contribution of each enzyme to the destruction of a limited number of cytosolic proteins.

11. CONCLUSION

In a sense, the title of this chapter is misleading since most of the discussion has focused on macropain, a 20 S protease that neither prefers ubiquitinated proteins as substrates nor requires ATP hydrolysis for activity. Our preferential coverage of the smaller enzyme was dictated in part by a lack of information on megapain, which was only recently discovered, and in part by our suspicion that the two enzymes share subunits. Thus, it is likely that many properties of macropain will be characteristic of megapain as well. Indeed, the similarities discussed suggest that this will be the case. We also suspect that each enzyme plays a major role in the degradation of cytosolic proteins.

REFERENCES

1. Hershko, A., Ciechanover, A., Heller, H., Haas, A. L., and Rose, I. A., 1980, Proposed role of ATP in protein breakdown: Conjugation of proteins with multiple chains of the polypeptide of ATP-dependent proteolysis, *Proc. Natl. Acad. Sci. U.S.A.* **77**: 1783–1786.
2. Chin, D. T., Kuehl, L., and Rechsteiner, M., 1982, Conjugation of ubiquitin to denatured hemoglobin is proportional to the rate of hemoglobin degradation in HeLa cells, *Proc. Natl. Acad. Sci. U.S.A.* **79**: 5857–5861.
3. Hershko, A., Eytan, E., Ciechanover, A., and Haas, A. L., 1982, Immunochemical analysis of the turnover of ubiquitin–protein conjugates in intact cells, *J. Biol. Chem.* **257**: 13964–13970.

4. Gregori, L., Marriott, D., West, C. M., and Chau, V., 1985, Specific recognition of calmodulin from *Dictyostelium discoideum* by the ATP, ubiquitin-dependent degradative pathway, *J. Biol. Chem.* **260:** 5232–5235.

5. Shanklin, J., Jabben, M., and Vierstra, R. D., 1987, Red light-induced formation of ubiquitin–phytochrome conjugates: Identification of possible intermediates of phytochrome degradation, *Proc. Natl. Acad. Sci. U.S.A.* **84:** 359–363.

6. Bachmair, A., Finley, D., and Varshavsky, A., 1986, In vivo half-life of a protein is a function of its amino-terminal residue, *Science* **234:** 179–186.

7. Finley, D., Ciechanover, A., and Varshavsky, A., 1984, Thermolability of ubiquitin-activating enzyme from the mammalian cell cycle mutant ts85, *Cell* **37:** 43–55.

8. Ciechanover, A., Finley, D., and Varshavsky, A., 1984, Ubiquitin dependence of selective protein degradation demonstrated in the mammalian cell cycle mutant ts85, *Cell* **37:** 57–66.

9. Munro, S., and Pelham, H. R. B., 1984, Use of peptide tagging to detect proteins expressed from cloned genes: Deletion mapping functional domains from *Drosophila* hsp 70, *EMBO J.* **3:** 3087–3993.

10. Carlson, N., Rogers, S., and Rechsteiner, M., 1987, Microinjection of ubiquitin: Changes in protein degradation in HeLa cells subjected to heat-shock, *J. Cell Biol.* **104:** 547–555.

11. Haas, A. L., and Rose, I. A., 1981, Hemin inhibits ATP-dependent ubiquitin-dependent proteolysis: Role of hemin in regulating ubiquitin conjugate degradation, *Proc. Natl. Acad. Sci. U.S.A.* **78:** 6845–6848.

12. Hough, R., and Rechsteiner, M., 1986, Ubiquitin–lysozyme conjugates: Purification and susceptibility to proteolysis, *J. Biol. Chem.* **261:** 2391–2399.

13. Hough, R., Pratt, G., and Rechsteiner, M., 1986, Ubiquitin–lysozyme conjugates: Identification and characterization of an ATP-dependent protease from rabbit reticulocyte lysates, *J. Biol. Chem.* **261:** 2400–2408.

14. Hershko, A., Leshinksy, E., Ganoth, D., and Heller, H., 1984, ATP-dependent degradation of ubiquitin–protein conjugates, *Proc. Natl. Acad. Sci. U.S.A.* **81:** 1619–1623.

15. Rechsteiner, M., Carlson, N., Chin, D., Hough, R., Rogers, S., Roof, D., and Rote, K., 1984, On the role of covalent protein modification and protein aggregation in intracellular proteolysis, in: *Protein Transport and Secretion* (D. L. Oxender, ed.), Alan R. Liss, New York, pp. 391–402.

16. Hershko, A., and Heller, H., 1985, Occurrence of a polyubiquitin structure in ubiquitin–protein conjugates, *Biochem. Biophys. Res. Commun.* **128:** 1079–1086.

17. Rechsteiner, M., 1985, Ubiquitin-dependent proteolysis in eucaryotic cells, *Curr. Topics Plant Biochem. Physiol.* **4:** 15–24.

18. Hershko, A., Heller, H., Eytan, E., and Reiss, Y., 1986, The protein substrate binding site of the ubiquitin–protein ligase system, *J. Biol. Chem.* **261:** 11992–11999.

19. Tanaka, K., Waxman, L., and Goldberg, A. L., 1983, ATP serves two distinct roles in protein degradation in reticulocytes, one requiring and one independent of ubiquitin, *J. Cell Biol.* **96:** 1580–1585.

20. Katznelson, R., and Kulka, R. G., 1983, Degradation of microinjected methylated and unmethylated proteins in hepatoma tissue culture cells, *J. Biol. Chem.* **258:** 9597–9599.

21. Chin, D. T., Carlson, N., Kuehl, L., and Rechsteiner, M., 1986, The degradation of guanidinated lysozyme in reticulocyte lysate, *J. Biol. Chem.* **261:** 3883–3890.

22. Hershko, A., Heller, H., Eytan, E., Kahij, G., and Rose, I. A., 1984, Role of the α-amino group of protein in ubiquitin-mediated protein breakdown, *Proc. Natl. Acad. Sci. U.S.A.* **81:** 7021–7025.

23. Glass, J. R., and Gerner, E. W., 1987, Spermidine mediates degradation of ornithine

decarboxylase by a non-lysosomal, ubiquitin-independent mechanism, *J. Cell. Physiol.* **130:** 133–141.

24. Tanaka, R. D., Li, A. C., Fogelman, A. M., and Edwards, P. A., 1986, Inhibition of lysosomal protein degradation inhibits the basal degradation of 3-hydroxy-3-methyl-glutaryl coenzyme A reductase, *J. Lipid Res.* **27:** 261–273.

25. Speiser, S., and Etlinger, J. D., 1982, ATP stimulates proteolysis in reticulocyte extracts by repressing an endogenous protease inhibitor, *Proc. Natl. Acad. Sci. U.S.A.* **80:** 3577–3580.

26. Murakami, K., and Etlinger, J. D., 1986, Endogenous inhibitor of nonlysosomal high molecular weight protease and calcium-dependent protease, *Proc. Natl. Acad. Sci. U.S.A.* **83:** 7588–7592.

27. Waxman, L., Fagan, J. M., and Goldberg, A. L., 1987, Demonstration of two distinct high molecular weight proteases in rabbit reticulocytes, one of which degrades ubiquitin conjugates, *J. Biol. Chem.* **262:** 2451–2457.

28. Fagan, J. M., Waxman, L., and Goldberg, A. L., 1987, Skeletal muscle and liver contain a soluble ATP + ubiquitin-dependent proteolytic system, *Biochem. J.* **243:** 335–343.

29. Simpson, M. V., 1953, The release of labeled amino acids from the proteins of rat liver slices, *J. Biol. Chem.* **201:** 143–154.

30. Ciechanover, A., Heller, H., Elias, S., Haas, A. L., and Hershko, A., 1980, ATP-dependent conjugation of reticulocyte proteins with the polypeptide required for protein degradation, *Proc. Natl. Acad. Sci. U.S.A.* **77:** 1365–1368.

31. Hough, R., and Rechsteiner, M., 1984, Effects of temperature on the degradation of proteins in rabbit reticulocyte lysates and after injection into HeLa cells, *Proc. Natl. Acad. Sci. U.S.A.* **81:** 90–94.

32. Hough, R., Pratt, G., and Rechsteiner, M., 1987, Purification of two high molecular weight proteases from rabbit reticulocyte lysate, *J. Biol Chem.* **262:** 8303–8313.

33. Achstetter, T., Emter, O., Ehmann, C., and Wolf, D. H., 1984, Proteolysis in eukaryotic cells: Identification of multiple proteolytic enzymes in yeast, *J. Biol. Chem.* **259:** 13334–13343.

34. Achstetter, T., Ehmann, G., Osaki, A., and Wolf, D. H., 1984, Proteolysis in eukaryotic cells: Proteinase yscE, a new yeast peptidase, *J. Biol. Chem.* **259:** 13344–13348.

35. Wolf, D. H., 1985, Proteinases, proteolysis and regulation in yeast, *Biochem. Soc. Trans.* **13:** 279–283.

36. Hase, J., Kobashi, K., Nakai, N., Mitsui, K., Iwata, K., and Takadera, T., 1980, The quaternary structure of carp muscle alkaline protease, *Biochim. Biophys. Acta* **611:** 205–213.

37. Etlinger, J. D., McMullen, H., Rieder, R. F., Ibrahim, A., Janeczko, R. A., and Marmorstein, S., 1985, Mechanisms and control of ATP-dependent proteolysis, in: *Intracellular Protein Catabolism* (E. A. Khairallah, J. S. Bond, and J. W. C. Bird, eds.), Alan R. Liss, New York, pp. 47–60.

38. Ishiura, S., and Sugita, H., 1986, Ingensin, a high-molecular-mass alkaline protease from rabbit reticulocyte, *J. Biochem.* **100:** 753–763.

39. Edmunds, T., and Pennington, R. J. T., 1982, A high-molecular weight peptide hydrolase in erythrocytes, *Int. J. Biochem.* **14:** 701–703.

40. McGuire, M. J., and DeMartino, G. N., 1986, Purification and characterization of a high molecular weight proteinase (macropain) from human erythrocytes, *Biochim. Biophys. Acta* **873:** 279–289.

41. Wilk, S., and Orlowski, M., 1983, Evidence that pituitary cation-sensitive neutral endopeptidase is a multicatalytic protease complex, *J. Neurochem.* **40:** 842–849.

42. Wilk, S., and Orlowski, M., 1980, Cation-sensitive neutral endopeptidase; Isolation and specificity of the bovine pituitary enzyme, *J. Neurochem.* **35:** 1172–1882.

43. Wilk, S., Pearce, S., and Orlowski, M., 1979, Identification and partial purification of a cation-sensitive neutral endopeptidase from bovine pituitaries, *Life Sci.* **24:** 457–464.
44. Blow, A. M. J., Van Heyningen, R., and Barrett, A. J., 1975, Metal-dependent proteinase of the lens: Assay, purification and properties of the bovine enzyme, *Biochem. J.* **145:** 591–599.
45. Ray, K., and Harris, H., 1985, Purification of neutral lens endopeptidase: Close similarity to a neutral proteinase in pituitary, *Proc. Natl. Acad. Sci. U.S.A.* **82:** 7545–7549.
46. Ray, K., and Harris, H., 1986, Comparative studies on lens neutral endopeptidase and pituitary neutral proteinase: Two closely similar enzymes, *FEBS Lett.* **194:** 91–95.
47. Dahlmann, B., Kuehn, L., and Reinauer, H., 1983, Identification of three high molecular mass cysteine proteinase from rat skeletal muscle, *FEBS Lett.* **160:** 243–247.
48. Dahlmann, B., Kuehn, L., Rutschmann, M., and Reinauer, H., 1985, Purification and characterization of a multicatalytic high-molecular-mass proteinase from rat skeletal muscle, *Biochem. J.* **228:** 161–170.
49. Dahlmann, B., Rutschmann, M., Kuehn, L., and Reinauer, H., 1985, Activation of the multicatalytic proteinase from rat skeletal muscle by fatty acids or sodium dodecyl sulphate, *Biochem. J.* **228:** 171–177.
50. Edmunds, T., and Pennington, R. J. T., 1985, A high molecular weight peptide hydrolase from rat skeletal muscle, in: *Intracellular Protein Catabolism* (E. A. Khairallah, J. S. Bond, and J. W. C. Bird, eds.), Alan R. Liss, New York, pp. 235–237.
51. Hardy, M. F., Mantle, D., Edmunds, T., and Pennington, R. J. T., 1981, A high-molecular-weight enzyme from skeletal muscle which hydrolyses chymotrypsin substrates, *Biochem. Soc. Trans.* **9:** 218–220.
52. Ismail, F., and Gevers, W., 1983, A high-molecular-weight cysteine endopeptidase from rat skeletal muscle, *Biochim. Biophys. Acta* **742:** 399–408.
53. Ishiura, S., Sano, M., Kamakura, K., and Sugita, H., 1985, Isolation of two forms of the high-molecular mass serine protease, ingensin, from porcine skeletal muscle, *FEBS Lett.* **189:** 119–123.
54. DeMartino, G. N., and Goldberg, A. L., 1979, Identification and partial purification of an ATP-stimulated alkaline protease in rat liver, *J. Biol. Chem.* **254:** 3712–3715.
55. Rose, I. A., Warms, J. V. B., and Hershko, A., 1979, A high molecular weight protease in liver cytosol, *J. Biol. Chem.* **254:** 8135–8138.
56. Sherman, M. T., Moran, M. C., Tuazon, F. B., and Stevens, Y.-W., 1983, Structure, dissociation, and proteolysis of mammalian steroid receptors, *J. Biol. Chem.* **258:** 10366–10377.
57. Rivett, A. J., 1985, Preferential degradation of the oxidatively modified form of glutamine synthetase by intracellular mammalian proteases, *J. Biol. Chem.* **260:** 300–305.
58. Rivett, A. J., 1985, Purification of a liver alkaline protease which degrades oxidatively modified glutamine synthetase, *J. Biol. Chem.* **260:** 12600–12606.
59. Tanaka, K., Kunio, I., Ichihara, A., Waxman, L., and Goldberg, A., 1986, A high molecular weight protease in the cytosol of rat liver. I. Purification, enzymological properties, and tissue distribution, *J. Biol. Chem.* **261:** 15197–15203.
60. Tanaka, K., Yoshimura, T., Ichihara, A., Kamayama, K., and Takagi, T., 1986, A high molecular weight protease in the cytosol of rat liver. II. Properties of the purified enzyme, *J. Biol. Chem.* **261:** 15204–15207.
61. Yamamoto, T., Nojima, M., Ishiura, S., and Sugita, H., 1986, Purification of the two forms of the high-molecular-weight neutral proteinase ingensin from rat liver, *Biochim. Biophys. Acta* **882:** 297–304.
62. Ishiura, S., Yamamoto, T., Nojima, M., and Sugita, H., 1986, Ingensin, a fatty acid-activated serine proteinase from rat liver cytosol, *Biochim. Biophys. Acta* **882:** 305–310.

63. Nojima, M., Ishiura, S., Yamamoto, T., Okuyama, T., Furuya, H., and Sugita, H., 1986, Purification and characterization of a high-molecular-weight protease, ingensin, from human placenta, *J. Biochem.* **99:** 1605–1611.

64. Ishiura, S., Nojima, M., Yamamoto, T., Fuchiwaki, T., Okuyama, T., Furuya, H., and Sugita, H., 1986, Effects of linoleic acid and cations on the activity of a novel high-molecular weight protease, ingensin, from human placenta, *Int. J. Biochem.* **18:** 765–769.

65. DeMartino, G. N., 1983, Identification of a high molecular weight alkaline protease in rat heart, *J. Mol. Cell. Cardiol.* **15:** 17–29.

66. Zolfaghari, R., Baker, C. R. F., Jr., Canizaro, P. C., Amirgholami, A., and Behal, F. J., 1987, A high-molecular-mass neutral endopeptidase-24.5 from human lung, *Biochem. J.* **241:** 129–135.

67. Gregory, M. R., and Notides, A. C., 1982, Characterization of two uterine proteases and their actions on the estrogen receptor, *Biochemistry* **21:** 6452–6458.

68. Iwata, K., Kohashi, K., and Hase, J., 1973, *Bull. Jpn. Soc. Sci. Fisheries* **39:** 1325–1337; cited in ref. 36.

69. Kopp, F., Steiner, R., Dahlmann, B., Kuehn, L., and Reinauer, H., 1986, Size and shape of the multicatalytic proteinase from rat skeletal muscle, *Biochim. Biophys. Acta* **872:** 253–260.

70. Barrett, A. J., 1977, Introduction to the history and classification of tissue proteinases, in: *Proteinases in Mammalian Cells and Tissues*, North-Holland, New York, pp. 1–55.

71. Sasaki, T., Kikuchi, T., Yumoto, N., Yoshimura, N., and Murachi, T., 1984, Comparative specificity and kinetic studies on porcine calpain I and calpain II with naturally occurring peptides and synthetic fluorogenic substrates, *J. Biol. Chem.* **259:** 12489–12494.

72. Sasaki, T., Kikuchi, T., Fukui, I., and Murachi, T., 1986, Inactivation of calpain I and calpain II by specifically oriented tripeptidyl chloromethyl ketones, *J. Biochem.* **99:** 173–179.

73. Gould, N. R., and Liener, I. E., 1965, Reaction of ficin with diisopropyphosphofluoridate. Evidence for contaminating inhibitor, *Biochemistry* **4:** 90–98.

74. Jansen, E. F., Nutting, M.-D. F., and Balls, A. K., 1948, The reversible inhibition of acetylesterase by diisopropyl fluorophosphate and tetraethyl pyrophosphate, *J. Biol. Chem.* **175:** 975–987.

75. Chaiken, I. M., and Smith, E. L., 1969, Reaction of a specific tyrosine residue of papain with diisopropylfluorophosphate, *J. Biol. Chem.* **244:** 4247–4250.

76. Ebata, M., and Yasunobu, K. T., 1963, Chymopapain III. The inhibition of chymopapain by diisopropylphosphorofluoridate, *Biochim. Biophys. Acta* **73:** 132–144.

77. Murachi, T., and Yasui, M., 1965, Alkylphosphorylation of stem bromelain by diisopropylphosphorofluoridate without inhibition of proteinase activity, *Biochemistry* **4:** 2275–2282.

78. Murachi, T., Inagami, T., and Yasui, M., 1965, Evidence for alkylphosphorylation of tyrosyl residues of stem bromelain by diisopropylphosphorofluoridate, *Biochemistry* **4:** 2815–2825.

79. Harper, J. W., Hemmi, K., and Powers, J. C., 1985, Reaction of serine proteases with substituted isocoumarins: Discovery of 3,4'-dichloroisocoumarin, a new general mechanism based serine protease inhibitor, *Biochemistry* **24:** 1831–1841.

80. Dahlmann, B., 1985, High M_r cysteine proteinases from rat skeletal muscle, *Biochem. Soc. Trans.* **13:** 1021–1023.

81. Rieder, R. F., Ibrahim, A., and Etlinger, J. D., 1985, A particle-associated ATP-dependent proteolytic activity in erythroleukemia cells, *J. Biol. Chem.* **260:** 2015–2018.

82. Waxman, L., Fagan, J. M., Tanaka, K., and Goldberg, A. L., 1985, A soluble ATP-dependent system for protein degradation from murine erythroleukemia cells, *J. Biol. Chem.* **260:** 11994–12000.

83. Watabe, S., and Kimura, T., 1985, ATP-dependent protease in bovine adrenal cortex: Tissue specificity, subcellular localization and partial characterization, *J. Biol. Chem.* **260:** 5511–5517.

84. Watabe, S., and Kimura, T., 1985, Adrenal cortex mitochondrial enzyme with ATP-dependent protease and protein-dependent ATPase activities: Purification and properties, *J. Biol. Chem.* **260:** 14498–14504.

85. Katayama-Fujimura, Y., Gottesman, S., and Maurizi, M. R., 1987, A multiple-component, ATP-dependent protease from *Escherichia coli*, *J. Biol. Chem.* **262:** 4477–4485.

86. Charette, M. F., Henderson, G. W., and Markovitz, A., 1981, ATP-hydrolysis-dependent protease activity of the *lon* (*capR*) protein of *Escherichia coli* K-12, *Proc. Natl. Acad. Sci. U.S.A.* **78:** 4728–4732.

87. Chung, C. H., and Goldberg, A. L., 1981, The product of the *lon* (*capR*) gene in *Escherichia coli* is the ATP-dependent protease, protease La, *Proc. Natl. Acad. Sci. U.S.A.* **78:** 4931–4935.

88. Phillips, T. T., VanBogelen, R., and Neidhardt, F., 1984, *Lon* gene product of *Escherichia coli* is a heat-shock protein, *J. Bacteriol.* **159:** 283–287.

89. Maurizi, M. R., 1987, Degradation *in vitro* of bacteriophage λ N protein by lon protease from *Escherichia coli*, *J. Biol Chem.* **262:** 2696–2703.

90. Goldberg, A. L., and Waxman, L., 1985, The role of ATP hydrolysis in the breakdown of proteins and peptides by protease La from *Escherichia coli*, *J. Biol. Chem.* **260:** 12029–12034.

91. Waxman, L., and Goldberg, A. L., 1986, Selectivity of intracellular proteolysis: Protein substrates activate the ATP-dependent protease (La), *Science* **232:** 500–503.

92. Menon, A. S., Waxman, L., and Goldberg, A. L., 1987, The energy utilized in protein breakdown by the ATP-dependent protease (La) from *Escherichia coli*, *J. Biol. Chem.* **262:** 722–726.

93. Reynolds, J. A., Johnson, E. A., and Tanford, C., 1985, Application of the principle of linked functions to ATP-driven ion pumps: Kinetics of activation by ATP, *Proc. Natl. Acad. Sci. U.S.A.* **82:** 3658–3661.

94. Pollard, T. D., and Craig, S. W., 1982, Mechanism of actin polymerization, *TIBS* **7:** 55–58.

95. Cleveland, D. W., 1982, Treadmilling of tubulin and actin, *Cell* **28:** 689–691.

96. Brenner, S. L., Mitchell, R. S., Marrical, S. W., Neuendorf, S. K., Schutte, B. C., and Cox, M. M., 1987, recA protein-promoted ATP hydrolysis occurs throughout recA nucleoprotein filaments, *J. Biol. Chem.* **262:** 4011–4016.

97. Edmunds, T., and Goldberg, A. L., 1986, Role of ATP hydrolysis in the degradation of proteins by protease La from *Escherichia coli*, *J. Cell Biochem.* **32:** 187–191.

98. Rothman, J. E., and Kornberg, R. D., 1986, An unfolding story of protein translocation, *Nature* **322:** 209–210.

99. Pollard, T. D., 1987, The myosin crossbridge problem, *Cell* **48:** 909–910.

100. Suzuki, K., 1987, Calcium activated neutral protease: Domain structure and activity regulation, *TIBS* **12:** 103–105.

101. Bålöw, R.-M., Tomkinson, B., Ragnarsson, U., and Zetterqvist, Ö., 1986, Purification, substrate specificity, and classification of tripeptidyl peptidase II, *J. Biol. Chem.* **261:** 2409–2417.

102. Schmid, H.-P., Akhayat, O., Martins de Sa, C., Puvion, F., Koehler, K., and Scherrer, K., 1984, The prosome: An ubiquitous morphologically distinct RNP particle associated

with repressed mRNPs and containing specific ScRNA and a characteristic set of proteins, *EMBO J.* **3**:29–34.

103. Shelton, E., Kuff, E. L., Maxwell, E. S., and Harrington, J. T., 1970, Cytoplasmic particles and aminoacyl transferase I activity, *J. Cell Biol.* **45**: 1–8.

104. Kleinschmidt, J. A., Hügle, B., Grund, C., and Franke, W. W., 1983, The 22 S cylinder particles of *X. laevis*. I. Biochemical and electron microscopic characterization, *Eur. J. Cell Biol.* **32**: 143–156.

105. Martins de Sa, C., Grossi de Sa, M.-F., Akyakat, O., Broders, F., Scherrer, K., Horsch, A., and Schmid, H.-P., 1986, Prosomes: Ubiquity and inter-species structural variation, *J. Mol. Biol.* **187**: 479–493.

106. Castaño, J. G., Ornberg, R., Koster, J. G., Tobian, J. A., and Zasloff, M., 1986, Eucaryotic pre-tRNA 5′ processing nuclease: Copurification with a complex cylindrical particle, *Cell* **46**: 377–387.

107. Vorgias, C. E., and Traub, P., 1986, Efficient degradation *in vitro* of all intermediate filament subunit proteins by the Ca^{2+}-activated neutral thiol proteinase from ehrlich ascites tumor cells and porcine kidney, *Biosci. Rep.* **6**: 57–64.

Chapter 5

Ubiquitin Carboxyl-Terminal Hydrolases

Irwin A. Rose

1. INTRODUCTION

There are a number of known ubiquitin (Ub) derivatives for which hydrolase activities are required to regenerate a functional Ub molecule. For example, removal of ubiquitin from ubiquitinated histones is extensive during metaphase and during periods of altered gene expression. One may expect that Ub, which is a very old protein in evolutionary terms, may be involved in post-translational modifications of other proteins, and hydrolases, possibly with allosteric control features, will be needed to control the degree of modification. Moreover, judging from the gene and mRNA structures that have been reported, some translated forms of Ub will consist of tandem repeats. This calls for Ub-specific peptidases in contrast to the hydrolases introduced above. In fact, fusion proteins involving Ub are rapidly hydrolyzed in yeast. In addition, a variety of conjugates of Ub can arise by reactions of Ub thiol esters of E1 and E2 with the simple thiols and amines that occur in cells. Hydrolysis is the only known way to repair these apparent mistakes.

Abbreviations used in this chapter: E1 = ubiquitin activating enzyme; E2 = ubiquitin transfer protein; E3 = E2-protein ubiquitin transferase; Ub = ubiquitin; Ubal = ubiquitin C-terminal aldehyde; DTT = dithiothreitol.

IRWIN A. ROSE • Institute for Cancer Research, Fox Chase Cancer Center, Philadelphia, Pennsylvannia 19111.

2. NOMENCLATURE

Assuming that the physiologically important hydrolytic enzymes will be specific for the Ub structure per se, we take the liberty of applying the inexact name "Ub hydrolase" to the enzymes that catalyze the reaction $UbX + H_2O \rightarrow Ub + XH$. The nature of the major substrate for each enzyme can be used to subdivide the hydrolases as follows: ULH, where X is a small, nonprotein ligand; $U^\varepsilon PH$, where X is a protein in amide linkage between Gly-76 of Ub and an N^ε lysine of the protein, an isopeptide linkage, and $U^\alpha PH$, where X is a protein in peptide linkage to Ub.

Other abbreviations that have been suggested are UCH (for Ub carboxyl-terminal hydrolase, where X is a nonprotein ligand) and UPL (Ub–protein lyase). We prefer "hydrolase" to "lyase" to avoid confusion with "ligase." (The term "ligase" seems appropriate for joining segments of polynucleotides without requiring specificity. The term "synthetase" has been used for ATP-dependent reactions leading to specific synthesis. We therefore recommend the abbreviations ULS, $U^\alpha PS$, and $U^\varepsilon PS$ for enzymes and reactions that give specific extension at the carboxyl terminal Gly of Ub.) The terms "deconjugation" and "disassembly" have been used to distinguish the decomposition of Ub–protein conjugates by Ub-hydrolase from the "decompositions" that result from cleavage within the protein. Decompositions may be thought of as destructive to domains.

3. DISCOVERY OF UBIQUITIN HYDROLASES

The independent and more rapid turnover of the Ub half of histone conjugates,[1-4] its decrease in metaphase chromatin,[2,5,6] and its disappearance from nondividing ts85 cells[7] at 39°C imply the existence of a Ub–histone hydrolase. The first in vitro study of this reaction was obtained by incubation of liver nucleoli as the enzyme source and liver chromatin as the source of Ub–histone substrate.[8] The products, Ub and H2A, were identified by gel electrophoresis. This may have been an inadequate identification in view of the release from liver lysosomes of an enzyme that cleaves between Arg-74 and Gly-75 of Ub[9] and Ub conjugates. Later studies by Anderson et al.[10] using purified Ub–H2A and a postmicrosomal liver fraction demonstrated an enzyme that acts at the isopeptide linkage (Ub-Gly-Gly-76-ε-NH-Lys-protein), because the Ub molecules isolated from the cleavage reaction produced glycylglycine upon incubation with trypsin.[10] By this time Hershko, Ciechanover, and co-workers had demonstrated the ATP-dependent formation of Ub conjugates, the release of the heat-stable peptide upon removal of ATP, and the ability of released Ub to form conjugates in a reconstituted reticulocyte system.[11] Hemin, at concentrations that inhibited ATP-dependent breakdown of proteins, also inhibited the ATP-independent hydrolysis of Ub conjugates.[12] This observation provides the first confirmation of the role of conjugates in protein breakdown and raised the question, still unanswered, as to whether hydrolases have a role in the ATP-dependent phase of protein breakdown.

3.1. Assay for Ubiquitin Hydrolases

Enzymes that produce Ub as their immediate product may be measured by specific assay for Ub. The first assay was based on the Ub requirement for PP_i:ATP exchange with Ub activating enzyme E1.[13] A more convenient method that may be applied in either a coupled or stepwise mode is an end-point assay[14] (see Chapter 1):

$$UbX \xrightarrow[X]{E} Ub \xrightarrow[[^3H]ATP]{E1} E1 \cdot [^3H]AMP-Ub$$

The Ub formed in the hydrolase reaction is converted to $E1[^3H]AMP-$ Ub with Ub activating enzyme in slight excess over Ub. Iodoacetamide-treated E1 is used to prevent further reactions of the enzyme and $[2,8^3H]ATP$ of high specific activity allows the formation of acid-insoluble radioactivity with AMP–Ub specific activities of about 10,000 cpm/pmole of Ub. The E1 is prepared from human erythrocytes by methods that remove most, but not all, hydrolase activity. Treatment with iodoacetamide inactivates remaining hydrolase activities and blocks the Ub accepting thiol group of the E1 so that the product of the assay will be the relatively stable E1·AMP–Ub. Sensitivity is a desirable feature of most hydrolase assays because of the limited amount of homogeneous substrate sample that might be available and because of the likelihood of inhibition by Ub produced in the reaction. These two considerations compromise an otherwise convenient assay using HPLC.[15] Product inhibition can be avoided in the coupled assay since Ub is converted to E·AMP–Ub as it is formed. When using the coupled assay, attention must be paid to the slow autocatalyzed hydrolysis of E1·AMP–Ub ($t_{1/2}$ = 10 min at 37°C). This effect is minimized by keeping the length of incubation within 10 min and by using sufficient iodoacetamide-treated E1 to assure an adequate rate of formation of the complex.

Liberation of labeled ligand, detected by electrophoresis[16] or conversion of radiolabeled ligand to an acid-soluble form,[17] has been used to assay hydrolases. Ideally, one would like a general substrate comparable to p-nitrophenylphosphate for the phosphatases. The ethyl ester of Ub has served such a role in studies by A. N. Mayer and K. D. Wilkinson (personal communication), and the labile acylphosphate bond of [^{3}H]AMP–Ub has made the compound useful for this purpose in the author's hands. The capacity of hydrolases to bind the C-terminal aldehyde form of Ub is used to determine the absolute concentration of active enzyme. (For synthesis and properties of Ubal, see Section 5.) In one approach, [^{125}I]Ubal is incubated briefly with enzyme and the complex separated from free Ubal on DE-52 or Sephacryl S-200. The Ubal, made from Ub of known specific activity (determined by comparing ^{3}H/^{125}I of the E1·AMP–Ub isolated on Sephacryl S-200), forms sufficiently tight complexes with a number of Ub hydrolases to permit their determination in molar terms by separation from the lower molecular weight Ubal. A method that does not depend on labeled Ubal has also been used. The hydrolase is titrated with a known concentration of Ubal, and the loss of hydrolase activity is measured.

4. NONPROTEIN LIGAND–UBIQUITIN HYDROLASES (ULH)

4.1. Discovery

It was observed in early studies with Ub activating enzyme[18] that the liberation of PP$_i$ in the reaction

$$2ATP + 2Ub + E1_{SH} \rightarrow E1_{S-Ub}^{AMP-Ub} + AMP + 2PP_i$$

was rapid to the extent of two enzyme equivalents. In the presence of thiols, such as dithiothreitol (DTT), the initial burst rate was followed by a much slower rate that varied in proportion to the concentration of DTT added. The continuing reaction was attributed to regeneration of free enzyme by transacylation of Ub from E1 to DTT in the medium. This could be looked on as a catalytic process in which both activation and transfer of Ub were properties of the activating enzyme.

A second phenomenon was soon recognized: DTT–Ub itself could be shown to be turning over; that is, utilization of ATP far exceeded the amount of Ub present. This additional cleavage of ATP stimulated by DTT depended on the presence of Ub and gave rise to AMP + PP$_i$, the products formed in the burst phase. Presumably, Ub was being recycled from DTT–Ub. Using [^{125}I]Ub and DTT there was no trace of [^{125}I]Ub retention by an Hg$^+$ column. This indicated that DTT–Ub did not accumulate in the original reaction mixture. In view of these results, we suspected that DTT–Ub was either inherently unstable or was being hydrolyzed by an enzymatic process.

To determine its chemical stability, DTT–Ub was synthesized by an alternative route. AMP–Ub was formed by E1 with [^{125}I]Ub and precipitated as such with 1 M TCA. Upon solubilization and treatment with DTT, DTT–Ub could be demonstrated by the Hg$^+$ column method. It was in fact quite stable, but only if 4 M urea present when the TCA pellet containing the AMP–Ub was solubilized. Therefore, we assumed that an acid-stable, urea-sensitive Ub thiolesterase contaminated the activating enzyme. Upon further purification, the Ub activating enzyme did not cause breakdown of ATP in excess of total Ub in the presence of DTT, or at least did so at a much slower rate. The recycling of Ub could be restored by addition of the less purified activating enzyme sample. The hydrolase activity toward AMP–Ub and DTT–Ub was then shown to be caused by a 30 kDa protein with notable stability to acid. It was later shown[19,20] that certain aliphatic amines are good acceptors for Ub molecules that have been activated by E1 in the presence of the Ub carrier proteins, E2s. These Ub amides are also good substrates for the hydrolase. The hydrolase probably uses a very nucleophilic thiol to form yet a third species of Ub thiol ester that differs from the others in its ease of hydrolysis (Scheme 1). As will be discussed later, the Ub that is transferred to a protein (R = lysyl residue in Scheme 1) due to action of E3 may be regenerated by action of hydrolases. Most of these are also SH enzymes, although at least one hydrolase has been found to be sensitive to PMSF.

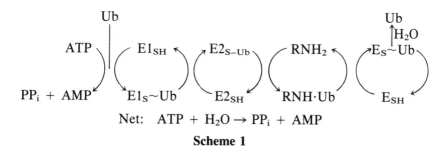

Net: ATP + H$_2$O → PP$_i$ + AMP

Scheme 1

4.2. Assays for Ubiquitin Carboxyl Ligand Hydrolases

The original assay using [^{32}P]ATP, E1, DTT, and pyrophosphatase and based on ^{32}P$_i$ formation is quite sensitive. However, it is limited to the V/K region of Ub–DTT dependence and therefore is not easily defined inasmuch as it depends on DTT concentration and three enzyme catalyzed steps. The general two-step assay (see Section 3.1), which is used frequently with Ub amide as substrate, relies on the increase in Ub with time as determined with iodacetamide-treated E1 and [^{3}H]ATP.

A direct assay with labeled ligand can be made sensitive and convenient if the free ligand is either acid soluble or of different charge than the substrate. Such an assay, not dependent on Ub formation, avoids the problem of a high Ub blank contributed by the enzyme. In our experience [^{3}H]AMP–Ub is a useful substrate because conversion of label from acid-insoluble to -soluble form is easily measured. [^{3}H]ATP, Ub, and iodoacetamide-treated E1 are incubated with a slight excess of E1 over Ub^{14} and TCA is added to liberate the AMP–Ub that is coprecipitated with added BSA for use as substrate. The hydrolase reaction is followed by production of TCA-soluble counts, which follows first-order kinetics to completion. A more stable radiolabeled Ub–X has been prepared using [^{3}H]NaBH$_4$ and the Ub amide of amino butyraldehyde.^{17} This substrate has not proved as useful as labeled AMP–Ub because of a high blank, less complete conversion to acid-soluble form, and a more involved synthesis.

4.3. Purification of ULHs

Isolation of homogeneous carboxy-terminal ligand hydrolase (ULH-A) of rabbit reticulocyte fraction II makes use of its acid stability and its affinity for a column of Ub–Sepharose.^{20} A large-scale preparation has been reported using human red cells that contain ~0.4 μM enzyme.^{20} Attempts to obtain large amounts of enzyme by this procedure have been troubled by failure of the enzyme to bind to the affinity column. This may result from competition with significant amounts of free Ub coming from the breakdown of endogenous conjugates or complexes. This possibility is supported by the observation (A. Hershko and H. Heller, personal communication) that a second DE-52 step following the acid precipitation step improved the yield of pure enzyme.

4.4. Specificity and Mechanistic Studies

Although the tests used in these studies have varied in many ways, it seems safe to say that all the UbX compounds tested, where X is a small ligand (e.g., AMP, DTT, $^\varepsilon$N-lysine, glycine methyl ester, spermidine, hydroxylamine, and ethyl ester), are similarly active as substrates for the hydrolase. However, the hydrolase showed a negligible rate (~0.1%) with Ub–isopeptide conjugates of such small proteins as yeast cytochrome c and egg white lysozyme[19] or Ub–thiol ester of denatured E1.[18] It has since been tested with Ub–H2A and found to be negative (I. A. Rose and R. L. Wixom, unpublished results). Values for k_{cat} and k_{cat}/K_m have been obtained for DTT–Ub ($\sim$24 sec^{-1} and $\sim$10^8 M^{-1} sec^{-1}) and Lys–Ub ($\sim$6 sec^{-1} and >6 $\times$ 10^7 M^{-1} sec^{-1}).[19] Maximum velocity with $^{\varepsilon}$N-Lys-Ub corresponds to $\sim$10 units/mg with a K_m of $\sim$10^{-7} M. The K_m of Ub–ethanol ester and the K_i of Ub were both $\sim$5 $\times$ 10^{-7} M.[15] In the hydrolysis of DTT–Ub, Des-Gly-Gly Ub, (Ub74) was found to be about 10$\times$ less inhibitory than Ub.[18] As shown in Section 5, there is evidence that Ub itself is a substrate for the hydrolase; that is, the carboxyl-terminal glycine of Ub interacts strongly with the enzyme, probably forming a covalent bond with the active site of the enzyme.

This wide range in substrate specificity is necessary to assure regeneration of Ub from conjugates that will result from the side reactions of E1 and E2 with cellular and environmental nucleophiles such as glutathione and spermidine. These reactions would divert all the cells' free Ub to useless forms in less than 60 sec. Any compound present in the cell that could trap the Ub in a form not susceptible to hydrolysis would present a considerable problem. For this reason perhaps, the specificity of Ub transfer from the E2–Ub to amines is restricted to terminal amino groups, that is, glycine and β-alanine but not alanine. Were the transfer specificity less sterically restrained than that of the hydrolases, the supply of the cell's Ub would be in danger.

Although the regeneration of Ub from nonfunctional forms is clearly significant, there is also the suggestion that the liberation of Ub from proteolytically processed protein conjugates would also be expected from this enzyme's ability to hydrolyze the Ub $^{\varepsilon}$N-amide of lysine methyl ester.[19] In that case, the amount of the enzyme is so great that under normal circumstances these terminal conjugates would scarcely be detectable at steady state. Estimating the hydrolase to be $\sim$6 $\times$ 10^{-7} M from its level in red cells[19] and $k_{cat}/K_m \simeq$ 10^8 M^{-1} sec^{-1} for Ub–$^{\varepsilon}$N lysine methyl ester, such a pool would have a half-life of 0.01 sec.

ULH-A is a thiol enzyme, sensitive to iodoacetamide protected by Ub.[18] The enzyme was shown to hydrolyze Ub hydroxamate formed in the presence of E1 plus E2 with hydroxylamine as the ultimate acceptor.[19] The observation that recycling of Ub slowly came to a stop and returned to the original rate upon addition of fresh hydrolase indicated either that the enzyme was slowly inactivated during hydrolysis of Ub hydroxamate or that hydroxylamine itself was a slow inactivator. Inactivation was not observed with hydroxylamine alone but required Ub. This suggested that hydrolase and Ub interact to form an activated complex, perhaps the Ub-enzyme derivative that is attacked by hydroxylamine to give inactive enzyme or complex. Although a thiol ester intermediate would be consistent with the sensitivity of the enzyme to thiol reagents, it would not explain the hydroxylamine effect since free Ub hydroxamate is a good substrate. The data could be explained by a mixed anhydride E–Ub intermediate that was attacked occasionally by hydroxylamine ($HONH_2$) to form an inactive enzyme acyl hydroxamate:

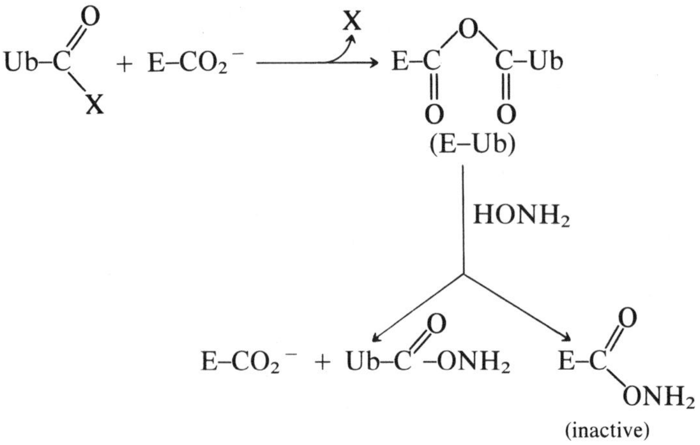

This mechanism can be ruled out by kinetic measurements. In a Ub-dependent inactivation study, an unusually low K_i value of ~1 mM was found for hydroxylamine. It was possible that this low K_i and the low rate of inactivation at saturation, ~10^{-2} sec^{-1} compared with k_{cat} ~20 sec^{-1}, might result if a step prior to formation of the anhydride were to become rate limiting at higher hydroxylamine concentrations. It was shown that even with very high concentrations of hydroxylamine, from 10 to 50× its apparent K_m. the enzyme was able to catalyze ~1200 hydrolytic turnovers for every molecule of enzyme inactivated. This result requires a more complicated model than the one shown below in which hydroxylamine reacts with an intermediate on the pathway of hydrolysis:

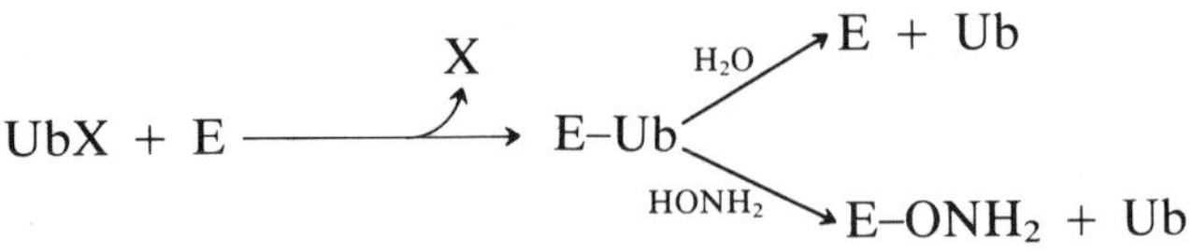

Since high levels of hydroxylamine did not markedly increase inactivation, competition between hydroxylamine and water, either as free or bound substrates, is ruled out. A satisfactory model to explain both the low value of K_m for $HONH_2$ and the lack of competition assumes that the main path acyl intermediate does not react directly with $HONH_2$; rather, a mixed anhydride is formed by spurious acyl transfer to a carboxyl group close to the active-site nucleophile. The reaction of the mixed anhydride with $HONH_2$ would be first order in $HONH_2$ and could be much more rapid than its reaction with water except that above 1 mM $HONH_2$, the acyl transfer becomes rate limiting at 10^{-3} times the rate of the competing hydrolysis of the intermediate. This Ub–acyl transfer must be readily reversible in the absence of a trap. In fact, one would be justified in concluding that Ub alone is as good a substrate of the enzyme as any of the Ub–X conjugates that have been studied: inactivation rate $\times$ (turnovers/inactivation) $= 10^{-2}$ sec^{-1} $\times$ 1200 $= 12$ sec^{-1}. This common rate seems unlikely to be determined by formation of the acyl-E intermediate, since the commitment to catalysis occurs at the diffusion rate, and the rate is independent of the leaving group X over a very wide range from AMP to HO^-.

Whether the common rate-limiting step for all substrates is acyl–enzyme hydrolysis or release of Ub may be determined by direct analysis for the acyl intermediate during steady state. It seems most likely that Ub release will be found to be the common rate-determining step since with $k_{cat}K_m = \sim10^8$ M^{-1} sec^{-1} and K_i of Ub $\sim10^{-7}$ M, then k_{off} would be 10 sec^{-1}. This value for k_{off} agrees with k_{cat} for most substrates, and it is likely that the noncovalent E·Ub contributes much more than the covalent E–Ub to the K_i equilibrium.

The presumption of a mixed anhydride led Pickart and Rose to test sodium borohydride as an inactivator.[17] Again, simply incubating the hydrolase with Ub and $NaBH_4$ resulted in inactivation. Several characteristics of the inactive product were best explained by a complex of enzyme with Ub aldehyde,

$$\text{Ub–}\underset{\underset{H}{|}}{C}\text{·E} \qquad \text{(Ubal enzyme complex)}$$

with $\overset{O}{\overset{\|}{}}$

or more likely,

$$\text{Ub–}\underset{\underset{H}{|}}{\overset{\overset{OH}{|}}{C}}\text{—S–E} \qquad \begin{array}{l}\text{(Ubal thiol hemiacetal}\\ \quad\text{enzyme complex)}\end{array}$$

1. Using [^{3}H]NaBH$_4$, with Ub and enzyme, tritium was found on a

calibrated Sepharose G75 column in the 30 kDa region, which also contained some active enzyme.

2. The inactive, labeled species after denaturing (heat, SDS, or acid) was not found at 30 kDa on gel electrophoresis but at ~8 kDa.

3. The recovered 8 kDa species did not behave like Ub but in fact caused inactivation when added to fresh enzyme in the absence of $NaBH_4$.

4. The inactivation was prevented by prior treatment of the presumed Ubal with $NaBH_4$ or sodium bisulfite ($NaHSO_3$).

5. With [^{3}H]$NaBH_4$ to reduce the presumed Ubal, followed by acid hydrolysis of the protein, the main tritiated product behaved like [^{3}H]ethanolamine by chromatography and produced [^{3}H]formaldehyde after periodate oxidation.

6. In the case of protection by bisulfite, inactivating Ubal could be regenerated following acid precipitation. This is consistent with an acid labile α-OH sulfonate, the product expected upon bisulfite addition to an aldehyde.

7. Chemical synthesis of Ubal by nonreductive methods, detailed in Section 5, has produced an inhibitor with the same properties as the Ubal obtained by one-step reduction on the enzyme.

8. A thiohemiacetal form of enzyme-bound Ubal is proposed since it would be insensitive to a second reduction by $NaBH_4$, and since the enzyme is inactivated by iodoacetamide ($t_{1/2} \simeq 10$ min at 0.5 mM iodoacetamide).[18]

9. The acid stability of the enzyme may be used to recover full hydrolase activity using $NaHSO_3$ in the buffer used to neutralize the Ubal–enzyme complex. Thus, it is clear that $NaBH_4$ reduction had no effect on the enzyme itself.

The reduction of Ub by $NaBH_4$ is not likely to occur from a simple E·Ub binary addition complex for the following reasons: (1) Ub74 at concentrations suitable for complex formation does not lead to inactivation with $NaBH_4$; (2) reduction of carboxyl functions by borohydride would be quite unusual; and (3) single-step reduction would be unexpected since aldehydes are much better substrates for reduction by $NaBH_4$ than are acids. Although a simple binary complex has not been excluded, it is more likely that the species is a Ub acyl intermediate of the enzymatic hydrolysis mechanism. For example,

$$\text{Ub–}\overset{O}{\overset{\|}{C}}\text{–OH} + {}^-\text{S–E} \rightarrow \text{Ub–}\overset{O}{\overset{\|}{C}} \sim \text{S–E} \xrightarrow{H^-} \text{Ub–}\underset{H}{\overset{OH}{\overset{|}{\underset{|}{C}}}}\text{–S–E}$$

Also, the great stability of the complex protects the potential aldehyde from reduction. The stability of the complex mitigates against a carboxylate nucleophile and a mixed anhydride as the primary intermediate because adducts between aldehydes and carboxyl groups are unknown even in the crystalline state. Other evidence for the high stability of the inhibited complex is (1) $K_{diss} < 10^{-12}$ M by titration is far less than $K_{diss} \simeq 10^{-7}$ M for Ub; (2) the reduced Ub–enzyme complex chromatographs as a sharp peak on extended separation by anion exchange as well as sizing columns; and (3) the inactivated complex is stable at high dilution. The complex is disrupted only by denaturation such as heating at 60°C for 10 min or by brief acid precipitation.

Wilkinson et al.[15] have made some further observations of mechanistic interest with reticulocyte ULH-A. Although the importance of its thiol group had been noted,[18] it was found with the ethyl ester of Ub that a time-dependent decrease in activity, due in part to product inhibition, was even greater unless DTT was present at high concentration. The origin of this effect may be more interesting than maintenance of the thiol function of the enzyme. Activation by high levels of DTT has not been seen when initial rates with low substrate levels are measured. This suggests that DTT may be relieving the inhibition by Ub, perhaps acting as a nucleophile to prevent the formation of side products that may develop from the mixed anhydride that was proposed to explain the hydroxylamine effect. A. N. Mayer and K. D. Wilkinson (personal communication) report the progressive loss of activity when catalysis is promoted at pH 5.5. Full activity is recovered when the pH is raised to 7.5. Using Ub–ethyl ester as substrate and HPLC analysis for Ub, these authors have analyzed pH dependence in terms of pK values of 5.2, 7.6, and 9.5. Maximum rate was at pH 8.5.

5. UBAL SYNTHESIS AND EVIDENCE FOR OTHER UBAL-SENSITIVE HYDROLASES

Ubal is readily made according to the method of Pickart and Rose[17] using sodium borohydride, hydrolase, and Ub. The Ubal is released upon mild heat denaturation of the complex after isolation, dialysis, and lyophilization. This is the most rapid procedure for preparing pure [^{125}I]Ubal of high specific activity from less than pure [^{125}I]Ub for use in identifying the other hydrolases. A second method that has allowed production of milligram amounts of Ubal (I. A. Rose, unpublished results) makes use of trypsin-catalyzed synthesis as reported by Wilkinson et al.[15] In this case, the C-terminal Gly-Gly of Ub is exchanged by Gly–amino propa-

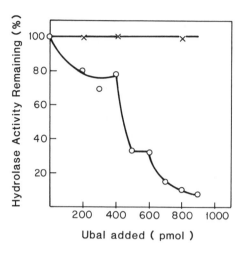

Figure 1. Differential sensitivity of hydrolases to Ubal. To rabbit reticulocyte fraction II (1 ml, 20 mg protein) was added the noted amount of Ubal. Samples were assayed for the initial rate of hydrolysis of [^{3}H]AMP–Ub ($\bigcirc$) and Ub–histone conjugates ($\times$).

nediol amide, which is readily prepared by reaction of Cbz glycine ethyl ester with 3-amino propane diol (both available) followed by reductive removal of Cbz by H_2 on Pd charcoal. The dialyzed, TCA-insoluble product, pre-Ubal, made in ~25% yield, is isolated on CM cellulose, pH 7.5, and reverse phase HPLC. It is converted to Ubal when needed by addition of HIO_4, followed, after an interval, by passage through DE-52 to remove the excess IO_4^-. The Ubal content is determined from the amount of hydrolase that it will inactivate. A second synthetic procedure (communicated to the author just prior to writing) was devised by Keith Wilkinson. In this method Des-Gly-Ub ethyl ester was converted to the hydrazide (by reaction with hydrazine), thence to the azide by reaction with $NaNO_2$, followed by reaction with 2-amino diethyl acetal. This method has the advantage that all the reagents are commercially available. The yield in the trypsin-catalyzed exchange reaction was ~48% using 2 M glycine ester, pH 6.8, 2 mg/ml trypsin, and 48°C for 12 hr.

When a solution of Ubal was added stepwise to reticulocyte fraction II, the disappearance of activity toward AMP–Ub established that a number of proteins, with different specific activities toward that substrate, were being titrated (Figure 1). Addition of the first 200 pmole of Ubal inhibited 20% of the total AMP–Ub hydrolase activity, whereas the next 200 pmole removed no more, presumably because another enzyme consumed the Ubal preferentially. Another 100 pmole removed about 45% of the remaining activity, which represents the enzyme of highest specific activity toward AMP–Ub hydrolyses, probably ULH-A. A second, low specific activity enzyme is titrated next. Since the assays are done after

100-fold dilution, it is likely that this represents inactivation and not inhibition. About 10% of the AMP–Ub hydrolase activity seems insensitive to Ubal.

Several proteins form Ubal complexes when whole fraction II is incubated with [^{125}I]Ub and NaBH$_4$. This is evident from the appearance on Sephacryl S-200 of peaks of radioactivity at ~150 kDa, >200 kDa, and at ~30 kDa. The same peaks appear if labeled Ubal is added directly to fraction II. They comprise 300–400 pmole /20 mg proteins of fraction II from 2 ml of reticulocytes. The high molecular weight complexes, >200 kDa, represent about half the total. The number of enzymes in this peak has not yet been determined. Hydrolase activity toward Ub–NH$_2$ is seen in the three regions that become labeled with Ubal. The majority migrates with an apparent M_r of 30 kDa.

6. OTHER HYDROLASES WITH MOLECULAR WEIGHT ~30 kDa AND GREATER

In addition to ULH-A described above, two other hydrolases able to split AMP–Ub have been obtained in pure form from reticulocyte extracts. ULH-B is obtained from fraction II as a second species in the ~30 kDa region on the Sephacryl S-200. The enzyme is insensitive to Ubal. It absorbs tightly to Ub–Sephadex 4B and is not eluted by 0.5 M KCl. A third hydrolase, ULH-C, was found in fraction I, the nonabsorbed effluent that is obtained when a dilute hemolyzate of reticulocytes is passed through DE-52 at pH 7.2. It was detected by [^{3}H]AMP–Ub hydrolysis. Hydrolase activity in fraction I was deemed likely since several studies have noted that some of the conjugates formed in the usual Ub/ATP/fraction II incubations did not readily disappear when ATP was withdrawn.[21,22] The AMP–Ub hydrolase activity in fraction I came as a single peak of apparent M_r = 30 kDa by filtration on Sephacryl S-200. The hydrolase activity was absorbed on passage through a Ub–Sepharose column and eluted with pH 9 buffer after washing the column with 0.5 M KCl. ULH-C does not appear in human erythrocytes, which contain ULH-A at about half the activity of rabbit reticulocyte fraction II.

A. N. Mayer and K. D. Wilkinson (personal communication) observed three regions of esterase activity from calf thymus cytosol using DEAE anion exchange (pH 7.5) and a KCl salt gradient. Four activities were identified when these peaks were further analyzed on Sephacryl S-300; three with M_r around 30 kDa and one at ~200 kDa. Of these, only one 30 kDa peak, which represented over 96% of the esterase activity, was absorbed on Ub–Sepharose 4B. All the enzymes were significantly inhibited by concentrations of Ubal well below the K_i value for Ub. The values $K_m = K_i$ of $UbOC_2H_5$ and Ub, respectively, were all in the micromolar range. All the esterase activities were completely lost after 15 min at 10 mM iodoacetamide, but not by either 10 mM EDTA or 1 mM PMSF.

7. UBIQUITIN–PROTEIN HYDROLASES

7.1. Assays

With a suitable Ub–protein conjugate in sufficient amount, direct assay for Ub produced by hydrolase is an immediately available procedure when one of the sensitive assays for Ub is used. Ub–H2A (A24) is available in pure form following the procedure of Hunter and Cary.[23] This substrate has been used in coupled assay with iodoacetamide-treated Ub activating enzyme and [³H]ATP to study the hydrolases of rabbit reticulocytes (I. A. Rose, R. L. Wixom, and M. Rigbi, unpublished results). However, greater sensitivity is achieved by the stepwise assay using longer incubation times. This assay then approaches the sensitivity of assays used by others with ¹²⁵I labeled Ub–H2A followed by polyacrylamide gel electrophoresis and radioautography. In lieu of other specific conjugates, a mixture of Ub–protein conjugates generated by incubation of fraction II with ATP from [¹²⁵I]Ub or [¹²⁵I]lysozyme provides a suitable mixture of substrates for the study of both the ATP-dependent proteases and ATP-independent hydrolases[21,22,24] using gel electrophoresis, radioautography, and determination of the distribution of label on the gel with time of incubation.

7.2. Ubiquitin–Histone Isopeptidases

An interest in the turnover of Ub associated with histones H2A and H2B has resulted in efforts to purify and characterize Ub–histone hydrolases from a variety of sources. Partial purification of such an activity (175-fold from calf thymus) has been reported after ion exchange, separation based on size, and elution from histone–Sepharose affinity columns.[16,25] The purification obtained by this last step was not what one would expect from a specific affinity interaction. Unfortunately, the enzyme did not interact with a Ub–Sepharose affinity column. The only published report of a molecular weight determination for a UPH assayed with Ub–H2A is that given for the cytosolic enzyme of cultured cells, 38 kDa. In its current state, this preparation was also shown to hydrolyze conjugates formed by rabbit reticulocyte fraction II.[25] [^{125}I]Ub was demonstrated to be the product of the hydrolysis of [^{125}I]Ub–H2A since it could be converted to conjugates by reticulocyte lysate.

The possibility of a histone specific Ub–H2A hydrolase is supported by observations of Goldknopf et al.[26,27] that show that Ub–H2A hydrolase activity of nuclei of mature chicken erythrocytes is negligible compared with that found in nuclei from immature, transcriptionally active cells. On the other hand the observation that the ubiquitin of nuclear histones is in rapid equilibrium with free ubiquitin in nondividing cells[28] suggests that the loss of ubiquitin may be at least in part due to a nonspecific process. In this case the control seen in the cell cycle may be a function of synthesis. Anderson et al.[10] extracted a rat liver Ub–H2A hydrolase that they demonstrated cleaved specifically at the isopeptide linkage since glycylglycine was produced by further treatment of the recovered Ub with trypsin. The enzyme was further localized to nucleoli in liver made hyperfunctional by partial hepatectomy and thioacetamide treatment, both of which lead to Ub-deficient nuclei.[8] Similarly, Anderson et al.[8] report an increase in enzyme activity associated with metaphase chromosomes with little change in the more abundant cytosolic enzyme in cultured cells. This also supports the existence of an Ub–hydrolase specific for uH2A and uH2B.

Reticulocyte fraction II has been studied as a possible source of a specific Ub–H2A hydrolase (I. A. Rose, R. L. Wixom, and M. Rigbi, unpublished results). A major source of activity was found in the high molecular weight region of a Sephacryl S-200 separation column where the "nonspecific" UPHs are also found. However, the two activities are due to different enzymes since the activity with Ub–H2A is not at all inhibited by Ubal (see Figure 1). On the other hand, in crude fraction II the hydrolysis of conjugates formed from lysozyme and methyl-Ub was completely inhibited by Ubal, and this resulted in the accumulation of endogenous protein conjugates[29] (see Section 7.4). Unlike many of the hydrolases of reticulocytes, the enzyme that acts on histone conjugates cannot be purified further by binding and elution from Ub–Sephadex 4B. In work on the Ub–histone hydrolase of rat nuclei, Anderson et al.[10] used PMSF liberally, indicating that their enzyme is probably not a serine hydrolase. The Ub–histone hydrolase of reticulocytes is sensitive to PMSF ($t_{1/2} < 10$ min with 1 mM PMSF).

7.3. Ubiquitin–$^\varepsilon$Protein Hydrolases of Reticulocytes

High molecular weight protein conjugates disappear if ATP is withdrawn from a reticulocyte incubation labeled with Ub.[11,12] In the absence of ATP, both synthesis and proteolytic breakdown of conjugates is halted, and the loss of conjugates must be due to hydrolases present in the reticulocyte fraction. This phenomenon is usually studied by the addition of conjugates generated in a preliminary incubation, isolated, and returned to a fraction II incubation or lysate with or without ATP.[21,22] In both reports cited, the disappearance of large conjugates was as rapid in the absence of ATP as in its presence. The hydrolases and proteases seem to compete for the conjugates since, with ATP present, conjugates derived from lysozyme produce as much lysozyme as acid-soluble radioactivity.

Hough et al.[24] examined the fractions of a DEAE–Fractogel column for a general Ub–protein hydrolase enzyme and found a major peak of activity that sedimented on glycerol gradient sedimentation with M_r ~200 kDa. A. N. Mayer and K. D. Wilkinson (personal communication, 1986) have noted the presence of a comparable enzyme in calf thymus and rat liver. In addition, they observed general Ub–protein hydrolase activity at 39 kDa. E. Leshinsky and A. Hershko have obtained $U^\varepsilon PH$ activities with M_r of 100, 300, and 400 kDa from rabbit reticulocyte fraction II (cited in ref. 29).

7.4. Role of Hydrolases in the Ubiquitin/ATP-Dependent Proteolytic Process

In addition to their importance for regeneration of Ub, hydrolases may have proofreading or correction roles.[11] It was supposed that proteins, whether native or denatured, may be conjugated in ways incompatible with the specificity requirements of the Ub-dependent protease. In this case, there must be a route, other than proteolysis, to restore the protein to its original state. If native, the protein would presumably resume normal function; if the protein remained, this would allow another opportunity for successful action by E3. Wilkinson and co-workers[30,31] have provided a structural model for the proofreading process. If Ub is able to interact with hydrophobic regions of a protein to which it is conjugated, it could result in conformational changes that make the protein a better substrate for proteolysis and possibly a poorer substrate for the hydrolases by virtue of the alteration in liganded protein, in Ub, or in both. Conjugates of Ub with native proteins would be less likely to develop these interactions.

Clues to the role of hydrolases in protein breakdown may be derived from effects of the hydrolase inhibitors, hemin and Ub aldehyde. The coincident effect of hemin on the ATP-independent disassembly and ATP-dependent degradation of high molecular weight conjugates[12] might be explained as a primary effect on the hydrolases giving rise to conjugates that inhibit proteolytic processing of properly conjugated proteins or as a direct effect of the hemin on the Ub-dependent protease. As shown by Hough and Rechsteiner,[22] the Ub conjugates obtained with labeled lysozyme in the presence of hemin are largely those that result in free lysozyme, not lysozyme fragments, when they are further incubated in fraction II without ATP. Therefore, the hemin effect, direct or indirect, seems to be on the protease. These authors also report that these conjugates are rapidly degraded to acid-soluble form when incubated with partially purified protease plus ATP in the absence of hemin. The absence of a lag in this rate suggests that inhibitory conjugates were not present. On the other hand, it is possible that in their reisolation these conjugates may have become inadvertently denatured and therefore become good substrates for the protease. The mechanism by which hemin inhibits hydrolases and proteases remains to be clarified. Hemin does not inhibit the action of ULH-A on $Ub-NH_2$ (I. A. Rose, unpublished results) so that the combination of hemin with Ub–X seems an unlikely general explanation, at least when X is a small ligand.

Ubal has been very useful in defining the role of hydrolases in Ub-dependent processes such as protein breakdown because of its apparent specificity as an inactivator for many important hydrolases. Ubal does not inhibit either protein conjugation of ATP-dependent breakdown as shown with $[^{125}I]Ub$ and $[^{125}I]$lysozyme.[29] By direct assay, Ub is seen to disappear rapidly, and a striking inhibition of lysozyme breakdown is observed if the Ub level is not maintained. The disappearance of Ub far exceeds any reasonable equivalence to enzyme in the system and must represent free intermediates or side products of the system. When very high levels of Ubal are used, about half of the Ub can be recovered by addition of purified ULH-A and therefore must be present as nonprotein conjugates form.

Very large conjugates, more abundant and even greater in variety than those seen at steady state without Ubal, accumulate in the Ubal-inhibited incubation under conditions in which lysozyme breakdown is uninhibited. Addition of performic acid oxidized RNase, a poorly degraded substrate, changed the pattern and greatly increased the rate of conjugate formation. A rapidly degraded protein, such as lysozyme, had much less effect on the accumulation of high molecular weight conjugates in the presence of Ubal. These data are consistent with a model in which the conjugates that have been observed from the earliest reports[11,32] on Ub/ATP-dependent protein degradation are largely intermediates in a "futile" cycle. They seem not to inhibit or participate in protein breakdown. In this view, the Ub–protein hydrolases with general specificity play an important role in the economy of the cell's Ub in monitoring incomplete specificity in the E3–protease system much as the Ub–ligand hydrolases are important in correcting the imperfections in specificity of the E1/E2 system. A high degree of specificity probably cannot be expected in a system that must serve the wide variety of sequences and structures represented in the cell's protein population.

8. UBIQUITIN–PROTEIN PEPTIDASES (U$^\alpha$PHs)

The generation of Ub from conjugates linked to an $^\alpha$N-terminal residue of a protein is required for the synthesis of Ub from the presumed polyprotein translation product of the poly-Ub message. The enzymes responsible for this may or may not be typical hydrolases. They may require ATP. The process may use enzymes that cleave at terminal Ub–Ub sites (exo-), internal Ub–Ub sites (endo-), or both. The C-terminal Ub unit is extended by a single amino acid or, in some cases, by a C-terminal polypeptide. Removal of C-terminal extensions will call for a Ub–peptide peptidase that may have unique specificity. Relevant to this, Bachmair et al.[33] encoded fusion proteins of Ub and β-galactosidase linked by different amino acids. In E. coli, the fusion proteins were not disassembled. In yeast, rapid cleavage of Ub from β-galactosidase occurred with any amino acid except proline. The responsible enzyme must be very active since fusion proteins were not seen in the yeast cell.

Ub–$^\alpha$N-conjugated proteins other than those presumed to arise by translation of the poly-Ub message are not yet known as products of metabolism. Hershko and Heller[34] have shown that fully lysine-guanylated lysozyme is degraded in the Ub/ATP reticulocyte system at approximately half the rate of nonguanylated lysozyme. The requirement for Ub, E1, and E2 suggested that $^\alpha$N-terminal ubiquitination is a required step. Blocking the $^\alpha$N terminal of the $^\varepsilon$N-guanylated lysozyme made it inactive in the system. The expected Ub–$^\alpha$N-lysozyme has not yet been observed, possibly because of a very active Ub protein peptidase in reticulocyte extracts, perhaps the same enzyme responsible for the rapid disassembly of Ub–fusion proteins in yeast observed by Bachmair et al.[33]

Efforts to characterize the Ub–$^\alpha$N peptide hydrolases will depend on the availability of suitable substrates. There is obvious interest in these enzymes both for the characterization of conjugates of the $^\alpha$N or $^\varepsilon$N type and for a more detailed characterization of conjugate structure. In addition, the study of Bachmair et al.[33] showing the presence of U$^\alpha$PH activity of broad specificity in yeast has encouraged the possibility that fusion of Ub with important peptides engineered for expression in E. coli would provide protection to these peptides for their large-scale preparation. These conjugates will then serve as the substrates for identification of hydrolases necessary for the liberation of the peptides desired.

9. CONCLUSION

Many questions regarding the hydrolases remain unanswered, and others have yet to be formulated. The shift in Ub–H2A/H2A ratio over the cell cycle implies a change in amount, control, or access of a specific hydrolase. The roles of hydrolases in proofreading remain to be analyzed in whole-cell systems. At the structural level, the hydrolases, with their high affinity for Ub and Ubal, should contribute to our understanding of the importance of specific regions of the Ub sequence as well as an appreciation of how specificity for particular protein ligands is obtained. A detailed structural comparison of the Ub–hydrolase and Ubal–hydrolase complexes should provide useful insight into the role of structural change in transition state stabilization. In terms of enzyme reaction mechanisms, we see evidence for covalent catalysis that may be the same in all Ubal-sensitive hydrolases. Those not inhibited by Ubal may represent another line of evolutionary development. Possibly, the two types of hydrolases will correlate with cysteine- and serine-based mechanisms. Failure of some hydrolases to be retained by a Ub-affinity column suggests that the Ub component of their natural substrates is unlike that of free Ub or else that their recognition of substrate depends on critical interactions with the Ub–carboxy-terminal ligand. Such ligand specific hydrolases may be important in regulating the level of functionally specific Ub–protein conjugates.

ACKNOWLEDGMENTS. The author thanks Drs. Keith D. Wilkinson and Avram Hershko for contributing information in advance of publication. My research on the hydrolases has been supported by National Institutes of Health grants GM-20940 and American Cancer Society grant BC-414, together with funds provided to the Institute for Cancer Research (CA-06927 and RR-05539) and an appropriation from the Commonwealth of Pennsylvania.

REFERENCES

1. Seale, R., 1981, Rapid turnover of the histone–ubiquitin conjugate protein A24, *Nucleic Acids Res.* **9**: 3151–3158.

2. Wu, R. S., Kohn, K. W., and Bonner, W. M., 1981, Metabolism of ubiquitinated histones, *J. Biol. Chem.* **256:** 5916–5920.

3. Atidia, J., and Kulka, R. G., 1982, Formation of conjugates of ^{125}I-labeled ubiquitin microinjected into cultured hepatoma cells, *FEBS Lett.* **142:** 72–76.

4. Matsumoto, Y.-I., Yasuda, H., Marunouchi, T., and Yamada, M. A., 1983, Decrease in μH2A (protein A24) of a mouse temperature sensitive mutant, *FEBS Lett.* **151:** 139–142.

5. Matsui, S.-I., Sandberg, A. A., Negoro, S., Seon, B.-K., and Goldstein, G., 1982, Isopeptidase: A novel eukaryotic enzyme that cleaves isopeptide bonds, *Proc. Natl. Acad. Sci. U.S.A.* **79:** 1535–1539.

6. Mueller, R. D., Yasuda, H., Hatch, C. L., Bonner, W. M., and Bradbury, E. M., 1985, Identification of ubiquitinated H2A and H2B in *Physarum polycephalum*: Disappearance of these proteins at metaphase and reappearance at anaphase, *J. Biol. Chem.* **260:** 5147–5153.

7. Mita, S., Yasuda, H., Marunouchi, T., Ishiko, S., and Yamada, M., 1980, A temperature-sensitive mutant of cultured mouse cells defective in chromosome condensation, *Exp. Cell Res.* **126:** 407–416.

8. Anderson, M. W., Ballal, N. R., Goldknopf, I. L., and Busch, H., 1981, Protein A24 lyase activity in nucleoli of thioacetamide-treated rat liver reduces histone 2A and ubiquitin from conjugated protein A24, *Biochemistry* **20:** 1100–1104.

9. Haas, A. L., Murphy, K. E., and Bright, P. M., 1985, The inactivation of ubiquitin accounts for the inability to demonstrate ATP-dependent proteolysis in liver extracts, *J. Biol. Chem.* **260:** 4694–4703.

10. Anderson, M. W., Goldknopf, I. L., and Busch, H., 1981, Protein A24 lyase is an isopeptidase, *FEBS Lett.* **132:** 210–214.

11. Hershko, A., Ciechanover, A., Heller, H., Haas, A. L., and Rose, I. A., 1980, Proposed role of ATP in protein breakdown: Conjugation of proteins with multiple chains of the polypeptide of ATP-dependent proteolysis, *Proc. Natl. Acad. Sci. U.S.A.* **77:** 1783–1786.

12. Haas, A. L., and Rose, I. A., 1981, Hemin inhibits ATP-dependent ubiquitin-dependent proteolysis: Role of hemin in regulating ubiquitin conjugate degradation, *Proc. Natl. Acad. Sci. U.S.A.* **78:** 6845–6848.

13. Ciechanover, A., Heller, H., Etzion-Katz, R., and Hershko, A., 1981, Activation of the heat-stable polypeptide of the ATP dependent proteolytic system, *Proc. Natl. Acad. Sci. U.S.A.* **78:** 761–765.

14. Rose, I. A., and Warms, J. V. B., 1987, A specific endpoint assay for ubiquitin, *Proc. Natl. Acad. Sci. U.S.A.* **84:** 1477–1481.

15. Wilkinson, K. D., Cox, M. J., Mayer, A. N., and Frey, T., 1986, Synthesis and characterization of ubiquitin ethyl ester, a new substrate for ubiquitin carboxyl-terminal hydrolase, *Biochemistry* **25:** 6644–6649.

16. Kanda, F., Matsui, S.-L., Sykes, D. E., and Sandberg, A. A., 1984, Affinity of chromatin structures for isopeptidase, *Biochem. Biophys. Res. Commun.* **122:** 1296–1306.

17. Pickart, C. M., and Rose, I. A., 1986, Mechanism of ubiquitin carboxyl-terminal hydrolase, *J. Biol. Chem.* **261:** 10210–10217.

18. Rose, I. A., and Warms, J. V. B., 1983, An enzyme with ubiquitin carboxy-terminal esterase activity from reticulocytes, *Biochemistry* **22:** 4234–4237.

19. Pickart, C. M., and Rose, I. A., 1985, Ubiquitin carboxyl-terminal hydrolase acts on ubiquitin carboxyl-terminal amides, *J. Biol. Chem.* **261:** 7903–7910.

20. Pickart, C. M., and Rose, I. A., 1985, Functional heterogeneity of ubiquitin carrier proteins, *J. Biol. Chem.* **260:** 1573–1581.

21. Hershko, A., Leshinsky, E., Ganoth, D., and Heller, H., 1984, ATP-dependent degradation of ubiquitin–protein conjugates, *Proc. Natl. Acad. Sci. U.S.A.* **81:** 1619–1623.
22. Hough, R., and Rechsteiner, M., 1986, Ubiquitin–lysozyme conjugates: Purification and susceptibility to proteolysis, *J. Biol. Chem.* **261:** 2391–2399.
23. Hunter, A. J., and Cary, P. D., 1985, Preparation of chromosomal protein A24 (μH2a) by denaturing gel filtration and preparation of its free nonhistone component ubiquitin by ion-exchange chromatography, *Anal. Biochem.* **150:** 394–402.
24. Hough, R., Pratt, G., and Rechsteiner, M., 1986, Ubiquitin–lysozyme conjugates: Identification and characterization of an ATP-dependent protease from rabbit reticulocyte lysates, *J. Biol. Chem.* **261:** 2400–2408.
25. Kanda, F., Sykes, D. E., Yasuda, H., Sandberg, A. A., and Matsui, S.-I., 1986, Substrate recognition of isopeptidase: Specific cleavage of the ε-(α-glycyl) lysine linkage in ubiquitin–protein conjugates, *Biochim. Biophys. Acta* **870:** 64–75.
26. Goldknopf, I. L., Cheng, S., Anderson, M. W., and Busch, H., 1981, Loss of endogenous nuclear protein A24 lyase activity during chicken erythropoiesis, *Biochem. Biophys. Res. Commun.* **100:** 1464–1470.
27. Goldknopf, I. L., Wilson, G., Ballal, N. R., and Busch, H., 1980, Chromatin conjugates protein A24 is cleaved and ubiquitin is lost during chicken embryo erythropoiesis, *J. Biol. Chem.* **255:** 10555–10558.
28. Wu, R. S., Kohn, K. W., and Bonner, W. M., 1981, Metabolism of ubiquitinated histones, *J. Biol. Chem.* **256:** 5916–5920.
29. Hershko, A., and Rose, I. A., 1987, Ubiquitin–aldehyde: A general inhibitor of ubiquitin-recycling processes, *Proc. Natl. Acad. Sci. U.S.A.* **84:** 1829–1833.
30. Wilkinson, K. D., and Mayer, A. N., 1986, Alcohol-induced conformational changes of ubiquitin, *Arch. Biochem. Biophys.* **250:** 390–399.
31. Cox, M. J., Haas, A. L., and Wilkinson, K. D., 1986, Role of ubiquitin conformations in the specificity of protein degradation: Iodinated derivatives with altered conformation and activities, *Arch. Biochem. Biophys.* **250:** 400–409.
32. Ciechanover, A., Heller, H., Elias, S., Haas, A. L., and Hershko, A., 1980, ATP-dependent conjugation of reticulocyte proteins with the polypeptide required for protein degradation, *Proc. Natl. Acad. Sci. U.S.A.* **77:** 1365–1368.
33. Bachmair, A., Finley, D., and Varshavsky, A., 1986, *In vivo* half-life of a protein is a function of its amino-terminal residue, *Science* **234:** 179–186.
34. Hershko, A., and Heller, H., 1985, Occurrence of a polyubiquitin structure in ubiquitin–protein conjugates, *Biochem. Biophys. Res. Commun.* **128:** 1079–1086.

Chapter 6

Ubiquitinated Histones and Chromatin

William M. Bonner, Christopher L. Hatch, and Roy S. Wu

1. INTRODUCTION

Ubiquitin (Ub) is a remarkably conserved eukaryotic protein found in the cytoplasm, in the nucleus, and on the cell surface. Whereas cytoplasmic Ub is known to play an important role in proteolysis, the function of nuclear Ub is still obscure. Presumably, ubiquitination of histones affects chromatin structure, but it should be stated at the outset that we do not know how this histone modification influences transcriptional or mitotic processes. We do know, however, some aspects of the structure and metabolism of histone–Ub conjugates, and these facts should provide clues to the role of Ub in chromatin dynamics.

2. STRUCTURE

The Ub adduct of histone H2A was originally found as a protein, named A24, the amount of which was greatly decreased in liver nucleoli isolated from rats after treatment with thioacetamide or after partial hepatectomy.[1-3] Since these treatments also led to high levels of rRNA syn-

WILLIAM M. BONNER, CHRISTOPHER L. HATCH, and ROY S. WU • Laboratory of Molecular Pharmacology, Division of Cancer Treatment, National Cancer Institute, National Institutes of Health, Bethesda, Maryland 20892.

thesis, it was surmised that protein A24 might be involved in the regulation of gene activity. Analysis of the purified protein showed that A24 had an amino acid composition similar to that of histone H2A,[4] and tryptic mapping showed that it contained histone H2A.[5] Sequencing the remaining portion[6] indicated its identity to Ub.[7] The Ub moiety was shown to be attached by an isopeptide bond through its carboxy-terminal glycine-76 to the ϵ-amino group of lysine-119 in histone H2A.[8] Most mammalian species contain four H2A variants, two predominant ones known as H2A.1 and H2A.2 and two minor ones known as H2A.X and H2A.Z;[9] all four have been shown to have Ub adducts. Known sequences of H2As from various species are presented in Figure 1.

The attachment site for Ub in H2A.1 is at lysine-119 following another lysine at position 118. This double lysine has been conserved in all the H2As that have been sequenced.[10] Mammalian H2A.X has not been sequenced through this region. Chicken H2A.F, sea urchin H2A.F/Z, and mammalian H2A.Z do have Lys-Lys at these positions.[11-13] H2A.Z, which is not phosphorylated like the other three mammalian H2A variants, is nevertheless ubiquitinated.[9] In mouse L1210 cells, each of the four variants seems to be ubiquitinated to the same or very similar extents, with approximately 11% of each variant in the ubiquitinated form. The Ub attachment site is also at the limit of trypsin digestion of H2A in chromatin, suggesting that the carboxy terminus of H2A may be more accessible to external molecules.[14]

Histone H2B can also be ubiquitinated, but much less information is available concerning those adducts.[15] In mouse L1210 cells there are two major H2B variants that differ only at position 76, a serine in H2B.2 for a glycine in H2B.1.[16] Both variants are ubiquitinated to the extent of 1–1.5%, much less than that of the H2A. The site of ubiquitination of H2B has been known to be carboxy terminal to methionine-62 from cyanogen bromide cleavage data.[15] Recently, Thorne et al.[17] have determined the exact site to be at lysine-120, which is invariant in all sequenced H2Bs (see Figure 2).

In intact nucleosomes, the ubiquitination site in H2A is also accessible to trypsin; however, none of the carboxy-terminal region of H2B is accessible to trypsin.[18] In addition, according to NMR studies, the carboxy-terminal region of H2B up to residue 114 is involved in formation of the H2A–H2B dimer.[19] These results suggest that compared to nucleosomal H2A the lysine residues in the carboxy-terminal regions of nucleosomal H2B may be less accessible to the enzymes involved in ubiquitination, providing one possible explanation for the lower extent of H2B ubiquitination.

No one has reported any functional or even behavioral difference

```
Bovine       H2A.1  -------SGRGKOGGKA --RAK-A--K--TRSSR AGLOFPVGRV HRLLRKGNYA E-RVGAGAPVY LAAVLEYLTA EILELAGNAA
                .2  -------.......... --...-.--.--S.... .......... .......... .-......... M......... ..........
Mouse           .1  -------.......... --...-.--.--..... .......... .......... .-......... .......... ..........
                .2  -------.......... --...-.--.--S.... .......... .......... N-......... M......... ..........
Chicken         .1  -------.......... --...-.--.--S.... .......... .......... .-......... .......... ..........
                .2  -------.........V --...-.--.--S.... .......... .......... .-......... M......... ..........
                .F  -----AGGKA..DS... --K..-.----VS..O. .........I ..H.KTRTTS HG....T.A.. S..I...... .V.......S
Yeast           .1  -------..-..-.... GSA..-.S--Q-S..AK ...T...... .....R.... Q-.I.S..... .T......A. ..........
                .2  -------..-..-.... GSA..-.S--Q-S..AK ...T...... .....R.... Q-.I.S..... .T......A. ..........
Tetrahymena     .1  -------.TT..-.... --KC.T.SS.OVS.A.  .........I S.F.KH.R.S .-.I.T..... ........A. .V........
                .2  -------.TT..-.... --KG.T.SS.QVS..A. .........I S.F.KN.R.S .-.I.T..... ........A. .V........

                                                                           119
                                                                            ↓
Bovine       H2A.1  RDNKKTRIIP RHLQLAIRND EELNKLLGKV TIAQGGVLPN IQAVLLP---KKT ESHHKAKGK-  (129)
                .2  .......... .......... .......... .......... .......---... .........-  (129)
Mouse           .1  .......... .......... ........R. .......... .......---... .........-  (129)
                .2  .......... .......... .......... .......... .......---... .........-  (129)
Chicken         .1  .......... .......... .......... .......... .......---... D..-...A.-  (129)
                .2  .......... .......... .......... .......... .......---... ...-...S.-  (128)
                .F  K.L.VK..T. ........G. ...DS.I-.A ...G...I.H .HKS.IG---..G OOK-T.----  (127)
Yeast           .1  .......... .......... D.......N. .......... .HQN...---..S AKAT..SQEL  (131)
                .2  .......... .......... D.......N. .......... .HQN...---..S AKTA..SQEL  (131)
Tetrahymena     .1  K.......V. ..IL...... ......MANT ...D...... .NPM...SKS... ..RGO.SQDI  (137)
                .2  K.......V. ..IL...... ......MANT ...D...... .NPM...SKT... SEAEH-----  (132)
```

Figure 1. Sequences of H2A. All sequences referenced to bovine. A space is skipped at every tenth residue of bovine sequence. Dots (···) mean the same as bovine H2A.1. Dashes (–––) mean a deletion owing to alignment in that sequence. Number in parentheses at the end of the sequence is the total number of amino acid residues in that sequence. Sequences or differences in sequences among variants were taken from Wu *et al.*[10] and references therein.

62
↓

		PEPAKSAPA	---P KKGS-----KKAVTK	AOKKDGKKRK	RSRKESYSVY	VYKVLKOVHP	DTGISSKAMG	IMNSFVNDIF
Bovine	H2B.1	-----						
Mouse	.2	-----						
	.1	-----						
	.2	-----						
	.3	-----						
Yeast	.1	SAKAE	KK..SK...EKK. AAKKTSTST------		...S KA..T..S.	I.......T..	..O.S.S	.I.....
	.2	SSAAE	KK..SK...EKK. AAKKTSTSV--		...S KV...T..S.	I.......T..	..O.S.S	.I.....
Tetrahymena		-----	.KK.PA.--E ..V-------..P.-	TE..N--.--	..--.TFAI.	IF....	.V...K.A.N	....I....S.
wheat		?????	?????????????????? ??????????????	??????????	?????T.KI.	IF....	.I.......S	....I....S.

120
↓

		ERIAGEASRL	AHYNKRSTIT	SREIQTAVRL	LLPGELAKHA	VSEGTKAVTK	YTSSK --	
Bovine	H2B.1						--	(125)
Mouse	.2	...Q.....					--	(125)
	.1						--	(125)
	.2	...S.....					--	(125)
	.3	...Q.....					--	(125)
Yeast	.1	...T...K.	.A...K...S A.........	I.........	...R.....	S..TQA	(130)
	.2	...T...K.	.A...K...S A.........	I.........	...R.....	S..TQA	(130)
Tetrahymena		...L.S.K.	VRF...R.LS ...V...K.	...R.....	I.........	FS..TN-	(119)
wheat		.KL...SAK.	.R...KP... ...S...V	V.........	...A.--	F..A.--	

Figure 2. Sequences of H2B. All sequences referenced to bovine. A space is skipped at every tenth residue of bovine sequence. Dots (···) mean the same as bovine H2B.1. Dashes (---) mean a deletion owing to alignment in that sequence. Question marks (???) mean unknown or not determined. Number in parentheses at the end of the sequence is the total number of amino acid residues in that sequence. Sequences or differences in sequences among variants were taken from Wu et al.[10] and references therein.

between nucleosomes with and without ubiquitinated H2A. Cross-linking of H2A with H2B by UV[20,21] or with H1 by carbodiimide[22] worked equally well for nucleosomes containing uH2A as for those containing H2A. HMG17 binding to mononucleosomes was unaffected by the presence of uH2A.[23] The DNA length in digested chromatin was found to be the same whether or not the H2A in the nucleosomes was ubiquitinated.[24,25]

In mammals, the percentage of histones present as ubiquitinated forms seems to be fairly constant. In rat cerebral cortex neurons, which are terminally differentiated and no longer divide after birth, 12–14% of the H2A and 1–2% of the H2B were ubiquitinated. These values are almost identical to those reported for mouse L1210 cells.[26] Zweidler reported that uH2A is a fairly constant fraction of total H2A in a wide variety of mouse tissues.[27] In nonmammalian species, evidence to date shows the presence of uH2A in chickens,[28] sea urchins,[29] *Drosophila,*[30] *Plodia,*[31] *Tetrahymena,*[32] and *Physarum.*[33] In these organisms, the Ub adducts are a small but significant fraction of the total H2A or H2B. For example, in *Physarum* 7% of the H2A and 6% of the H2B were ubiquitinated.[33] It is notable that even though the relative amounts of uH2A and uH2B differ significantly between mice and *Physarum,* the total percentage of ubiquitinated histone is similar in the two organisms.

Very small amounts of diubiquitinated and multiubiquitinated adducts of H2A have been detected in mammalian nuclei using anti-Ub antibodies (A. Varshavsky, unpublished results). Multiubiquitinated adducts of H2A are also present when H2A is synthesized *in vitro* in a rabbit reticulocyte extract (R. S. Wu and W. M. Bonner, unpublished results).

In contrast to other eukaryotes, yeast growing under optimal conditions seems to lack uH2A completely; it is undetectable by the techniques sensitive enough to measure 0.1% of the adduct in mammals (D. Finley, B. Bartel, and A. Varshavsky, personal communication). However, when the yeast are grown under mildly stressful but not lethal conditions of elevated temperature, a significant fraction of the H2A becomes ubiquitinated (D. Finley, B. Bartel, and A. Varshavsky, personal communication). This novel finding may lead to new insights regarding the role of histone ubiquitination.

3. METABOLISM

The metabolism of ubiquitinated histones has been studied both in dividing and nondividing cells.[34–40] There is a consensus that the Ub moiety of uH2A and uH2B is in a rapid equilibrium with a pool of free Ub in all cells. Several different approaches demonstrate turnover of the

Ub moiety and conservation of the H2A moiety in uH2A. Using mouse L1210 cells, Wu et al.[34] studied the decay kinetics of radioactively labeled uH2A in conjunction with tryptic peptide mapping of the uH2A. The decay kinetics were bimodal; tryptic analyses showed a rapid decay of the Ub peptides and no decay of the H2A peptides in uH2A. Radioactivity in the Ub peptides of uH2A decreased parallel to that in free Ub. The identical half-life of free Ub and the Ub moiety of uH2A indicates a rapid equilibrium between free Ub and H2A-bound Ub.

Using human HeLa cells, Seale[35] took advantage of the fact that Ub contains an amino-terminal methionine whereas histone H2A does not contain methionine. A series of pulse-chase studies using either lysine or methionine as the label led him to the conclusion that the Ub moiety of uH2A turns over at a higher rate than the H2A moiety. It was also suggested that newly synthesized Ub and newly synthesized H2A may be preferentially conjugated, and that both newly synthesized and preexisting H2A may be reconjugated to Ub following cleavage.

Trempe and Leffak[36] closely examined the conjugation of Ub to H2A in MSB-1 cells, a continuous line of chicken lymphocytes transformed by Marek's virus. Combining the techniques of peptide mapping, dense amino acid labeling, isopycnic centrifugation, and SDS–PAGE, these authors concluded that newly synthesized histone H2A molecules are not preferentially ubiquitinated. It was further suggested that newly synthesized H2A can be conjugated with either newly synthesized or preexisting Ub. Also, newly synthesized Ub may become conjugated to H2A molecules more than 2–3 hr after the histones have been deposited in the cell nucleus.

These findings on the metabolism of ubiquitinated histones have been extended to nondividing cells. Wu et al.[34] studied the formation of uH2A in unstimulated lymphocytes and in L1210 cells treated with inhibitors of DNA synthesis. Cells in the quiescent state (G_0) and cells treated with inhibitors of DNA synthesis synthesize greatly reduced amounts of certain S-phase histone variants.[41-44] Despite the inhibition of histone synthesis, substantial radioactivity was found in the Ub adducts of H2A. These results indicate that the formation of ubiquitinated histones as well as the synthesis of Ub are linked to neither DNA synthesis nor histone synthesis. Furthermore, the results with noncycling lymphocytes showed that cells need not traverse mitosis to effect turnover of ubiquitinated histones.

The rapid turnover of the Ub moiety in uH2A and uH2B in nondividing as well as dividing cells provides a possible explanation for many observations that the relative amount of ubiquitinated histones in different cells changes under diffferent physiological states. Decreases in the amount of ubiquitinated histone have been observed in rat liver cells stimulated to hypertrophy by thioacetamide[2] or partial hepatectomy,[1] in the maturation of chicken erythrocytes,[28] and in the temperature-sensitive mutant cell ts85 (derived from the FM3A mouse mammary carcinoma cell line) at the nonpermissive temperature.[45] Increases in the amount of ubiquitinated histone have been reported during bullfrog red cell maturation[46] and in Ehrlich ascites tumor cells after treatment with 1-methyl-1-nitrosourea and 1,3-bis(2-chlorethyl)-1-nitrosourea.[47] Furthermore, the levels of ubiquitinated histones have been shown to vary with the position in the cell cycle[33,34,37-39,48-51] and with physiological stress.[40,52]

3.1. Cell Cycle Studies

Many groups have studied the behavior of ubiquitinated histones during the cell cycle. Using synchronized HeLa cells, Goldknopf et al.[48] first demonstrated that Ub labeling and uH2A formation were largely interphase events, rising to a maximum rate in early G_1 phase and declining to a minimum during cell division. Matsui et al.[50] reported the complete disappearance of ubiquitinated H2A from metaphase chromosomes in synchronized hamster DON cells. They also demonstrated that reubiquitination of histone H2A during the mitosis–G_1 transition could happen under conditions of complete inhibition of protein synthesis. This latter finding supports other reports that preexisting H2A and Ub can be reconjugated.[36] From their finding, Matsui et al.[50] suggested that the ubiquitination of histone may play a role in the condensation and decondensation of metaphase chromatin and that at least two enzymes may be involved in the ubiquitination cycle: one to release Ub from uH2A at the G_2–mitosis transition and another to catalyze the addition of Ub to H2A at the mitosis–G_1 transition.

Wu et al.[34] confirmed and extended the findings of Matsui et al.[50] by showing that all Ub adducts of H2A and H2B variants present in the interphase nucleus are absent from metaphase chromosomes isolated from CHO cells. Furthermore, they suggested that the period during which ubiquitinated histones were absent in CHO cells was quite short, a finding derived from the observation that mitotic cells, as opposed to metaphase chromosomes, contained ubiquitinated histone in amounts that varied from experiment to experiment. Cytological analysis revealed that these cells, synchronized and isolated by the procedure of Tobey and Ley,[51] were composed of a mixture of cells in late prophase to anaphase.

Mueller et al.[33] studied the cell cycle behavior of uH2A.1, uH2A.Z, and uH2B in Physarum polycephalum, a macroplasmodial mold in which all the nuclei divide synchronously within 2–3 min. The 9 hr nuclear division cycle is very precise; early prophase lasts 15 min, prophase 5 min, metaphase 7 min, and anaphase 3 min. The ubiquitinated histones were present during prophase, anaphase, and telophase, as well as during S, G_1, and G_2 phases of the cell cycle, but disappeared precisely at metaphase. These observations were similar to those reported in mammalian cells[34,48,49] and confirm those of Matsui et al.[50] that deubiquitination and ubiquitination occur during chromosome condensation and decondensation, respectively.

However, these observations do not necessarily support the hypothesis that deubiquitination is the trigger for chromosome condensation as proposed by Matsui et al.[50] This hypothesis requires that only a small subset of H2A and H2B molecules be ubiquitinated but that these ubiquitinated forms be uniformly distributed throughout the chromosomes. The first condition is readily satisfied by the low ratios of uH2A and uH2B to H2A and H2B. However, the second criterion has not been demonstrated. In addition, evidence from a temperature-sensitive cell cycle mutant argues against a causal role for deubiquitination in chromosome condensation. This mutant, ts85, loses all its uH2A and uH2B and arrests in early to middle G_2 when the cells are brought to the nonpermissive temperature; and yet, ts85 chromatin remains dispersed at the nonpermissive temperature.[29,44,51-54]

3.2. Other Metabolic Studies

Several other investigators[37-40,55-64] have taken different approaches to the study of ubiquitination. A useful approach has been to study the enzymes in the Ub conjugation and deconjugation pathways. A hydrolase or isopeptidase has been identified and partially purified.[59,60] This enzyme specifically recognizes and cleaves the ε-(α-glycyl)lysine bond in Ub–protein conjugates.[58] One study suggested that the level of isopeptidase activity associated with metaphase chromatin may be tenfold higher than that associated with interphase chromatin.[61] Anderson et al.[56] have reported increased levels of isopeptidase after treating rats with thioacetamide, coupled with a decrease of uH2A in nucleoli from these treated animals.

The Ub ligase system for histones is composed of three types of enzymes[65,66] (discussed in detail in Chapter 3) and requires ATP.[45,64] Briefly, Ub carrier proteins (E2s) acquire a Ub moiety in an ATP-dependent reaction catalyzed by an activating enzyme (E1) and then donate that moiety to protein amino groups in a reaction that usually requires a third enzyme (E3). A thermolabile Ub activating enzyme (E1) has been reported as the temperature-sensitive lesion in the ts85 cell line.[55,64] Ubiquitinated H2A (and presumably H2B) disappears from these cells during incubation at the nonpermissive temperature.

In reticulocyte extracts, there are five carrier proteins.[65] Of these five, two (with approximate molecular masses of 20 and 14 KDa), are able to donate their Ub moieties to histones H2A and H2B and to cytochrome c in a reaction that does not require enzyme E3. In these cases, the major products have molecular masses expected for monoubiquitinated adducts. Recently, a protein analogous to the 20 kDa mammalian protein has been identified in yeast as the product of the *RAD6* gene.[66] This yeast protein also mediates the *in vitro* ubiquitination of histone H2B in an E3-independent reaction. The *RAD6* mutants are extremely vulnerable to a variety of DNA damaging agents and are deficient in a variety of cellular functions including DNA repair, induced mutagenesis, and sporulation.[67] This finding implicates the ubiquitination of histones as mediated by the *RAD6* protein in the alterations of chromatin structure required to perform these cellular functions. Notably, ubiquitinated histone adducts are not detectable in yeast except when they are exposed to stressful (but not lethal) conditions such as elevated temperature[36] (D. Finley, B. Bartel, and A. Varshavsky, personal communication).

Another approach places [^{125}I]Ub inside cells by microinjection or permeabilization in order to study the intracellular location, stability, and molecular weight distribution of the [^{125}I]Ub conjugates.[37-40] Atidia and Kulka,[37] Carlson and Rechsteiner,[39] and Carlson *et al.*[40] have introduced [^{125}I]Ub into cells by the red-cell-mediated microinjection technique described by Schlegel and Rechsteiner.[68] These authors found that the injected Ub molecules were rapidly conjugated to cellular proteins. Carlson and Rechsteiner[39] found that the injected Ub was present as high molecular weight conjugates in both the nucleus and the cytoplasm, as Ub-histone conjugates in the nucleus, and as free Ub in the cytoplasm. Under normal physiological conditions, 10% of the injected Ub was linked to histones, 40% was in high molecular weight conjugates, and the remainder was free Ub. There was, however, a clear reduction in the amount of Ub-histone conjugates in mitotic cells relative to interphase cells. Furthermore, if the injected cells were subjected to heat shock, the level of high molecular weight conjugates almost doubled as the free Ub pool and the level of Ub-histone conjugates decreased dramatically.[40]

Raboy *et al.*,[38] using permeabilized cells, compared the pattern of [^{125}I]Ub conjugation in interphase and metaphase cells and reported a specific decrease in the ubiquitination of histones in permeabilized metaphase cells. These additional studies all demonstrate the dynamic nature of the Ub process and are in agreement with previous results. However, none has as yet provided a clear indication of the role of histone ubiquitination.

4. HISTONE CONJUGATES AND TRANSCRIPTION

The association of ubiquitinated nucleosomes with particular stretches of chromatin is studied by determining which DNA sequences and proteins are present in various forms of mononucleosomes.[24,25] Mononucleosomes are first resolved on low ionic strength polyacrylamide deoxyribonucleoprotein (DNP) gels. Then two different second-dimension gels are used to analyze either the proteins or the DNA fragments present in the various regions of the DNP gels. The DNA fragments resolved in the two-dimensional gel system can be transferred to a hybridization support matrix and probed for various DNA sequences.[69] If a DNA sequence comigrates in the DNP gel with certain proteins, it may suggest, but does not prove, that the DNA sequence in question was present in a nucleosome with those particular proteins.

By this technique a class of mononucleosome was identified that contained 146 base pairs of DNA with one molecule of uH2A; a minor class of mononucleosomes possibly with two molecules of uH2A was also present. When various satellite and highly repetitive DNAs were used as probes to study the relation of ubiquitinated nucleosomes to nontranscribed DNA sequences, uH2A was identified as a component of nucleosomes comigrating on DNP gels with a highly repetitive human DNA sequence.[69]

Levinger and Varshavsky[30] and Levinger[70] analyzed the protein composition of nucleosomes comigrating on DNP gels with several different *Drosophila* satellite DNA sequences. Nucleosomes that comigrated with a 1.672 density simple satellite sequence (AATAT) were found not to contain Ub but did contain D1, a specific ~55 kDa nonhistone protein previously shown by immunocytochemical approaches to reside in those regions of the polytene chromosomes enriched in (A + T)-DNA.[30,71] The nucleosomes that comigrated with a 1.705 density simple satellite sequence (AAGAG) lacked D1 but contained uH2A, some with one and some with two molecules per nucleosome.

Levinger and Varshavsky[30] also analyzed the composition of mononucleosomes comigrating with *Drosophila* hsp-70 heat-shock and copia gene sequences. Although only 20–25% of the bulk DNA comigrated on DNP gels with monoubiquitinated nucleosomes, approximately 50% of the hsp-70 and copia DNA sequences were found in this region. Similarly, while only about 5% of the bulk DNA comigrated with diubiquitinated nucleosomes, 10–15% of the DNA hybridizable to the hsp-70 heat-shock gene probe was found in this region. In heat-shocked cells, there was complete loss of DNA sequences hybridizable to the hsp-70 heat-shock gene probe.

In contrast, rehybridization of the same blots with the copia gene probe indicated that heat shock did not significantly alter the amount or mobility on DNP gels of DNA from this gene. It was suggested that ubiquitinated nucleosomes may be associated with moderate rates of transcription, but that higher rates of transcription lead to a loss of recognizable nucleosomal structure.

Barsoum and Varshavsky[72] analyzed the composition of nucleosomes comigrating with DNA sequences from various regions of the 31 kilobase, transcriptionally active, amplified DHFR gene locus in mouse cells. The first DHFR exon, which is approximately 200 bp in length, preferentially comigrated with diubiquitinated nucleosomes at a tenfold higher extent than the DNA sequences just upstream or downstream. Equal amounts of DNA sequences from the first exon were found to comigrate with nucleosomes containing zero, one, or two uH2As even though the latter two forms were only 10 and 2%, respectively, of the bulk nucleosomes.

The above results showed that transcriptionally active DNA sequences are enriched in nucleosomes that migrate more slowly than the bulk of nucleosomes on DNP gels. Since ubiquitinated nucleosomes were present in these regions, transcriptionally active regions of the genome appear to be enriched in ubiquitinated nucleosomes. An alternative interpretation, less likely but equally compatible with the data, is that other mononucleosomes, which lack uH2A, nevertheless comigrate with the ubiquitinated mononucleosomes owing to either unusual conformational properties or the presence of variant protein components other than uH2A. The following recent experiments raise the possibility that the latter interpretation may be correct.

Huang et al.[73] studied the composition of nucleosomes comigrating with DNA sequences from the active immunoglobulin kappa chain gene locus of mouse plasmacytoma (MPC11 or G403) cells. After staphylococcal nuclease digestion, they obtained two soluble chromatin fractions, S1 and S2, comprising 5 and 80%, respectively, of the total nuclear DNA. An insoluble fraction, P, comprised the remaining 15% of the nuclear DNA. Sequences either within the gene itself or 10 kb downstream were enriched at least threefold in the S1 fraction. By contrast, nontranscribed sequences such as satellite DNA, the β-globin gene, or most $[d(C-A)]_n$ repetitive elements were significantly depleted from the S1 fraction and were almost entirely represented in the S2 and P fractions. All core histone subtypes as well as the ubiquitinated forms of H2A and H2B were present in all three chromatin fractions at equivalent levels. However, the S1 fraction was about 50-fold depleted of H1 and 2.5–4-fold enriched in HMG-14 and HMG-17 as compared to the S2 fraction. Analysis of the S1 fraction on DNP gels revealed that kappa gene sequences comigrated with mononucleosomes that appeared to contain HMG-14 and/or HMG-17 and the ubiquitinated forms of H2A and H2B. These nucleosomes also appeared to contain the minor histone variant H2A.X.

Ub can be removed from nucleosomes with isopeptidase.[61] If the DNA sequences of interest were bound to ubiquitinated nucleosomes, they should comigrate with nonubiquitinated nucleosomes after isopeptidase treatment. When this experiment was carried out so that 80–90% of the uH2A and uH2B molecules were converted to their nonubiquitinated forms, the kappa chain gene sequences were unchanged in their electrophoretic mobility. This result indicates that nucleosomal components other than Ub were responsible for the slower migration of kappa chain gene sequences on DNP gels. The H2A variant, H2A.X, was found preferentially in nucleosomes comigrating with these sequences. The authors suggest that nucleosomes containing this variant may be bound to the kappa gene, but just as with uH2A, comigration of any particular protein with any particular sequence does not prove their association. However, unless the kappa chain gene nucleosomes contain ubiquitinated histones that are unusually resistant to isopeptidase cleavage, this form of histone modification does not seem to play a direct role in the structure and function of this gene. These findings admit the possibility that the probed DNA sequences are bound to nucleosomes that happen to comigrate with ubiquitinated nucleosomes but that are not themselves ubiquitinated.

In light of these recent results, whether Ub is involved in transcription remains a completely open question. That ubiquitinated histones are found on some satellite DNAs and not on others creates even more confusion concerning the role of Ub in chromatin. Thus, even though substantial information exists concerning the behavior of Ub in chromatin, little insight into its functional role has yet been achieved.

REFERENCES

1. Ballal, N. R., and Busch, H., 1973, Two dimensional gel electrophoresis of acid-soluble nucleolar proteins of Walker 256 carcinosarcoma, *Cancer Res.* **33:** 2737–2743.
2. Ballal, N. R., Goldknopf, I. L., Goldberg, D. A., and Busch, H., 1974, The dynamic state of liver nucleolar proteins as reflected by their changes during administration of thioacetamide, *Life Sci.* **14:** 1835–1845.

3. Ballal, N. R., Kang, Y.-S. J., Olson, N. O. J., and Busch, H., 1975, Changes in nucleolar proteins and their phosphorylation patterns during liver regeneration, *J. Biol. Chem.* **250:** 5921–5925.

4. Goldknopf, I. L., Taylor, C. W., Baum, R. M., Yeoman, L. C., Olson, M. O. J., Prestayko, A. W., and Busch, H., 1975, Isolation and characterization of protein A24, a "histone–non-histone chromosomal protein," *J. Biol. Chem.* **250:** 7182–7187.

5. Goldknopf, I. L., and Busch, H., 1975, Remarkable similarities of peptide fingerprints of histone 2A and nonhistone chromosomal protein A24, *Biochem. Biophys. Res. Commun.* **65:** 951–960.

6. Olson, M. O. J., Goldknopf, I. L., Guetzow, K. A., James, G. T., Hawkins, T. C., Mays-Rothberg, C. J., and Busch, H., 1976, The NH_2- and COOH-terminal amino-acid sequence of nuclear protein A24, *J. Biol. Chem.* **251:** 5901–5903.

7. Hunt, L. T., and Dayhoff, M. O., 1977, Amino-terminal sequence and identity of ubiquitin and the nonhistone component of nuclear protein A24, *Biochem. Biophys. Res. Commun.* **74:** 650–655.

8. Goldknopf, I. L., and Busch, H., 1978, Modification of nuclear proteins: The ubiquitin-histone 2A conjugate, in: *The Cell Nucleus*, Vol. VI, Academic Press, New York, pp. 149–180.

9. West, M. H. P., and Bonner, W. M., 1980, Histone 2A: A heteromorphous family of eight protein species, *Biochemistry* **19:** 3238–3245.

10. Wu, R. S., Panusz, H., Hatch, C. L., and Bonner, W. M., 1986, Histones, *CRC Crit. Rev.* **20:** 201–263.

11. Harvey, R. P., Whiting, J. A., Coles, L. S., Kreig, P. A., and Wells, J. R. E., 1983, H2A.F: An extremely variant histone H2A sequence expressed in the chicken embryo, *Proc. Natl. Acad. Sci. U.S.A.* **80:** 2819–2823.

12. Ernst, S. G., Miller, H., Brenner, C. A., Nocente-McGrath, C., Francis, S., and McIssac, R., 1987, Characterization of a cDNA clone coding for a sea urchin histone H2A variant related to the H2A.F/Z histone protein in vertebrates, *Nucleic Acids Res.* **15:** 4629–4643.

13. Hatch, C. L., and Bonner, W. M., 1988, Sequence of cDNAs for mammalian H2A.Z, an evolutionarily diverged but highly conserved basal histone H2A isoprotein species, *Nucleic Acids Res.* (in press).

14. Bohm, L., Crane-Robinson, C., and Sautiere, P., 1980, Proteolytic digestion studies of chromatin core–histone structure: Identification of a limit peptide of histone H2A, *Eur. J. Biochem.* **106:** 525–530.

15. West, M. H. P., and Bonner, W. M., 1981, Histone 2B can be modified by the addition of ubiquitin, *Nucleic Acids Res.* **10:** 4671–4680.

16. Franklin, S. G., and Zweidler, A., 1977, Nonallelic variants of histone 2A, 2B and 3 in mammals, *Nature* **266:** 273–275.

17. Thorne, A. W., Sautiere, P., Briand, G., and Crane-Robinson, C., 1987, The structure of ubiquitinated histone H2B, *EMBO J.* **6:** 1005–1010.

18. Bohm, L., Briand, G., Sautiere, P., and Crane-Robinson, C., 1982, Proteolytic digestion studies of chromatin core–histone structure: Identification of the limit peptides from histone H2B, *Eur. J. Biochem.* **123:** 299–303.

19. Moss, T., Cary, P. D., Aberchrombie, B. D., Crane-Robinson, C., and Bradley, E. M., 1976, A pH-dependent interaction between histones H2A and H2B involving secondary and tertiary folding, *Eur. J. Biochem.* **71:** 337–350.

20. Martinson, H. G., True, R., Lau, C. K., and Mehrabian, M., 1982, Histone–histone interactions with chromatin, Preliminary location of multiple contact sites between histones 2A, 2B, and 4, *Biochemistry* **18:** 1075–1082.

21. DeLange, R. J., Williams, L. C., and Martinson, H. G., 1979, Identification of interacting amino acids at the histone 2A–2B binding site, *Biochemistry* **18:** 1942–1946.
22. Bonner, W. M., and Stedman, J., 1979, Histone 1 is proximal to histone 2A and A24, *Proc. Natl. Acad. Sci. U.S.A.* **76:** 2190–2194.
23. Swerdlow, P. S., and Varshavsky, A., 1983, Affinity of HMG17 for a mononucleosome is not influenced by the presence of ubiquitin–H2A semihistone but strongly depends on DNA fragment size, *Nucleic Acids Res.* **11:** 387–401.
24. Albright, S. C., Wiseman, J. M., Lange, R. A., and Garrard, W. T., 1980, Subunit structures of different electrophoretic forms of nucleosomes, *J. Biol. Chem.* **255:** 3673–3684.
25. Levinger, L., and Varshavsky, A., 1980, High-resolution fractionation of nucleosomes: Minor particles, "whiskers," and separation of mononucleosomes containing and lacking A24 semihistone, *Proc. Natl. Acad. Sci. U.S.A.* **77:** 3244–3248.
26. Pina, B., and Suau, P., 1985, Core histone variants and ubiquitinated histones 2A and 2B of rat cerebral cortex neurons, *Biochem. Biophys. Res. Commun.* **133:** 505–510.
27. Zweidler, A., 1976, Complexity and variability of the histone complement, *Life Sci. Res. Rep.* **4:** 187–196.
28. Goldknopf, I. L., Wilson, G., Ballal, N. R., and Busch, H., 1980, Chromatin conjugate protein A24 is cleaved and ubiquitin is lost during chicken embryo erythropoiesis, *J. Biol. Chem.* **255:** 10555–10558.
29. Wu, R. S., Nishioka, D., and Bonner, W. M., 1982, Differential conservation of histone 2A variants between mammals and sea urchins, *J. Cell Biol.* **93:** 426–431.
30. Levinger, L., and Varshavsky, A., 1982, Selective arrangement of ubiquitinated and D1 protein-containing nucleosomes within the *Drosophila* genome, *Cell* **28:** 375–385.
31. Pataryas, T. A., Sekeri-Pataryas, K. T., Bonner, W. M., and Marinou, V. A., 1984, Histone variants of the insect *Plodia interpunctella* during metamorphosis, *Comp. Biochem. Physiol.* **77B:** 749–753.
32. Fusauchi, Y., and Iwai, K., 1985, *Tetrahymena* ubiquitin–histone conjugate uH2A. Isolation and structural analysis, *J. Biochem.* **97:** 1467–1476.
33. Mueller, R. D., Yasuda, H., Hatch, C. L., Bonner, W. M., and Bradbury, E. M., 1985, Identification of ubiquitinated H2A and H2B in *Physarum polycephalum:* Disappearance of these proteins at metaphase and reappearance at anaphase, *J. Biol. Chem.* **260:** 5147–5153.
34. Wu, R. S., Kohn, K. W., and Bonner, W. M., 1981, Metabolism of ubiquitinated histones, *J. Biol. Chem.* **256:** 5916–5920.
35. Seale, R., 1981, Rapid turnover of the histone–ubiquitin conjugate, protein A24, *Nucleic Acids Res.* **9:** 3151–3158.
36. Trempe, J., and Leffak, M., 1982, Assembly of semihistone A24, *Nucleic Acids Res.* **10:** 5467–5481.
37. Atidia, J., and Kulka, R. G., 1982, Formation of conjugates by ^{125}I-labeled ubiquitin microinjected into cultured hepatoma cells, *FEBS Lett.* **142:** 72–76.
38. Raboy, B., Parag, H. A., and Kulka, R. G., 1986, Conjugation of ^{125}I-ubiquitin to cellular proteins in permeabilized mammalian cells: Comparison of mitotic and interphase cells, *EMBO J.* **5:** 863–869.
39. Carlson, N., and Rechsteiner, M., 1987, Microinjection of ubiquitin: Intracellular distribution and metabolism in HeLa cells maintained under normal physiological conditions, *J. Cell Biol.* **104:** 537–546.
40. Carlson, N., Rogers, S., and Rechsteiner, M., 1987, Microinjection of ubiquitin: Changes in protein degradation in HeLa cells subjected to heat-shock, *J. Cell Biol.* **104:** 547–555.

41. Wu, R. S., and Bonner, W. M., 1981, Separation of basal histone synthesis from S-phase histone synthesis in dividing cells, *Cell* **27:** 321–330.
42. Wu, R. S., Tsai, S., and Bonner, W. M., 1982, Patterns of histone variant synthesis can distinguish G_0 from G_1 cells, *Cell* **31:** 367–374.
43. Sariban, E., Wu, R. S., Erickson, L. C., and Bonner, W. M., 1985, Interrelationships of protein and DNA syntheses during replication of mammalian cells, *Mol. Cell. Biol.* **5:** 1279–1286.
44. Wu, R. S., and Bonner, W. M., 1985, Mechanism for differential sensitivity of the chromosome and growth cycles of mammalian cells to the rate of protein synthesis, *Mol. Cell. Biol.* **5:** 2959–2966.
45. Matsumoto, Y.-I., Yasuda, H., Marunonuchi, T., and Yamada, M.-A., 1983, Decrease in uH2A (protein A24) of a mouse temperature sensitive mutant, *FEBS Lett.* **151:** 139–142.
46. Shirada, T., Okihama, Y., Murata, C., and Shukuya, R., 1981, Occurrence of H1°-like protein and protein A24 in the chromatin of bullfrog erythrocytes lacking histone 5, *J. Biol. Chem.* **256:** 10577–10582.
47. Dornish, J. M., and Smith-Kielland, I., 1981, Increase in protein A24 following treatment of Ehrlich ascites tumor cells with 1-methyl-1-nitrosourea and 1,3-bis(2-chloroethyl)-1-nitrosourea, *FEBS Lett.* **136:** 41–44.
48. Goldknopf, I. L., Sudhakar, S., Rosenbaum, F., and Busch, H., 1980, Timing of ubiquitin synthesis and conjugation into protein A24 during the HeLa cell cycle, *Biochem. Biophys. Res. Commun.* **95:** 1253–1260.
49. Goldknopf, I. L., French, M. F., Dashal, Y., and Busch, H., 1978, A reciprocal relationship between contents of free ubiquitin and protein A24—its conjugate with histone 2A in chromatin fractions obtained by the DNAse II, Mg^{++} procedure, *Biochem. Biophys. Res. Commun.* **84:** 786–793.
50. Matsui, S.-I., Seon, B., and Sandberg, A., 1979, Disappearance of a structural chromatin protein A24 in mitosis: Implications for molecular basis of chromatin condensation, *Proc. Natl. Acad. Sci. U.S.A.* **76:** 6386–6390.
51. Tobey, R. A., and Ley, K. D., 1971, Isoleucine-mediated regulation of genome replication in various mammalian cell lines, *Cancer Res.* **31:** 46–51.
52. Mita, S., Yasuda, H., Marunouchi, T., Ishiko, S., and Yamada, M., 1980, A temperature-sensitive mutant of cultured mouse cells defective in chromosome condensation, *Exp. Cell Res.* **126:** 407–416.
53. Yasuda, H., Matsumoto, Y., Mita, S., Marunouchi, T., and Yamada, M., 1981, A mouse temperature sensitive mutant defective in H1 histone phosphorylation is defective in deoxyribonucleic acid synthesis and chromosome condensation, *Biochemistry* **20:** 4414–4419.
54. Ciechanover, A., Finley, D., and Varshavsky, A., 1984, Ubiquitin dependence of selective protein degradation demonstrated in the mammalian cell cycle mutant ts85, *Cell* **37:** 57–66.
55. Ciechanover, A., Finley, D., and Varshavsky, A., 1985, Mammalian cell cycle mutant defective in intracellular protein degradation and ubiquitin–protein conjugation, in: *Intracellular Protein Catabolism,* Alan R. Liss, New York, pp. 17–31.
56. Anderson, M. N., Ballal, N. R., Goldknopf, I. L., and Busch, H., 1981, Protein A24 lyase activity in nucleoli of thioacetamide-treated rat liver reduces histone 2A and ubiquitin from conjugated protein A24, *Biochemistry* **20:** 1100–1104.
57. Goldknopf, I. L., Cheng, S., Anderson, M. W., and Busch, H., 1981, Loss of endogenous nuclear protein A24 lyase activity during chicken erythropoiesis, *Biochem. Biophys. Res. Commun.* **4:** 1464–1470.

58. Anderson, M. W., Goldknopf, Z. L., and Busch, H., 1981, Protein A24 lyase is an isopeptidase, *FEBS Lett.* **132:** 210–214.
59. Kanda, F., Matsui, S.-L., Sykes, D. E., and Sandberg, A. A., 1984, Affinity of chromatin structures for isopeptidase, *Biochem. Biophys. Res. Commun.* **122:** 1296–1306.
60. Kanda, F., Slykes, D. E., Yasuda, H., Sandberg, A. A., and Matsui, S.-I., 1986, Substrate recognition of isopeptidase: Specific cleavage of the ε-(α-glycyl) lysine linkage in ubiquitin–protein conjugates, *Biochim. Biophys. Acta* **870:** 64–75.
61. Matsui, S.-I., Sandberg, A. A., Negoro, S., Seon, B.-K., and Goldstein, G., 1982, Isopeptidase: A novel eukaryotic enzyme that cleaves isopeptide bonds, *Proc. Natl. Acad. Sci. U.S.A.* **79:** 1535–1539.
62. Hershko, A., and Ciechanover, A., 1982, Mechanism of intracellular protein breakdown, *Annu. Rev. Biochem.* **51:** 335–364.
63. Finley, D., and Varshavsky, A., 1985, The ubiquitin system functions and mechanism, *TIBS* **Sept.:** 343–347.
64. Finley, D., Ciechanover, A., and Varshavsky, A., 1984, Thermolability of ubiquitin-activating enzyme from the mammalian cell cycle mutant ts85, *Cell* **37:** 43–55.
65. Pickart, C. M., and Rose, I. A., 1985, Functional heterogeneity of ubiquitin carrier proteins, *J. Biol. Chem.* **260:** 1573–1581.
66. Jentsch, S., McGrath, J. P., and Varshavsky, A., 1987, The DNA repair gene *RAD6* encodes a ubiquitin-conjugating enzyme, *Nature* **329:** 131–134.
67. Haynes, R. H., and Kunz, B. A., 1981, in: *The Molecular Biology of the Yeast Saccharomyces cerevisiae: Life Cycle and Inheritance* (J. Strathern, E. Jones, and J. Broac, eds.), Cold Spring Harbor Laboratory, Cold Spring Harbor, NY, pp. 371–414.
68. Schlegel, R. A., and Rechsteiner, M., 1978, Red cell mediated microinjection of macromolecules into mammalian cells, *Methods Cell Biol.* **20:** 341–354.
69. Levinger, L., Barsoum, J., and Varshavsky, A., 1981, Two dimensional hybridization mapping of nucleosomes. Comparison of DNA and protein patterns, *J. Mol. Biol.* **146:** 287–304.
70. Levinger, L., 1985, Nucleosomal structure of two *Drosophila melanogaster* simple satellites, *J. Biol. Chem.* **260:** 11799–11804.
71. Rodriquez-Alfageme, C., Rudkin, G. T., and Cohen, L. K., 1980, Isolation properties and cellular distribution of D1, a chromosomal protein of *Drosophila, Chromosoma* **78:** 1–15.
72. Barsoum, J., and Varshavsky, A., 1985, Preferential localization of variant nucleosomes near the 5'-end of the mouse dihydrofolate reductase gene, *J. Biol. Chem.* **260:** 7688–7697.
73. Huang, S.-Y., Barnard, M. B., Xu, M., Matsui, S. I., Rose, S. M., and Garrard, W. T., 1986, The active immunoglobulin kappa chain gene is packaged by non-ubiquitin-conjugated nucleosomes, *Proc. Natl. Acad. Sci. U.S.A.* **83:** 3738–3742.

Chapter 7

Immunochemical Probes of Ubiquitin Pool Dynamics

Arthur L. Haas

1. INTRODUCTION

The mechanism originally proposed for the ubiquitin (Ub)-dependent degradative pathway[1] can be represented by the following minimum kinetic model[2]:

$$\text{protein} + \text{Ub} \underset{k_{dis}}{\overset{k_{form}}{\rightleftharpoons}} \text{protein–Ub} \xrightarrow{k_{deg}} \text{amino acids} + \text{Ub}$$

in which k_{form} is the bulk rate constant for Ub–protein ligation, k_{dis} the rate constant for disassembly of Ub–protein conjugates, and k_{deg} the rate constant for the net multistep process by which conjugates are degraded to free amino acids with the liberation of functional Ub. Since degradation is a multistep process probably catalyzed by several distinct proteases, k_{deg} refers only to the rate-limiting step, which could vary with different protein substrates. The contribution of ATP·Mg in the model is grouped into the term for k_{form} since the intracellular concentration of the nucleotide is saturating with respect to the conjugation reaction.[3] A second

---

Arthur L. Haas • Department of Biochemistry, The Medical College of Wisconsin, Milwaukee, Wisconsin 53226.

ATP·Mg term can be grouped into k_{deg} for cases where the ATP-dependent protease[4,5] of the pathway is rate limiting.* For simplicity, the model neglects the characteristic processive nature of Ub ligation by which multiubiquitinated proteins are formed, suggesting strong positive cooperation.[1] Nonetheless, the model illustrates the salient features of the pathway: the catalytic role of Ub in protein degradation and the partitioning of Ub conjugates between alternate fates of disassembly and degradation.

In the original model, short-lived and abnormal proteins were proposed to undergo an initial ATP-coupled conjugation to Ub. This posttranslational modification then committed the protein to degradation, presumably by Ub-specific proteases, to yield the constituent amino acids and free functional Ub. In addition, a "correcting enzyme" was proposed as a rescue pathway to cleave Ub from incorrectly conjugated target proteins not intended for degradation. This latter process has more recently been termed disassembly.[5]

Several fundamental questions immediately arise concerning the model. How does Ub conjugation distinguish short-lived and abnormal proteins from other potential substrates within the intracellular milieu? How is the "correcting enzyme" able to discriminate between those conjugates committed to degradation and those formed by errors in ligation specificity? If Ub conjugation is a signal for degradation, how does one account for proteolytically stable Ub adducts such as uH2A? Finally, what are the intracellular concentrations of free and conjugated Ub? The last question is critical to arguments of substrate specificity with respect to rates of conjugation since Ub activation is kinetically removed from the protein ligation step in the overall conjugation mechanism.[6,7] Therefore, if intracellular free Ub is subsaturating, then activation and not ligation becomes rate limiting, the consequence of which is the loss of substrate specificity with respect to rates of target protein ligation. Most of our knowledge about this degradative pathway has been gleaned from cell-free studies that may not accurately represent functional aspects of the system within an intact cell. If the system is specific for proteins of abnormal structure as proposed, then how does the presence of denatured proteins generated during preparation of the cell extract affect *in vitro* function? More troublesome is whether potential regulatory processes critical to the function of the pathway are disrupted during extraction? For example, model studies with exogenous conjugates *in vitro* demonstrate that rates of conjugate disassembly are kinetically significant compared to rates of degradation.[4,5] The observed nonproductive cycling sug-

---

* If the ATP-dependent protease is not the rate-limiting step of degradation for a given protein, then the contribution of ATP in k_{deg} will be undetected.

gests poor specificity for the target proteins subject to conjugation and is not consistent with the notion of Ub ligation as an absolute signal for degradation. However, if rates of disassembly are negligible within the cell, it is possible that some regulatory link between degradation and disassembly has been disrupted during cell extraction.

Many of these questions can be examined by directly probing Ub pools within intact cells. However, the unique nature of the Ub ligation reaction makes this post-translational modification inaccessible to the usual experimental approaches. Unlike protein phosphorylation, which appears to be functionally analogous to ubiquitination, cells cannot be pulsed with appropriately labeled precursors. Therefore, several alternative approaches have been developed.[1,8,9] The least invasive of these methods has involved microinjection of [^{125}I]Ub into cultured cells.[9,10] Although recent evidence suggests Ub pools are not dramatically altered by this manipulation,[11] perturbation of cell function by microinjection can never be totally precluded. More important, [^{125}I]Ub is not functionally identical to the native polypeptide in supporting protein degradation[12] or in the steady-state distribution of conjugates.[13] Finally, the study of Ub function within intact tissues is not possible by microinjection. My group has adopted immunological methods for the direct detection and quantitation of Ub pools. The principal advantage of this approach is that both cultured cells and intact tissues can be examined by a variety of methods without prior experimental manipulation. In this chapter, I examine the different experimental methods afforded by this approach and discuss recent observations concerning the dynamics of Ub pools.

2. IMMUNOCHEMICAL METHODS

2.1. Preparation of Polyclonal Antibodies

The highly conserved primary sequence and low molecular weight of Ub present challenges for the production of antibodies to the polypeptide. These problems are compounded by the existence of both free and conjugated forms of the protein. However, one can define four criteria that an ideal immunoprobe of Ub should satisfy: (1) the capacity to discriminate between free and conjugated Ub; (2) the ability to react with both native and SDS-denatured polypeptide; (3) a high titer and binding affinity for Ub; and (4) a convenient and reproducible method of production. The first two requirements relate to the detection and independent quantitation of free and conjugated Ub and to the immunoisolation of both forms of Ub from various samples. The last two criteria define an immunological reagent of general utility and sensitivity. Most of the anticipated applications for such immunoprobes involved solid phase methods having samples immobilized on nitrocellulose. Therefore, we chose to optimize production of polyclonal antisera rather than to develop monoclonal cell lines, since monoclonal antibodies generally prove less satisfactory in solid phase applications. Our recent studies[13,14] demonstrate that polyclonal antibodies to Ub that satisfy these four criteria can be generated reproducibly.

Since Ub is an inherently poor immunogen, its antigenicity must be enhanced. The usual approach is to cross-link the desired protein to some highly immunogenic carrier, and screening of several carrier proteins has shown that bovine γ-globulin yields satisfactory antibody titers in rabbits.[8,13,14] Only intact Ub should be used for immunization. Because of rapid proteolytic cleavage of the carboxyl-terminal glycine dipeptide from Ub during isolation from cells containing lysosomes, it is best to use erythroid cells for the purification of the polypeptide.[14] Outdated human blood is a convenient source for the isolation of large quantities of Ub[15] and provides two advantages over bovine erythrocytes. First, erythrocytes can be obtained from a blood bank in the form of packed cells, which eliminates initial centrifugation. Second, human hemoglobin has a lower denaturation temperature than the bovine protein, which results in a more efficient purification and improved yield of Ub.

Since the protocols for antibody production have been published previously,[13,14] the general approaches are only summarized here. Pure Ub is cross-linked to bovine γ-globulin using glutaraldehyde. When using Ub-bovine γ-globulin as the immunogen, two factors appear important in determining the ultimate specificity for either free or conjugated Ub: the molar ratio of Ub/globulin and heat denaturation in the presence of SDS prior to immunization. Of the two factors, heat denaturation is the more critical. If the Ub–globulin is prepared at a molar ratio of 18 and injected directly into rabbits using complete Freund's adjuvant, then antisera specific for free Ub are consistently obtained.[14] In contrast, if rabbits are challenged with Ub–globulin cross-linked at a ratio of 4 and boiled in the presence of SDS prior to immunization, then antisera specific for conjugated Ub are routinely obtained.[13] Adequate titers for both types of antibodies are reached after four to five injections spaced at two week intervals. Thereafter, periodic booster injections a week prior to blood collection is sufficient to maintain titers.

By using Ub affinity columns, both types of antibody can readily be purified with better than 80% yield.[13] Overnight batch adsorption of serum to the Ub-linked matrix provides the most efficient method of isolation. The bound antibodies can then be eluted with 0.1 M glycine–HCl, pH

2.8.[13] Affinity isolation of the conjugate-specific antibodies is especially critical since the resulting probe consistently gives a lower nonspecific background staining of nitrocellulose-immobilized samples. This maneuver greatly enhances both the signal-to-noise ratio and limit of detection for conjugated polypeptide. Isolation of conjugate-specific antibodies on free Ub affinity columns at first would appear inconsistent with the specificity of these reagents; however, as discussed in Section 2.3, such antibodies exhibit a low affinity cross-reaction with free polypeptide. Therefore, a Ub column containing ~5 mg Ub/ml bed volume provides sufficient capacity to bind both types of antibody.

2.2. Solution Phase Quantitation of Free Ubiquitin

Free Ub is readily quantitated by solution phase radioimmunoassay (RIA) in which bound ligand is precipitated with 10% polyethylene glycol.[14] Typical data from such a modified Farr assay are illustrated in Figure 1. These antisera consistently have an interesting sensitivity for modification at the carboxyl terminus of Ub. Cleavage of the carboxyl-terminal glycine dipeptide from Ub by the action of trypsin[16] or a lysosomal cathepsin[14] results in diminished binding affinity. That the slopes of the RIA curves are nearly identical after such selective proteolysis demonstrates that the affinity and not the identity of the epitope has been altered. We have found that the degree of cross-reaction by partially cleaved Ub with respect to the native molecule is linearly proportional to the extent of carboxyl-terminal modification (Figure 1). This property

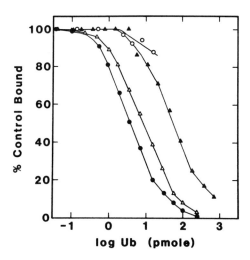

Figure 1. Solution phase radioimmunoassay for free Ub. The radioimmunoassay was conducted as described in refs. 13 and 14 using either native Ub (●), uH2A (○), 50% trypsin-inactivated Ub[16] (△), or 100% trypsin-inactivated Ub (▲). The content of uH2A is expressed in Ub equivalents.

has been exploited to correlate data from functional assays of Ub activating enzyme-catalyzed ATP:PP$_i$ exchange with the extent of proteolytic modification occurring during isolation of the polypeptide. With such correlative data, plant Ub was shown to be functionally equivalent to human Ub[17] in spite of three amino acid changes in the primary sequence.[18] Conjugated Ub responds similarly to completely inactivated polypeptide in the RIA, probably as a consequence of masking the carboxyl-terminal glycine by isopeptide bond formation.[13] Therefore, over a limited range of values it is possible to assay free polypeptide to the exclusion of conjugated Ub.

2.3. Solid Phase Quantitation of Conjugated Ubiquitin

The absolute quantitation of conjugated Ub is achieved with a solid phase dot-blot assay in which samples immobilized on nitrocellulose filters are detected after sequential incubation with conjugate-specific antibody and [^{125}I]protein A. Only the operational characteristics of this assay are reviewed since details of the method have been described previously.[13]

Figure 2A illustrates that when bound antibody is quantitated by γ counting of the associated [^{125}I]protein A, a degree of selectivity between free and conjugated Ub is observed.[13] Detection of conjugated Ub is independent of the nature of the target protein since uH2A shows reactivity equivalent to a heterogeneous mixture of reticulocyte Ub conjugates.

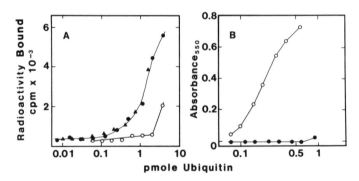

Figure 2. Solid phase dot-blot assay for conjugated Ub. A solid phase assay for conjugated Ub was conducted as described in ref. 13. (A) Immunospecific signal was quantitated by direct counting of bound [^{125}I]protein A for samples of free Ub (○), uH2A (●), and a heterogeneous mixture of reticulocyte Ub conjugates (▲). (B) Immunospecific signal was quantitated by densitometry following autoradiography for samples of free Ub (●) or a heterogeneous mixture of reticulocyte conjugates (○).

Other studies have shown that the apparent specificity of the antibody is based on a difference in binding affinity for conjugated versus free Ub. Although we have not directly mapped the antigenic determinant(s), it is likely that the observed specificity results from differing degrees of exposure for the epitope(s) between conjugated and free Ub. If a binding curve such as that in Figure 2A is obtained with native samples not exposed to either heat or SDS, specificity between the two forms of Ub is retained although the signal for conjugated Ub is attenuated by about 50%. In contrast, no change in the signal for free Ub is observed. Ub is known to retain a certain degree of tertiary structure after SDS denaturation, which accounts for its migration on SDS–PAGE at a molecular weight less than that predicted by amino acid composition.[19] This partial denaturation in the presence of SDS appears to result in a somewhat greater exposure of the epitope(s) for conjugated Ub. Consistent with this interpretation is our observation that samples completely denatured by the autoclaving procedure of Swerdlow *et al.*[20] exhibit no specificity between free and conjugated polypeptide.

The specificity of the assay for conjugated Ub can be amplified, and the signal-to-noise ratio can be greatly improved by densitometric quantitation after autoradiography of the dot-blot filters.[13] A representative example of such an assay is shown in Figure 2B. Conjugated Ub is typically quantitated over a range of 0.08–0.5 pmole with a standard error of <5% of the mean. With care, the limit of detection has been as low as 5 fmole of conjugated Ub. Such extreme sensitivity depends on the binding affinity of the antibodies and the intrinsic level of nonspecific background staining. Except for samples in which conjugated Ub represents a very small fraction of total protein, it is rarely necessary to operate at the limit of detection for the assay. Fortunately, in most cells Ub represents a considerable fraction of total intracellular protein (see Section 3.1).

Experience has shown that nitrocellulose exhibits a finite binding capacity critically dependent on the degree of hydration. Within the first hour of hydration in distilled water, the binding capacity of nitrocellulose for bovine serum albumin (BSA) more than doubles. Greater than 90% of maximum binding capacity is reached within 4 hr. The ease of hydration decreases with pore size, and nitrocellulose having an average pore diameter of 0.2 μm provides the best compromise between binding capacity and ease of hydration. With any new sample it is advisable to conduct a dilution series to determine empirically the dilution required to be within the linear binding range of the nitrocellulose. Absolute quantitation is achieved by comparison to a series of conjugate standards applied to each blot. The standards are a dilution series composed of a heterogeneous mixture of reticulocyte conjugates prepared as described previously.[13]

Inclusion of the standards on the same blot as the samples provides an internal control for any variation in staining among assays conducted at different times. In addition, it is advisable to include a single free Ub standard equivalent to the highest conjugate standard to serve as a negative control. Of the two principal methods currently used for detecting specifically bound antibodies, enzyme-linked second antibodies or [^{125}I]protein A, the latter is preferable for three reasons. First, sensitivity is far greater by autoradiography than any of the colorimetric methods and ultimately is limited only by the level of nonspecific background signal. Second, adjusting the exposure time readily allows one to optimize the signal for greatest sensitivity. Third, with a broad range of conjugate standards, merely by altering the exposure time one can simultaneously quantitate a collection of samples having very different levels of conjugated Ub. Both antibody and [^{125}I]protein A staining solutions employ BSA as a carrier protein to minimize nonspecific adsorption. We have found some commercial preparations of BSA to contain low levels of contaminating Ub presumably arising by erythrocyte lysis during blood collection. The effect of such contaminating Ub is not to increase nonspecific binding, since BSA is in great excess, but rather to decrease the sensitivity of the assay.

2.4. Visualization of Ubiquitin Conjugates after Electrophoretic Transfer

The immunostaining procedures used for dot-blot assays can also be used to visualize Ub conjugates after electrophoretic transfer to nitrocellulose from polyacrylamide gels.[13] This procedure can provide information on the size distribution of Ub conjugates. As with the dot-blot method, it is important not to exceed the binding capacity of the nitrocellulose. Observation of a constant pattern of conjugate distribution with successive dilutions of a given sample is an indication that the protein content for all regions of the gel lanes is within the binding capacity of the filter. Even when optimized, the distribution of conjugates revealed by SDS–PAGE shows extreme size heterogeneity. This typically is manifested as a smeared pattern on Western blots[13] and has also been noted with conjugates formed *in vitro* to [^{125}I]Ub.[1,15,17] Some of this smearing results from comigration of non-Ub proteins in the sample and can be alleviated by immunoisolating the conjugates prior to electrophoresis (see Section 2.5). A variation of this last approach can be exploited for the identification of specific target proteins subject to ubiquitination. Samples are first screened with antibody against the target protein. The resulting immunoisolates are then resolved in replicate by SDS–PAGE and electrophoretically transferred to nitrocellulose. Replicate sets of samples can then be stained with antibodies against either target protein or Ub. Bands detected by both probes provide good evidence for Ub–target protein adducts.

Another application that holds significant promise involves the sensitive detection of ubiquitinated histones. The immunostaining procedures used for Western blots are capable of detecting uH2A present in samples comprising relatively few cells compared to more conventional means.[13] As a quantitative method, immunodetection has been used successfully to follow the developmentally related decline in uH2A that accompanies the fusion and terminal differentiation of primary chicken myoblast cultures.[21] One obstacle to the facile quantitation of ubiquitinated histones by this method is the incomplete electrophoretic transfer of these proteins after resolution by SDS–PAGE.[21] Methanol present in the transfer buffer apparently extracts SDS from histones and leads to their precipitation within the gel. If methanol is replaced with 0.01% SDS, histones are completely cleared from the gel.[21] We have found though that short exposure of nitrocellulose to SDS above a concentration of 0.1% or longer exposures to lower concentrations (as would occur during transfer) significantly compromises the binding capacity of the filter.

In Western blot applications, free Ub is also detected using the conjugate-specific antibody.[13] This appears inconsistent with the data of Figure 2, demonstrating marked specificity between the two forms of polypeptide. However, the detection of free Ub on Western blots reflects an absolute quantity of the polypeptide that would be predicted based on the dot-blot data. Nonetheless, the signal for free Ub relative to conjugates is greatly attenuated by the difference in binding affinity. By using the autoclaving technique,[20] the signal for free Ub can be greatly enhanced and this allows such Western blots to serve as a sensitive means of quantitating free polypeptide in samples not suitable for or below the limit of detection by RIA.[2,13]

2.5. Immunoisolation of Ubiquitin Conjugates

There are many applications, such as pulse-chase studies, for which the immunoisolation of Ub conjugates are required. This approach has been used previously to demonstrate a marked increase in Ub-bound label owing to enhanced conjugation of abnormal proteins in reticulocytes following amino acid analog incorporation.[8] In these studies, antisera were selected that showed no appreciable cross-reaction with free Ub. In non-erythroid cells recent observations demonstrate Ub to be a relatively short-lived protein.[2] Therefore, it becomes essential to correct for the contribution of label present in the Ub moiety of conjugates. In these cases it is preferable to immunoisolate total Ub, then use pool quantitation to correct for label present in Ub to determine the net label from the target protein (see Section 3.4).

That the conjugate-specific antibodies cross-react with free Ub indicates that a sufficiently high concentration of antibody should result in the immunoisolation of total Ub from samples. In practice, we have found that affinity-purified antibody covalently linked to solid supports at a concentration of 250 μg/ml bed volume (1400 pmole Ub equivalents/ml) provides an efficient means of isolating total Ub pools comprising 10% or less of the column capacity. Since the antibody reacts with native Ub, it is not necessary to boil samples in the presence of SDS prior to immunoisolation as was used in previous reticulocyte studies.[8] Inclusion of 0.1% (v/v) Triton X-100 in the samples obviates nonspecific adsorption to the column matrix. Such antibody affinity columns can be used several times if care is taken to neutralize the columns quickly following sample elution with 0.1 M glycine–HCl, pH 2.8, containing 0.1% Triton to prevent nonspecific readsorption to the matrix.

2.6. Immunohistochemical Localization of Ubiquitin Conjugates

A double-antibody screening procedure was described in Section 2.4 for the identification of target proteins to which Ub is ligated. Another potentially useful approach, providing the target protein is well localized at the microscopic level, involves using the conjugate-specific antibody as a histochemical or cytochemical probe. At first the potential detection of free Ub may make this seem impractical. However, the immunospecific identification of conjugated Ub in tissue sections is fundamentally identical to the immunospecific quantitation of conjugated Ub in a solid phase dot-blot assay (Section 2.3). Therefore, the density of free and conjugated Ub per unit area in a tissue section need only be within a range where comparable densities in the dot-blot assay predict specificity for conjugated polypeptide. Danny Riley and James Bain (Department of Anatomy and Cell Biology, Medical College of Wisconsin), with whom we have had an ongoing collaboration, have shown that the conjugate-specific antibodies can successfully detect conjugated Ub to the exclusion of free polypeptide at both the light and electron microscopic levels.[22] The method has been especially informative in examining Ub ligation to muscle cell proteins.

While the densities per unit area of free and conjugated Ub predict conjugate-specific immunohistochemical localization, this alone is not adequate proof. Compelling evidence for the specific detection of conjugated Ub is shown in Figure 3. Blood smears from rabbit erythrocytes and

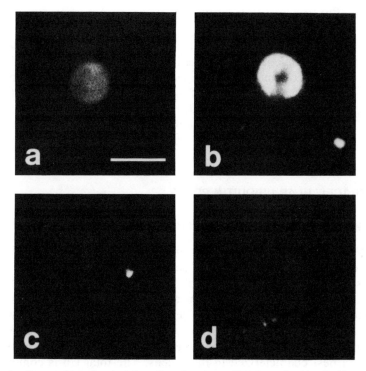

Figure 3. Immunohistochemical detection of conjugated Ub in erythroid cells. Normal and reticulocyte-rich rabbit blood samples were collected and washed with phosphate-buffered saline.[14] Subsequent steps were as described in ref. 22. Briefly, smears of cells were made on glass slides and fixed with paraformaldehyde. Slides were then incubated with affinity-purified polyclonal antibodies against SDS-denatured Ub and bound antibody then detected by incubation with fluorescein-labeled, affinity-purified goat anti-rabbit IgG. (a) An erythrocyte stained with conjugate-specific antibody. (b) A reticulocyte stained with conjugate-specific antibody. (c) Identical to (b) except that the reticulocytes were depleted of ATP by incubation with 20 mM 2-deoxyglucose and 0.2 mM 2,4-dinitrophenol for 90 min prior to preparation of the smear. (d) A reticulocyte control stained in the absence of conjugate-specific antibody. The intensely fluorescent small spots in (b) and (c) are phenylhydrazine particles incompletely rinsed from the cells. Bar = 10 μm.

phenylhydrazine-induced reticulocytes were probed with conjugate-specific antibody and detected with fluorescein-labeled goat anti-rabbit IgG.[22] The difference in signal between the two cell types is qualitatively proportional to the content of conjugated Ub.* Reticulocytes (panel b) con-

---

* Considerable population heterogeneity in the conjugate signal was noted for erythrocytes,[22] suggesting that an initial substrate-limiting decline in the conjugate pool for these cells[13] is followed by a progressive loss in ability to ligate Ub as the erythrocytes continue to age.

sistently give a stronger signal than erythrocytes (panel a). This is consistent with a smaller pool of conjugated Ub in erythrocytes even though the content of total Ub is comparable for both cell types.[13] When the pool of reticulocyte conjugates is depleted to free polypeptide by incubation with 2-deoxyglucose and 2,4-dinitrophenol, the immunospecific signal decreases to that of a control smear from which the conjugate-specific antibody was omitted (panels c and d, respectively). Elimination of signal on ATP depletion provides strong evidence for the specific detection of conjugates to the exclusion of free Ub.

3. INTRACELLULAR UBIQUITIN POOLS

3.1. Ubiquitin Pool Sizes in Cultured Cell Lines

Since the development of the immunoassay methods described in Section 2, the Ub pools present in several cultured cell lines and intact tissues have been quantitated. Table I summarizes values for total Ub present per 10^6 cells, the equivalent copy number per cell, and the percentage of total Ub conjugated for selected cultured cell lines. Data for rabbit reticulocytes and the corresponding mature erythrocytes are included for comparison since the ATP,Ub-dependent proteolytic system has been best characterized in these cells. The total Ub is comparable among the cell types listed in Table I and averages 90 pmole/10^6 cells with a range of about threefold to either side. However, these values are only approximations since in IMR-90, and probably other cells, the pool of total Ub changes with the number of population doublings and fluctuates in response to external effectors.[2]

Qualitatively, the Ub content in cultured cell lines is roughly proportional to cell volume. Using conservative estimates of cytoplasmic volume, the intracellular concentration of total Ub appears to be within the range of 10–20 μM.[13] Erythroid cells, the smallest in the table, are an exception to this size correlation and have Ub concentrations five- to tenfold higher. The K_d for Ub activation in reticulocytes is 0.6 μM[3]; therefore, in erythroid cells the rate of conjugation is saturating with respect to Ub. If one assumes that Ub activation within the cultured cells also proceeds with a submicromolar K_d, then the rate of Ub ligation is not substrate limiting with respect to the polypeptide even when the fraction of total Ub conjugated is considered in calculating the concentration of free polypeptide. The affinity of activating enzyme (E1) for Ub is identical in erythroid cells and rabbit liver (A. L. Haas, unpublished results), the only other E1 characterized to date. It is reasonable to assume that

Table I. Ubiquitin Pools in Selected Cell Lines

	Total Ub (pmole/10^6 cells)	Copies per cell ($\times\ 10^7$)	Percentage conjugated	Reference
Rabbit reticulocyte	40	2.4	83	13
Rabbit erythrocyte	38	2.3	31	13
Friend erythroleukemia	47	2.8	52	13
Human lung fibroblast (IMR-90)	135	8.1	60	13
African green monkey kidney (CV-1)	295	18	74	13
Rabbit lens epithelium (primary)	46	2.7	40	27
Bovine lens epithelium (primary)	93	5.6	67	27
Mouse fibroblast (L929)	88	5.3	31	76
Human lung carcinoma (A-549)	31	1.9	45	76

this enzyme will be nearly identical in other cells as well since the system is so highly conserved in other respects.

Unlike total Ub, no obvious correlation is apparent between either the identity or physical size of any cell line in Table I and the percentage of Ub conjugated. The fractional level of conjugation appears to be an intrinsic property of a particular cell type at a given population doubling. The importance of this "set point" is suggested by the existence of a homeostatic mechanism within cells to maintain this balance between the free and conjugated pools under normal conditions[2] (see Section 3.4). Longitudinal studies with IMR-90 have shown that the fractional level of conjugation is not immutable, however. With increasing population doublings the fractional level of conjugation progressively declines as the total Ub content increases. As a consequence, a relatively constant pool of conjugated Ub is maintained.[2] The significance of these progressive, age-related changes in Ub pools is uncertain at present, but it may relate to programmed cell senescence. Indeed, inability of senescent, Phase III cultures to carry out Ub–protein ligation could account for both the cessation of cell division and the observed accumulation of abnormal proteins that apparently results from impaired degradation.[23] An interesting parallel may exist between senescent cells and ts85, a mammary carcinoma cell line characterized by a temperature-sensitive mutation in Ub activation.[24] Both cell cycle arrest and abnormal protein accumulation occur in ts85 at the nonpermissive temperature, presumably because of a block in Ub ligation.[24,25]

The intracellular concentration of Ub is surprisingly high for any protein and approaches that encountered for many metabolic intermediates. Ub represents 0.5% by weight of nonhemoglobin protein in erythroid cells[13,26] and 0.2% by weight of total protein in IMR-90 fibroblasts.[2] The lowest ratio found to date is for the primary rabbit lens epithelial cells for which Ub represents 0.04% of total protein.[27] The other entries of the table range between the latter two values. Clearly, Ub is a major protein constituent of cells as reflected in the high average copy number of $\sim 6 \times 10^7$ per cell (Table I). From the Ub content and percentage conjugated, one can estimate a limiting value for the relative fraction of total intracellular proteins ligated. For example, the total protein content of confluent IMR-90 cultures averages 0.53 ± 0.03 mg/10^6 cells.[2] If one assumes the entire pool of cellular protein is susceptible to Ub ligation and an average cellular protein has a molecular weight of 50 kDa, then about 1% of the protein molecules could contain a single Ub based on the data of Table I. Such an estimate is fraught with uncertainty since some proteins may not co-compartmentalize with the Ub ligation system. Also, the processive nature of Ub conjugation to target proteins would favor multi- rather than monoubiquitination.[1] A more reliable estimate is to calculate the fraction of total lysines ubiquitinated. If one again assumes all the IMR-90 proteins are subject to ubiquitination and that lysine constitutes 1/20th of all amino acid residues, then for an average residue molecular weight of 109 Da the total protein pool corresponds to 0.25 µmole equivalents of lysine. From the level of conjugated Ub in the cultures, this represents a lower limit of 0.05% of total lysine residues conjugated. Even if one assumes that only one-fourth of total protein is susceptible to conjugation, only a very small fraction of potential sites are ubiquitinated at any time.

3.2. Ubiquitin Pools within Intact Tissue

Immunological methods are the only currently available means of assessing how Ub pools in cultured cell lines compare to those in intact tissues. However, in dealing with tissue samples, the acute proteolytic sensitivity of the Ub carboxyl terminus and the inherent rates of conjugate disassembly/degradation within homogenates become important considerations. The carboxyl-terminal glycine dipeptides of free and conjugated Ub are susceptible to cleavage at Arg-74 by the action of a lysosomal cathepsin released during initial homogenization.[14] The rapid inactivation of Ub at Arg-74 is consistent with the exposure of this region to bulk solvent based on the 2.8 Å crystal structure of the protein[28] and accounts for both the early ambiguity in carboxyl-terminal sequence for the polypeptide[16] and the general inability to observe ATP/Ub-dependent proteolysis in nonerythroid cell extracts.[14] The cathepsin has an extremely high turnover number and submicromolar K_m for free Ub at neutral pH. Even with mild homogenization of rabbit liver, both free and conjugated Ub are quantitatively inactivated within 2 min at 4°C (A. L. Haas and P. M. Bright, unpublished results). An analogous cleavage by a neutral protease has been observed in extracts of etiolated oat shoots[17] that do not possess lysosomes and suggests that any protease with a trypsin-like specificity is capable of this limited proteolytic modification. This artifact and the inherent metabolism of Ub conjugates can be circumvented with tissue culture samples by harvesting cells directly into SDS sample buffer.[2,13,27] However, this approach is precluded with tissue samples and requires the presence of a battery of protease inhibitors within the homogenization buffer.[13,22] Even the inclusion of protease inhibitors is not sufficient to stabilize Ub pools in tissues having a high lysosomal content. For this reason we have yet to obtain consistently satisfactory pool values for liver samples. However, in tissues having a relatively low lysosomal content the approach can be very satisfactory.

In rabbit lens epithelial cells the total content of Ub is comparable to that found in the corresponding primary tissue cultures[27]; Ub constitutes 0.03–0.04% of total protein in both tissue-derived and cultured cells, well within the range found in Table I. However, the fractional level of conjugation is eightfold greater in the primary cultures than in the tissue samples. The basis for this difference is presently unknown. Eye lens tissues exhibit a progressive developmental/age-related decline in total Ub content among epithelial, core, and cortex cells.[27] The fractional level of conjugation is found to be inversely related to the total Ub within these three cell layers, the result of which is that the absolute conjugate pool size remains relatively fixed.[27] Although the changes in total Ub and fractional conjugation observed in lens tissue are opposite to those found with increased population doublings in IMR-90 cells (above), the overall impression is of compensatory processes that seek to maintain a constant pool of conjugated Ub. About 95% of the eye lens proteins turn over with exceedingly long half-lives and are subject to considerable cummulative photooxidative damage. Preliminary observations of Ub pool changes in lens suggest that the age-related accumulation in lens of abnormal, photooxidized proteins that leads to cataractous opacities may result in part from a compromised Ub ligation system.[27]

Skeletal muscle continues to be an important system for examining aspects of protein degradation, particularly with respect to the differential rates of myofibrillar protein turnover and the pathological atrophy that accompanies a negative nitrogen balance, disuse, or disease. In addition, certain conditions such as disuse atrophy are characterized by the selective enhanced degradation of myofibrillar rather than sarcoplasmic proteins. Therefore, it is an obvious tissue in which to examine Ub pools and their role in selective protein turnover. Muscle also has the advantage that the potential target proteins of most interest with respect to muscle atrophy, the myofibrillar components, occur in a three-dimensional supramolecular assembly that can be examined immunocytochemically for localizing Ub conjugates to complement parallel pool studies.[22] For these reasons, Riley and I have begun to collect baseline data on Ub pools within normal and atrophic skeletal muscles in rats. Table II lists values for intracellular Ub pools in selected normal rat hind limb skeletal muscles.[22] In tissue samples, normalization to protein content corrects for minor variability in homogenization efficiency. Among the four different muscles, Ub content is in the range found with cultured cell lines and represents 0.02–0.04% of total protein by weight. If the Ub content is instead normalized to gross muscle volume, then the lower limit for the intracellular concentration of Ub is approximately 2 μM. If we assume that the sarcoplasmic volume of muscle is about 25% of gross tissue volume, the concentration of Ub falls within the range found in the cultured cell lines of Table I.

For the muscles listed in Table II, there is no correlation between the size of the total Ub pool and the gross muscle weight. The smallest muscle, soleus, has nearly twice the total Ub found in the other three muscles although the latter differ by fourfold in gross mass. The total Ub pool does correlate with muscle fiber type composition. Soleus is composed exclusively of oxidative muscle fibers (slow oxidative and fast oxidative-glycolytic), whereas the other three are mixed-fiber type muscles composed of both oxidative and glycolytic (fast glycolytic) fibers. This

Table II. Ubiquitin Pools in Selected Rat Skeletal Muscles[a]

	Weight (mg)	Total Ub (pmole/mg protein)	Percentage conjugated
Soleus	143 ± 5	61.2 ± 4.0	24.5 ± 2.2
Extensor digitorum longus	153 ± 4	36.6 ± 3.5	25.1 ± 3.3
Plantaris	288 ± 11	32.7 ± 1.5	28.4 ± 0.9
Gastrocnemius	622 ± 39	39.5 ± 2.4	25.3 ± 2.0

[a] Data are taken from ref. 22 and represent the mean ± SE for seven age- and weight-matched Sprague Dawley rats.

correlation was also observed with adductor longus, a muscle comparable to soleus in both oxidative fiber composition and total intracellular Ub pool.[22] Thus, the total pool of Ub appears to be muscle-type specific and depends on the ratio of glycolytic to oxidative fibers. Although Ub content is fiber-type specific, the fractional level of conjugation among the four muscles in Table II is identical within experimental error. This observation suggests the fractional level of conjugation within muscle is tissue-type specific. Although muscle-specific differences in Ub pools have been consistently observed, the absolute pool sizes and fractional conjugation appear to vary with the age of the animals, reminiscent of the results with IMR-90 cultures.

The conclusions from Table II have been confirmed independently by immunohistochemical methods as is shown in Figure 4.[22] Cross sections of soleus muscle fibers consistently exhibit uniform staining for conjugated Ub (panel a). A serial control section from which the Ub antibody was omitted shows only peripheral endogenous autofluorescence (panel b). A section of extensor digitorum longus (EDL), a mixed-fiber type muscle, reveals marked fiber-specific differences in Ub conjugate levels (panel c) that cannot be attributed to a higher endogenous background fluorescence.[22] In serial sections of EDL, the localization of differentially elevated levels of conjugated Ub to oxidative fibers has been confirmed by myofibrillar ATPase fiber typing.[22] These conclusions, which are predicated on the ability to detect conjugated Ub specifically in tissue sections, are supported by the results of Figure 3. Using Ub pool values from Table II, one can calculate the density of Ub per unit area in the muscle sections by assuming a uniform distribution of the polypeptide and a reasonable penetration depth for the reagents into the sections. Based on these calculations, the density of Ub per μm^2 is within a range where only conjugated Ub is detected in dot-blot assays of either native or SDS-denatured samples[13,22] (Figure 2B). In addition, by using immunogold detection at the electron microscopic level, the density of immunospecific gold particles per μm^2 also closely agrees with that predicted for conjugated Ub from similar calculations.[22] Therefore, the muscle-type specific differences in total Ub in Table II reflect the relative content of oxidative versus glycolytic fibers.

The basis for this fiber-type specific difference is unknown at present but it may correlate with a greater oxidative stress within fibers exhibiting elevated conjugate pools, particularly since oxidative damage appears to predispose proteins to enhanced conjugation.[9,13] It is significant that oxidative muscle has a higher intrinsic rate of protein turnover than glycolytic muscle.[29,30] Therefore, the larger absolute pool of conjugated Ub within oxidative fibers correlates with their higher reported rate of protein

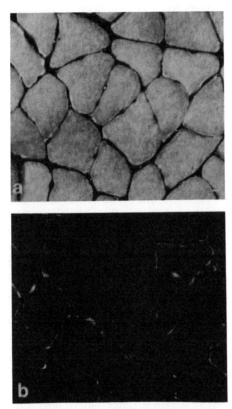

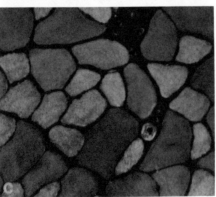

Figure 4. Immunohistochemical detection of Ub conjugates in rat muscle cross sections. Sections were prepared and immunostained using conjugate-specific antibody as described in ref. 22. Immunospecifically bound antibody was detected by subsequent incubation with fluorescein-labeled affinity-purified goat anti-rabbit IgG.[22] (a) Cross-section of soleus showing uniform fiber staining. (b) Control serial section to that of (a) stained in the absence of conjugate-specific antibody. (c) Cross-section of extensor digitorum longus showing fiber-type specific levels of Ub conjugates. Bar = 10 μm.

degradation. Such a correlation does not *a priori* demonstrate a causal relation between the muscle-type specific rates of protein turnover and conjugate pool size. However, enhanced cytochemical localization of Ub conjugates to Z lines, containing several relatively short-lived components, is consistent with this conclusion for specific target proteins.[22] In addition, localized enhanced conjugation has been observed at sites of myofibrillar disruption in spontaneously atrophic muscle fibers (D. A. Riley and A. L. Haas, unpublished results). Conclusive proof will require demonstration that the extent of conjugation to specific target proteins parallels their rates of turnover.

3.3. Subcellular Localization of Ubiquitin Conjugates

Since the Ub–protein ligation system is currently believed to be exclusively cytoplasmic, it has been assumed that only soluble proteins are subject to Ub ligation. This view has required modification based on recent evidence for covalent Ub adducts to membrane-bound proteins such as the lymphocyte homing receptor,[31] the platelet-derived growth factor receptor,[32] the synaptosomal high-affinity sodium-dependent choline transporter,[33] and unidentified proteins of reticulocytes,[13] hepatoma cells,[10] and HeLa cells.[11] Because little is known about the subcellular distribution of Ub conjugates, these adducts may represent a considerable fraction of the total conjugate pool. Conjugate-specific antibodies are potentially useful probes for identifying such adducts. Ubiquitination of membrane-bound proteins appears to serve two different roles: regulation of function[31–33] (discussed elsewhere in this volume) and commitment to degradation.[13,34]

In erythroid cells conjugation to membrane-bound proteins probably mediates the normal maturational degradation of proteins not required in the circulating red cell. The stromal and ribosomal pellets of reticulocyte lysates contain approximately 25 and 2%, respectively, of the total conjugate pool by immunoquantitation.[13] In contrast, particulate-bound conjugates are largely absent in erythrocytes. It has long been known that ribosomes and many stromal proteins, mostly of mitochondrial origin, are lost during the terminal maturation of reticulocytes. Ubiquitin-mediated degradation of these proteins is proposed to account for their programmed loss,[13,34] which is reflected in the shift in conjugate subcellular distribution between reticulocytes and erythrocytes.[13] Rapoport and co-workers have reported evidence for the *in vitro* degradation of stromal proteins by the ATP,Ub-dependent pathway.[34,35] Disintegration of the mitochondrial membrane by the action of a lipoxygenase may trigger the degradation of sequestered mitrochondrial enzymes.[36] Oxidative damage may also play a role in initiating the degradation of proteins sensitive to this modification. Soluble proteins and those present on the outer mitochondrial membrane may be degraded directly, as with the hexokinase isozymes Ia and Ib. Ub participates in the selective degradation of the soluble Ib isozyme of hexokinase; the slower rate of isozyme Ia degradation, the sole form of enzyme retained in erythrocytes, is believed due to the partial protection afforded by binding to mitochondria prior to destruction of the organelle.[37] These observations parallel data for hexokinase II, a K^+- and ligand-stabilized mitochondria-bound enzyme that undergoes enhanced ATP-dependent degradation in ascites cells in the absence of K^+ or after dissociation from the mitochondrial membrane by EDTA.[38] More recent data indicate that dog erythrocyte Na,K-ATPase is also degraded by an ATP-dependent process in rabbit reticulocyte extracts, although direct participation of Ub was not tested.[39]

The subcellular distribution of IMR-90 Ub conjugates with respect to molecular weight and location is typical of that found in cultured cell lines (Figure 5). The distribution of conjugates is markedly skewed to high molecular weight, and although specific bands can be discerned, for the most part the pattern of Ub conjugates is very heterogeneous as is consistently observed.[1,10,13,40] Two predominant conjugate bands of 102 and 56 kDa are routinely observed in cells, with the exception of Friend erythroleukemia.[13] Few conjugates are typically found below 30 kDa.[13] Whether this reflects a size dependence for protein conjugation or the rapid cleavage of Ub from proteolytic fragments via the action of isopeptidase is uncertain. However, that few low molecular weight conjugates are ever observed is consistent with model studies demonstrating an inverse relation between polypeptide size and relative extent of ATP-dependent versus ATP-independent degradation in reticulocytes.[41] Other than uH2A, few of the conjugate bands localize to a single subcellular fraction. The broad distribution of conjugate species in Figure 5 is not an artifact of fractionation since pellets were rigorously washed to remove contaminants present in the included volume. A similar broad distribution of conjugates over several subcellular fractions has also been noted in other cultured cell lines after microinjection of $[^{125}I]$Ub.[11,40] One possibility for the dispersed distribution of intracellular conjugates, particularly of high molecular weight, is that these intermediates represent precipitated protein aggregates. In reticulocytes, such aggregates are preferentially degraded by an energy-dependent process[42] that is presumably Ub mediated. Alternatively, Ub has been shown to undergo a conformational change in solutions of low dielectric constant.[43] Perhaps, a fraction of conjugated Ub becomes anchored by the intercalation of the hydrophobic portion of the polypeptide within membranes.

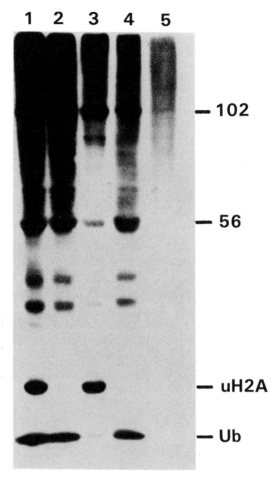

Figure 5. Subcellular fractionation of IMR-90 Ub conjugates. Cultures of IMR-90 were grown to confluency and then harvested by scraping from plates. The cells were then disrupted into Tris-Cl, pH 7.5, containing isotonic sucrose and a mixture of protease inhibitors. Fractionation was similar to that reported in ref. 13 for erythroid cells. Lane 1: Original cell extract. Lane 2: 8000 g supernatant. Lane 3: 8000 g pellet. Lane 4: 10^5 g supernatant. Lane 5: 10^5 g pellet. Free Ub, uH2A, and the prominent 102 and 56 kDa bands are marked to the right.

3.4. The Dynamics of Ubiquitin Pools

One can consider intracellular conjugate pools either with respect to total protein or with respect to specific targets for Ub ligation. This distinction is important since quite different conclusions can be drawn depending on one's perspective. For example, it is possible for the total conjugate pool to be disproportionate to the rate of degradation through the Ub pathway if only a small subpopulation of conjugates are true degradative intermediates. In contrast, the levels of conjugates to specific proteins within this labile pool may be proportional to their individual rates of degradation.

Studies in erythroid cells indicate that a substantial fraction of intracellular conjugates are proteolytically stable[13]; that is, for this subpopulation of conjugates $k_{dis} > k_{deg}$. This conclusion is based on the marked disproportionality between rates of degradation[13] and the sizes of conjugate pools between reticulocytes and erythrocytes.[44] Earlier reticulocyte studies involving immunoisolation of pulsed Ub conjugates indicated that perhaps 50% of the labeled conjugates were not degraded.[8] Similar results were also reported for ascites cells.[8] The low rate of Ub synthesis in reticulocytes, and other control studies, ruled out the possibility that this apparent stable pool resulted simply from label present in the Ub moiety of the conjugates.

Reticulocytes and mature erythrocytes differ principally in their content of Ub conjugates of stromal proteins. These particulate-bound conjugates constitute about 20% of the total Ub pool of reticulocytes and are ligated to a subpopulation of proteins undergoing maturation-linked degradation.[13,34-36] The distribution of nonstromal (cytosolic) conjugates is identical between reticulocytes and erythrocytes, although the pool size of the latter is decreased by half. Since the concentration of conjugates is not directly proportional to the degradative flux through the Ub-dependent pathway, this suggests the existence of stable conjugates. Moreover, this observation argues against Ub ligation as an irreversible signal event for commitment to degradation per se and suggests that features of the target proteins other than the presence of an attached Ub dictate the relative magnitudes of k_{deg} versus k_{dis}.[13] In contrast to this pool of stable conjugates, the concentrations of conjugates to labile proteins are consistent with their overall rates of degradation. In reticulocytes, label is observed to chase from amino acid analog-containing conjugates at a rate identical to the degradation of these target proteins.[8] In addition, when [^{125}I]hemoglobin was microinjected into HeLa cells, the pool of globin conjugates that formed following in situ denaturation of hemoglobin with phenylhydrazine was also proportional to the rate of degradation.[9] Comparable results were seen in the phenylhydrazine induction of a similar set of conjugates in erythrocytes[13]; the concentration of these conjugates paralleled a reported increase in protein degradation.[45] The latter observations have been questioned in the report of an ATP-independent activity that degrades denatured globin.[46] Goldberg and co-workers correctly point out that Ub conjugation cannot be equated directly with degradation.[46] However, the rates of ATP,Ub-independent degradation of model substrates were considerably less than those involving Ub-mediated processes.[46] Also, the structure of phenylhydrazine-denatured globin is probably so severely compromised that degradation by several pathways may occur simultaneously.

In reticulocytes where Ub is very long lived,[13] the rate of Ub synthesis is negligible.[8] However, the polypeptide is a relatively short-lived protein in other cells, having a half-life of 28–31 hr in IMR-90[2] and 10 hr in both HeLa[9] and L1210[47] cell lines. The physical properties of Ub do not fit the typical profile of high molecular weight and acidic isoelectric point for short-lived proteins.[48] Since Ub is stable in erythroid cells, the rapid turnover is probably not a consequence of its role in degradation.[2,13] Nor is it likely owing to lysosomal autophagy in nutritionally balanced cultures.[2] Approximately 5% of the turnover rate can be accounted for by secretion of Ub from IMR-90[2] at approximately 2% of the internal pool per day (A. L. Haas and P. M. Bright, unpublished results). However, some specific mechanism must exist within cells to account for the majority of Ub turnover. The intracellular levels of other short-lived proteins are usually under dynamic regulation.[48] For these key metabolic enzymes, rapid turnover is a selective advantage because the half-life determines the time required for a protein to shift between different steady states.[49] Initial evidence indicates Ub levels are also subject to regulation. Ub has been identified as a heat-shock protein in chicken embryo fibroblasts,[50] the gene for which has been found to contain an upstream heat-shock promoter.[51] In IMR-90 and other cells, total cellular Ub increases in response to a serum factor.[2] Ub induction in IMR-90 results from a twofold increase in the rate of synthesis while the half-life for degradation of the polypeptide remains constant.[2] Other conditions triggering Ub induction will undoubtedly be discovered in the future.

Because of these differences in the rate of Ub turnover, reticulocytes do not reflect the dynamics of Ub pools observed in normal cells where the balance between free and conjugated polypeptide is tightly regulated.[2] During both the transient induction of Ub in IMR-90 by the serum factor and the loss of the polypeptide following serum withdrawal, free and conjugated Ub fluctuate in concert so that the fractional level of conjugation remains constant. No change in the uH2A pool was noted under either condition, suggesting independent regulation of nuclear and cytoplasmic conjugation. This constant set point for the fractional level of cytoplasmic conjugation within a cell is unexpected, but not peculiar to the IMR-90 cell line. Similar behavior has been found in other cultured cell lines and intact tissue as well.[2,22,52] Rat soleus muscle undergoing suspension-induced disuse atrophy shows an analogous constant set point during Ub induction: Ub synthesis is induced while the fractional level of conjugation remains unaltered.[52] Therefore, cells appear to possess a mechanism for maintaining the balance between free and conjugated Ub. At present the temporal relation of these changes is unclear. That is, does the conjugate pool increase in response to induction of total Ub? Or does

Ub induction result from an increase in the pool of conjugates, signaled in turn by a corresponding transient decrease in free Ub? A rapid decline in free Ub during its enhanced conjugation to thermally denatured proteins has been speculated to trigger synthesis of stress proteins during heat shock.[53]

Since intracellular free Ub concentrations are saturating with respect to E1, the homeostasis between Ub pools probably does not reflect a velocity–substrate dependence for the activation step. However, one cannot discount the possibility that Ub is divided between active and inactive forms owing to some post-translational modification, such as phosphorylation,[54] that renders activation subsaturating. This seems unlikely since conjugation would then be limited by the activation step and show no target protein specificity with respect to the rate of ligation. Potential alternative hypotheses require that the factors defining the set point be known. From the minimum kinetic model presented in Section 1, the pool of conjugated Ub can be represented by the following equation[2]:

$$[\text{protein-Ub}] = \frac{k_{\text{form}} [\text{protein}]}{k_{\text{dis}} + k_{\text{deg}}} \tag{1}$$

In deriving the expression, the concentration of Ub was assumed saturating with respect to conjugation; this reduces the three rate constants to first-order terms. The equation for the steady-state concentration of conjugates predicts regulation by (1) an increase in the pool of proteins susceptible to conjugation, (2) an increase in k_{form}, or (3) a decrease in either k_{deg} or k_{dis}. The size distribution of IMR-90 conjugates is invariant during induction.[2] Therefore, it is unlikely that a subpopulation of kinetically favored substrates is induced since this would result in the appearance of specific conjugates and an altered size distribution. On the other hand, an increase in k_{form}, representing an up-regulation of the rate of ligation, is consistent with the invariant size distribution of conjugates during induction. If IMR-90 contains a stable pool of conjugates, that is, if the bulk rate of disassembly is faster than degradation, then a decline in k_{dis} could account for regulation of Ub pools. Free Ub is a competitive inhibitor of Ub carboxyl-terminal hydrolase[55] at concentrations observed intracellularly.[2,13] If the functionally distinct isopeptidase(s)[4,5,56] possesses a binding site for Ub similar to that of the hydrolase, then free Ub could serve as a feedback inhibitor of conjugate disassembly. By this model, the rate of disassembly would be sensitive to fluctuations in free Ub. The final alternative, down-regulation in the rate of degradation during Ub induction, seems contradictory and appears to be precluded by studies showing an unchanged rate of total protein degradation between conditioned and fed IMR-90 cultures[57] during a time frame in which Ub induction is maximal.[2] However, the question remains unresolved because the latter study only examined the fate of long-lived proteins. If IMR-90 contains a stable pool of conjugates to long-lived proteins, for which the rate of disassembly greatly exceeds that of degradation then small changes in the turnover of the long-lived pool through the Ub pathway might have gone undetected. Any potential effect should be most sensitive to the rate of short-lived protein degradation for which $k_{deg} > k_{dis}$.

In Equation (1), the relative magnitudes for k_{dis} versus k_{deg} are unknown, and they depend on whether nonerythroid cells contain a significant pool of proteolytically stable conjugates. As cited earlier, pulse-chase studies with ascites cells suggest approximately half the conjugate pool to be stable to degradation.[8] In addition, microinjected [^{125}I]Ub rapidly equilibrates with the conjugate pool in cultured cell lines.[11,40] Preliminary double-label, pulse studies in IMR-90 indicate that the isotopic ratio for the conjugate pool is intermediate between that of total protein and the short-lived pool (A. L. Haas and P. M. Bright, unpublished results). Nonerythroid cells therefore also appear to contain a pool of conjugates to long-lived proteins that exhibit rapid disassembly in comparison to degradation. The dynamics of the cytoplasmic conjugate pool is thus analogous to the nuclear uH2A pool, which similarly exhibits a rapid steady-state between formation and disassembly[47,58] and which the level of ubiquitinated histones is regulated by modulation of the opposing processes.[59]

Significant conjugation to long-lived proteins appears to contradict studies in which it was concluded that conjugation occurred only to short-lived proteins, based on immunoisolation of the resulting conjugate pool.[25] However, brief pulses selectively, but not exclusively, label short-lived proteins. More critical to unambiguous interpretation in such antibody-based applications is the presence of label in the Ub moiety of conjugates. This is a consequence of the rapid rate of Ub synthesis. The pool of poly-Ub precursor[60] in cells appears to be negligible, so that increases in the Ub pool occur directly from *de novo* synthesis.[2] The rate of Ub synthesis is substantial in IMR-90,[2] L1210,[47] bovine parathyroid,[61] bovine pituitary,[62] rat hypothalamus,[63] and both human and mouse pituitary tumors.[63] Rapid rates of Ub synthesis are not surprising in view of the substantial pool and short half-life of this protein within cells. Therefore, a considerable fraction of label in immunoisolated conjugates can potentially derive from Ub. Based on pool quantitation, conjugates isolated from IMR-90 after a 1 hr pulse in the presence of [^{3}H]leucine contain 50–60% of the label in Ub (A. L. Haas and P. M. Bright, unpublished results). For L929 cultures in which Ub turnover is more rapid, the majority of isotope in conjugates results from the labeled Ub (A. L. Haas and P. M. Bright, unpublished results). Unfortunately, even abundant amino acids, such as leucine, are poorly incorporated into the target proteins in conjugates; therefore, alternative approaches using less abundant tracers such as tryptophan, which does not occur in Ub,[19,64] are precluded. In summary, careful corrections based on pool quantitation are required for unambiguous interpretation of such data.

An area for future research is the response of Ub pools to conditions of stress such as heat shock, amino acid analog incorporation, administration of protein-modifying agents, and tissue anoxia. In the present kinetic model, stress conditions lead to a rapid increase in the [protein] term, arising from the generation of proteins having abnormal conformations. Pulse studies with reticulocytes,[8] ascites cells,[8] and ts85[25] have shown coordinate increases in immunoisolated Ub-associated label and rates of degradation during amino acid analog incorporation. These increases have been interpreted as evidence for elevated conjugate pools. In the absence of direct measurements on pool sizes, such conclusions are untenable since it is equally possible that abnormal proteins compete out the endogenous, stable population of conjugated proteins to become a greater fraction of a constant, total conjugate pool. The latter alternative appears to occur in IMR-90 during thialysine and canavanine incorporation; the absolute pool of Ub conjugates remains constant or decreases slightly (A. L. Haas and P. M. Bright, unpublished results). This is analogous to the phenylhydrazine-induced denaturation of hemoglobin that results in the appearance of globin conjugates and enhanced degradation but has no effect on the absolute concentration of total conjugates.[9,13] Contradictory observations are also found in the limited data available for the heat-shock response. Under conditions in which heat-shock proteins (including Ub) are induced, indirect evidence suggests little change in the absolute conjugate pool within chicken embryo fibroblasts.[50] A similar result is obtained during heat shock of IMR-90.* In contrast, a substantial increase in the fractional level of conjugation is observed in HeLa cells during brief thermal stress.[65] These cell line specific responses of Ub pools to heat shock may relate to differences in the temperature required for induction of heat-shock proteins versus that for significant protein denaturation.

Rapid enhanced ubiquitination of specific target proteins, presumably

---

* A slow induction of Ub is also observed during amino acid analog incorporation, consistent with the role of Ub as a general stress-inducible protein (A. L. Haas and P. M. Bright, unpublished results).

also owing to an increase in the [protein] term, has been observed under physiological conditions. The best characterized example is phytochrome, a constitutive protein central to a number of regulatory processes in plants. Phytochrome undergoes photoinduced enhanced turnover that may initiate the cascade of metabolic changes.[66] A significant induction of Ub conjugates to phytochrome accompanies this increased degradation.[66] The effect of photoactivation on the structure of phytochrome is uncertain, although it must involve a substantial conformational change since the protein forms insoluble aggregates.[67] Whether enhanced conjugation occurs to soluble phytochrome or to the precipitated form can be examined immunohistochemically.

Collaborative studies with Winston Kao (Department of Ophthalmology, University of Cincinnati College of Medicine) have yielded analogous results with prolyl hydroxylase, a stable enzyme of confluent fibroblast cultures.[68] If cultures are replated to subconfluent densities, collagen synthesis ceases and prolyl hydroxylase is rapidly degraded.[68] Coincidently, there is a marked increase in Ub ligation to the enzyme (W. W.-Y. Kao and A. L. Haas, unpublished results). The "trigger event" for enhanced conjugation could be one of several regulatory processes known to modulate prolyl hydroxylase. Such induction in individual subpopulations of the conjugate pool are subtle. In the case of phytochrome and prolyl hydroxylase, the induced conjugates, which could not be detected above the background of the stable endogenous conjugates, were only observed after enrichment by double-antibody screening procedures. Finally, the localized increases in ubiquitination to sites of myofibrillar disruption during spontaneous atrophy represent a form of enhanced conjugation at the histochemical level (see Section 3.2).

3.5. The Role of Ubiquitin–Protein Ligation: A Hypothesis

To account for stable conjugates in erythroid cells, we previously proposed that the role of Ub ligation in energy-dependent degradation was not signal recognition per se, but rather served to destabilize target proteins and yield proteolytically labile open structures.[13] Features determining the relative magnitude of k_{dis} versus k_{deg} would then depend on target protein sensitivity to Ub-induced unfolding. From the current understanding of Ub pool dynamics it is possible to expand this hypothesis to the following general model of Ub function.

That a variety of exogenous, extracellular proteins and endogenous long- and short-lived proteins can undergo Ub conjugation *in vitro* argues against the presence of a consensus ligation recognition sequence analogous to those specifying sites for other forms of post-translational modification. Indeed, preliminary reports suggest the active site of isopeptide ligase may be tailored to bind the polypeptide chains nonspecifically,[69] although in some cases with high affinity.[70] Therefore, steric accessibility of lysine side chains and amino termini probably dictates the sites of ubiquitination. This conclusion is consistent with the specific monoubiquitination of lysine-rich histones[71,72] and *Dictyostelium* calmodulin.[73] In the latter case, the Lys-115 subject to Ub ligation is exposed to solvent and is highly mobile by NMR.[74] Because the conjugate pool is subject to rapid disassembly compared to degradation, Ub is continuously ligated to and released from numerous protein substrates. We propose that following initial ubiquitination the partitioning between disassembly and degradation for a given conjugate is determined by the relative rates of Ub-induced protein unfolding and exposure of new sites for ligation versus that of disassembly. Therefore, the specificity of Ub-mediated proteolysis resides not in ligation but in the partitioning between disassembly and degradation. Such partitioning would depend on the inherent conformational stability of the target protein. The dynamics of the Ub pools in such a model would reflect a constant sampling of cellular proteins for conformationally labile species arising by spontaneous thermal denaturation, errors of synthesis, or induced instability by specific trigger events. Whether Ub serves a passive[13] or active[12] role in promoting such protein unfolding remains to be tested.

This model for Ub function is consistent with a considerable body of literature indicating that features of protein structure and stability determine *in vivo* half-life.[48] If ligation is nonspecific, then long-lived proteins would be subject to slow, Ub-mediated proteolysis since a small fraction of such conjugates would always partition toward degradation. The last point reconciles two recent observations: greater than 90% of short-lived and abnormal protein degradation is Ub mediated,[25] and ATP-dependent proteolysis accounts for approximately 90% of the degradative rate for both short- and long-lived proteins.[75]

Metabolic control is invariably designed around regulating the balance between rates of opposing reactions because of the exquisite sensitivity afforded by such a mechanism. Branch-point and committed step enzymes are prime examples of such control. That Ub ligation and disassembly exist in a similar dynamic balance suggests that proteolysis via the Ub-dependent pathway can be modulated by altering total intracellular Ub. Since all conjugates would exhibit a finite rate of partitioning toward degradation, up-regulation in the pool of such intermediates would result in a net increase in their rate of proteolysis even though the ratio of partitioning between the opposing steps remained unchanged. Similar modulation of conjugate pools could potentially serve additional regulatory functions distinct from degradation by displacing equilibria between protein conformers having different biological activity. For example, in the following scheme,

$$\text{protein}_{\text{active}} \rightleftharpoons \text{protein}_{\text{inactive}} \underset{k_{\text{dis}}}{\overset{k_{\text{form}}}{\rightleftharpoons}} \text{protein}_{\text{inactive}} - \text{Ub}$$

Ub ligation could regulate protein function by altering the equilibrium between active and inactive states. An increase in total Ub would result in an increased conjugate pool and thereby shift the equilibria toward the stable protein$_{\text{inactive}}$–Ub form.

Our model suggests that Ub-mediated degradation does not require (but does not exclude) the participation of Ub-specific proteases since ligation-induced unfolding would generate open structures susceptible to general cellular proteases. Once such unfolding occurred, subsequent degradation may not require the continued presence of ligated Ub. Alternatively, Ub might be rapidly cleaved from the initial proteolytic fragments by the action of isopeptidases. The model predicts Ub conjugates to be inherently more labile to proteases than the native, unmodified target protein. This prediction agrees with observations of Hough et al. that in vitro conjugates to lysozyme are more susceptible than native lysozyme to basal ATP-independent degradation and to various exogenous proteases.[56] Therefore, the frequently conflicting observations on the substrate specificity of the Ub-mediated pathway may be accounted for in part by the relative rates of proteolysis of a target protein versus its corresponding conjugate.

If this post-translational modification merely promotes protein unfolding, then what functional advantage does Ub ligation serve within the cell? This question is particularly relevant since prokaryotes easily degrade denatured proteins in the absence of Ub-dependent pathways. "Denaturation" is an ambiguous description of protein conformation especially since it often connotes an extended random coil more commonly found in urea. Under physiological conditions, such extended polypeptides are thermodynamically unstable; therefore, the conformation of a "denatured" protein need not deviate significantly from that of the native, minimum free-energy conformation. Since peptide bond cleavage of both native and nonnative proteins is directed only to external accessible sites, only limited degradation of most cellular proteins may occur in the absence of mechanisms for exposing additional sequestered sites. Proteins are frequently observed to tolerate one or more such endoproteolytic cleavages and still maintain a quasinative conformation and activity; selective cleavage of ribonuclease is a classic example of such a case, as is the more recent results obtained with Ub.[14,16] Therefore, the function of Ub ligation would be to ensure the rapid degradation of such proteins by promoting their unfolding, eliminating the accumulation of partially degraded polypeptides within the cell. In contrast, prokaryotes are not faced with this problem since "nicked" proteins can be diluted by cell division.

4. CONCLUSION

Ub–protein ligation is a novel post-translational modification that has proved difficult to study within intact cells and tissues. These experimental limitations can be circumvented by immunological methods available for the quantitation, identification, and cytochemical localization of intracellular Ub adducts. The dynamics of Ub pools suggests that Ub ligation is analogous to phosphorylation in several respects: (1) the steady-state levels of both modifications depend on the dynamic balance between rates of adduct formation and scission; (2) both modifications appear to be relatively nonspecific with respect to the total protein population; and (3) in the short term only a subpopulation of modifications generate productive events (degradation in the case of Ub ligation). In addition, the regulation of Ub pools is significantly more involved than the passive tagging of proteins for degradation originally proposed. Remaining questions relate to the occurrence and characteristics of stable conjugate pools within cells, the relative magnitudes of disassembly versus degradation for specific conjugates, the nature of events that predispose proteins to enhanced conjugation and degradation, the potential for nondegradative Ub ligation in regulation, and the dynamics of Ub pools under normal and stress conditions.

ACKNOWLEDGMENT. Unpublished observations of the author were supported by United States Public Health Service grant GM 34009. The contribution of Ms. Patricia Bright to these unpublished studies is gratefully acknowledged.

REFERENCES

1. Hershko, A., Ciechanover, A., Heller, H., Haas, A. L., and Rose, I. A., 1980, Proposed role of ATP in protein breakdown: Conjugation of proteins with multiple chains of the polypeptide of ATP-dependent proteolysis, *Proc. Natl. Acad. Sci. U.S.A.* **77:** 1783–1786.

2. Haas, A. L., and Bright, P. M., 1987, The dynamics of ubiquitin pools in cultured human lung fibroblasts, *J. Biol. Chem.* **262:** 345–351.
3. Haas, A. L., and Rose, I. A., 1982, The dynamics of ubiquitin activating enzyme: A kinetic and equilibrium analysis, *J. Biol. Chem.* **257:** 10329–10337.
4. Hershko, A., Leshinsky, E., Ganoth, D., and Heller, H., 1984, ATP-dependent degradation of ubiquitin–protein conjugates, *Proc. Natl. Acad. Sci. U.S.A.* **81:** 1619–1623.
5. Hough, R., and Rechsteiner, M., 1986, Ubiquitin–lysozyme conjugates: Purification and susceptibility to proteolysis, *J. Biol. Chem.* **261:** 2391–2399.
6. Haas, A. L., Warms, J. V. B., Hershko, A., and Rose, I. A., 1982, Ubiquitin-activating enzyme: Mechanism and role in ubiquitin–protein conjugation, *J. Biol. Chem.* **257:** 2543–2548.
7. Hershko, A., Heller, H., Elias, S., and Ciechanover, A., 1983, Components of the ubiquitin–protein ligase system: Resolution, affinity purification, and role in protein breakdown, *J. Biol. Chem.* **258:** 8206–8214.
8. Hershko, A., Eytan, E., Ciechanover, A., and Haas, A. L., 1982, Immunochemical analysis of the turnover of ubiquitin–protein conjugates in intact cells, *J. Biol. Chem.* **257:** 13964–13970.
9. Chin, D. T., Kuehl, L., and Rechsteiner, M., 1982, Conjugation of ubiquitin to denatured hemoglobin is proportional to the rate of hemoglobin degradation, *Proc. Natl. Acad. Sci. U.S.A.* **79:** 5857–5861.
10. Atidia, J., and Kulka, R. G., 1982, Formation of conjugates by ^{125}I-labelled ubiquitin microinjected into cultured hepatoma cells, *FEBS Lett.* **142:** 72–76.
11. Carlson, N., and Rechsteiner, M., 1987, Microinjection of ubiquitin: Intracellular distribution of and metabolism in HeLa cells maintained under normal physiological conditions, *J. Cell Biol.* **104:** 537–546.
12. Cox, M. J., Haas, A. L., and Wilkinson, K. D., 1986, Role of ubiquitin conformations in the specificity of protein degradation: Iodinated derivatives with altered conformations and activities, *Arch. Biochem. Biophys.* **250:** 400–409.
13. Haas, A. L., and Bright, P. M., 1985, The immunochemical detection and quantitation of intracellular ubiquitin–protein conjugates, *J. Biol. Chem.* **260:** 12464–12473.
14. Haas, A. L., Murphy, K. E., and Bright, P. M., 1985, The inactivation of ubiquitin accounts for the inability to demonstrate ATP,ubiquitin-dependent proteolysis in liver extracts, *J. Biol. Chem.* **260:** 4694–4703.
15. Haas, A. L., and Wilkinson, K. D., 1985, The large scale purification of ubiquitin from human erythrocytes, *Prep. Biochem.* **15:** 49–60.
16. Wilkinson, K. D., and Audhya, T. K., 1981, Stimulation of ATP-dependent proteolysis requires ubiquitin with the COOH-terminal sequence Arg-Gly-Gly, *J. Biol. Chem.* **256:** 9235–9241.
17. Vierstra, R. D., Langan, S. M., and Haas, A. L., 1985, Purification and initial characterization of ubiquitin from the higher plant, *Avena sativa, J. Biol. Chem.* **260:** 12015–12021.
18. Vierstra, R. D., Langan, S. M., and Schaller, G. E., 1886, Complete amino acid sequence of ubiquitin from the higher plant *Avena sativa, Biochemistry* **25:** 3105–3108.
19. Wilkinson, K. D., Urban, M. K., and Haas, A. L., 1980, Ubiquitin is the ATP-dependent proteolysis factor I of rabbit reticulocytes, *J. Biol. Chem.* **255:** 7529–7532.
20. Swerdlow, P. S., Finley, D., and Varshavsky, A., 1986, Enhancement of immunoblot sensitivity by heating of hydrated filters, *Anal. Biochem.* **156:** 147–153.
21. Wunsch, A. N., Haas, A. L., and Lough, J., 1987, Synthesis and ubiquitination of histones during myogenesis, *Dev. Biol.* **119:** 85–93.
22. Riley, D. A., Bain, J. L. W., Ellis, S., and Haas, A. L., 1987, The quantitation and

immunocytochemical localization of ubiquitin conjugates within rat red and white skeletal muscles, *J. Histochem. Cytochem.* (in press).

23. Bradley, M. O., Hayflick, L., and Schimke, R. T., 1976, Protein degradation in human fibroblasts (WI-38): Effects of aging, viral transformation, and amino acid analogs, *J. Biol. Chem.* **251:** 3521–3529.

24. Finley, D., Ciechanover, A., and Varshavsky, A., 1984, Thermolability of ubiquitin-activating enzyme from the mammalian cell cycle mutant ts85, *Cell* **37:** 43–55.

25. Ciechanover, A., Finley, D., and Varshavsky, A., 1984, Ubiquitin dependence of selective protein degradation in the mammalian cell cycle mutant ts85, *Cell* **37:** 57–66.

26. Ciechanover, A., Elias, S., Heller, H., Ferber, S., and Hershko, A., 1980, Characterization of the heat-stable polypeptide of the ATP-dependent proteolytic system from reticulocytes, *J. Biol. Chem.* **255:** 7525–7528.

27. Jahngen, J., Haas, A. L., Ciechanover, A., Blondin, J., Eisenhauer, D., and Taylor, A., 1986, The eye has an active ATP-dependent ubiquitin–lens protein conjugation system, *J. Biol. Chem.* **261:** 13760–13767.

28. Vijay-Kumar, S., Bugg, C. E., Wilkinson, K. D., and Cook, W. J., 1985, Three-dimensional structure of ubiquitin at 2.8 Å resolution, *Proc. Natl. Acad. Sci. U.S.A.* **82:** 3582–3585.

29. Li, J. B., and Goldberg, A. L., 1976, Effects of food deprivation on protein synthesis and degradation in rat skeletal muscles, *Am. J. Physiol.* **231:** 441–448.

30. Obinata, T., 1973, Dynamics of protein turnover in vertebrate skeletal muscle, *Muscle Nerve* **4:** 456–488.

31. Siegelman, M., Bond, M. W., Gallatin, W. M., St. John, T., Smith, H. T., Fried, V. A., and Weissman, I. L., 1986, Cell surface molecule associated with lymphocyte homing is a ubiquitinated branch-chain glycoprotein, *Science* **231:** 823–829.

32. Yarden, Y., Escobedo, J. A., Kuang, W.-J., Yang-Feng, T. L., Daniel, T. O., Tremble, P. M., Chen, E. Y., Ando, M. E., Harkins, R. N., Francke, U., Fried, V. A., Ullrich, A., and Williams, L. T., 1986, Structure of the receptor for platelet-derived growth factor helps define a family of closely related growth factor receptors, *Nature* **323:** 226–232.

33. Meyer, E. M., West, C., and Chau, V., 1986, Antibodies directed against ubiquitin inhibit high affinity [³H]choline uptake in rat cerebral cortical synaptosomes, *J. Biol. Chem.* **261:** 14365–14368.

34. Rapoport, S., Dubiel, W., and Muller, M., 1985, Proteolysis of mitochondria in reticulocytes during maturation is ubiquitin-dependent and is accompanied by a high rate of ATP hydrolysis, *FEBS Lett.* **180:** 249–252.

35. Dubiel, W., Muller, M., and Rapoport, S., 1986, Kinetics of ¹²⁵I-ubiquitin conjugation with and liberation from rabbit reticulocyte stroma, *FEBS Lett.* **194:** 50–55.

36. Dubiel, W., Muller, M., and Rapoport, S., 1981, Proteolysis of reticulocyte mitochondria is preceded by the attack of lipoxygenase, *Biochem. Int.* **3:** 165–171.

37. Magnani, M., Stocchi, V., Chiarantini, L., Serafini, G., Dacha, M., and Fornaini, G., 1986, Rabbit red blood cell hexokinase: Decay mechanism during reticulocyte maturation, *J. Biol. Chem.* **261:** 8327–8333.

38. Rose, I. A., and Warms, J. V. B., 1982, Stability of hexokinase II *in vitro* and in ascites tumor cells, *Arch. Biochem. Biophys.* **213:** 625–634.

39. Inaba, M., and Maedo, Y., 1986, Na,K-ATPase in dog red cells: Immunological identification and maturation-associated degradation by the proteolytic system, *J. Biol. Chem.* **261:** 16099–16105.

40. Raboy, B., Parag, H. A., and Kulka, R. G., 1986, Conjugation of [¹²⁵I]ubiquitin to cellular proteins in permeabilized mammalian cells: Comparison of mitotic and interphase cells, *EMBO J.* **5:** 863–869.

41. McKay, M. J., and Hipkiss, A. R., 1982, ATP-independent proteolysis of globin cyanogen bromide peptides in rabbit reticulocyte cell-free extracts, *Eur. J. Biochem.* **125:** 567–573.
42. Daniels, R. S., McKay, M. J., Worthington, V. C., and Hipkiss, A. R., 1982, Effects of ATP and cell development on the metabolism of high molecular weight aggregates of abnormal proteins in rabbit reticulocytes and cell-free extracts, *Biochim. Biophys. Acta* **717:** 220–227.
43. Wilkinson, K. D., and Mayer, A. N., 1986, Alcohol-induced conformational changes of ubiquitin, *Arch. Biochem. Biophys.* **250:** 390–399.
44. Boches, F. S., and Goldberg, A. L., 1982, Role for the adenosine triphosphate-dependent proteolytic pathway in reticulocyte maturation, *Science* **215:** 978–980.
45. Goldberg, A. L., and Boches, F. S., 1982, Oxidized proteins in erythrocytes are rapidly degraded by the adenosine triphosphate-dependent proteolytic system, *Science* **215:** 1107–1109.
46. Fagan, J. M., Waxman, L., and Goldberg, A. L., 1986, Red blood cells contain a pathway for the degradation of oxidant-damaged hemoglobin that does not require ATP or ubiquitin, *J. Biol. Chem.* **261:** 5705–5713.
47. Wu, R. S., Kohn, K. W., and Bonner, W. M., 1981, Metabolism of ubiquitinated histones, *J. Biol. Chem.* **256:** 5916–5920.
48. Goldberg, A. L., and St. John, A. C., 1976, Intracellular protein degradation in mammalian and bacterial cells, *Annu. Rev. Biochem.* **45:** 747–803.
49. Schimke, R. T., 1973, Control of enzyme levels in mammalian tissues, *Adv. Enzymol.* **37:** 135–187.
50. Bond, U., and Schlesinger, M. J., 1985, Ubiquitin is a heat shock protein in chicken embryo fibroblasts, *Mol. Cell. Biol.* **5:** 949–956.
51. Bond, U., and Schlesinger, M. J., 1986, The chicken ubiquitin gene contains a heat shock promoter and expresses an unstable mRNA in heat-shocked cells, *Mol. Cell. Biol.* **6:** 4602–4610.
52. Riley, D. A., Bain, J. L., and Haas, A. L., 1986, Increased ubiquitin conjugation of proteins during skeletal muscle atrophy, *J. Cell Biol.* **103:** 401a.
53. Munro, S., and Pelham, H., 1985, What turns on heat shock genes? *Nature* **317:** 477–478.
54. Levy-Wilson, B., Denker, M. S., and Ito, E., 1983, Isolation, characterization, and post-synthetic modification of *Tetrahymena* high mobility group proteins, *Biochemistry* **22:** 1715–1721.
55. Rose, I. A., and Warms, J. V. B., 1983, An enzyme with ubiquitin carboxyl-terminal esterase activity from reticulocytes, *Biochemistry* **22:** 4234–4237.
56. Hough, R., Pratt, G., and Rechsteiner, M., 1986, Ubiquitin–lysozyme conjugates: Identification and characterization of an ATP-dependent protease from rabbit reticulocyte lysates, *J. Biol. Chem.* **261:** 2400–2408.
57. Slot, L. A., Lauridsen, A.-M. B., and Hendil, K. B., 1986, Intracellular protein degradation in serum-deprived human fibroblasts, *Biochem. J.* **237:** 491–498.
58. Seale, R. L., 1981, Rapid turnover of the histone–ubiquitin conjugate, A24, *Nucleic Acids Res.* **9:** 3151–3158.
59. Matsui, S.-I., Sandberg, A. A., Negoro, S., Seon, B. K., and Goldstein, G., 1982, Isopeptidase: A novel eukaryotic enzyme that cleaves isopeptide bonds, *Proc. Natl. Acad. Sci. U.S.A.* **79:** 1535–1539.
60. Ozkaynak, E., Finley, D., and Varshavsky, A., 1984, The yeast ubiquitin gene: Head-to-tail repeats encoding a polyubiquitin precursor protein, *Nature* **312:** 663–666.
61. Hamilton, J. W., and Rouse, J. B., 1980, The biosynthesis of ubiquitin by parathyroid gland, *Biochem. Biophys. Res. Commun.* **96:** 114–120.

62. Seidah, N. G., Crine, P., Benjannet, S., Scherrer, H., and Chretien, M., 1978, Isolation and partial characterization of a biosynthetic N-terminal methionyl peptide of bovine pars intermedia: Relationship to ubiquitin, *Biochem. Biophys. Res. Commun.* **80:** 600–608.

63. Scherrer, H., Seidah, N. G., Benjannet, S., Crine, P., Lis, M., and Chretien, M., Biosynthesis of a ubiquitin-related peptide in rat brain and in human and mouse pituitary tumors, *Biochem. Biophys. Res. Commun.* **84:** 874–885.

64. Schlesinger, D. H., Goldstein, G., and Niall, H. D., 1975, The complete amino acid sequence of ubiquitin, an adenylate cyclase stimulating polypeptide probably universal in living cells, *Biochemistry* **14:** 2214–2218.

65. Carlson, N., Rogers, S., and Rechsteiner, M., 1987, Microinjection of ubiquitin: Changes in protein degradation in HeLa cells subjected to heat-shock, *J. Cell Biol.* **104:** 547–555.

66. Shanklin, J., Jabben, M., and Vierstra, R. D., 1987, Red light induced formation of ubiquitin–phytochrome conjugates: Identification of possible intermediates of phytochrome degradation, *Proc. Natl. Acad. Sci. U.S.A.* **84:** 359–363.

67. McCurdy, D. W., and Pratt, L. H., 1986, Kinetics of intracellular redistribution of phytochrome in *Avena* coleoptiles after its photoconversion to the active, far-red absorbing form, *Planta* **167:** 330–336.

68. Hebda, P. A., Ebert, J., Chou, K.-L., Shields, M., and Kao, W.-Y., 1983, The association between prolyl hydroxylase metabolism and cell growth in cultured L-929 fibroblasts, *Biochim. Biophys. Acta* **758:** 128–134.

69. Breslow, E., Daniel, R., Ohba, R., and Tate, S., 1986, Inhibition of ubiquitin-dependent proteolysis by non-ubiquitinatable proteins, *J. Biol. Chem.* **261:** 6530–6535.

70. Hershko, A., Heller, H., Eytan, E., and Reiss, Y., 1986, The protein substrate binding site of the ubiquitin–protein ligase system, *J. Biol. Chem.* **261:** 11992–11999.

71. Goldknopf, I. L., and Busch, H., 1977, Isopeptide linkage between nonhistone and histone 2A polypeptides of chromosomal conjugate–protein A24, *Proc. Natl. Acad. Sci. U.S.A.* **74:** 864–868.

72. West, M. H. P., and Bonner, W. M., 1980, Histone 2B can be modified by the attachment of ubiquitin, *Nucleic Acids Res.* **8:** 4671–4680.

73. Gregori, L., Marriott, D., West, C. M., and Chau, V., 1985, Specific recognition of calmodulin from *Dictyostelium discoideum* by the ATP,ubiquitin-dependent degradative pathway, *J. Biol. Chem.* **260:** 5232–5235.

74. Seamon, K. B., 1980, Calcium- and magnesium-dependent conformational states of calmodulin as determined by nuclear magnetic resonance, *Biochemistry* **19:** 207–215.

75. Gronostajski, R. M., Pardee, A. B., and Goldberg, A. L., 1985, The ATP dependence of the degradation of short- and long-lived proteins in growing fibroblasts, *J. Biol. Chem.* **260:** 3344–3349.

76. Haas, A. L., Ahrens, P., Bright, P. M., and Ankel, H., 1987, Interferon induces a 15 kilodalton protein exhibiting marked homology to ubiquitin, *J. Biol. Chem.* **262:** 11315–11323.

Chapter 8

Protein Breakdown and the Heat-Shock Response

Stephen A. Goff, Richard Voellmy, and Alfred L. Goldberg

1. INTRODUCTION

A primary function of protein breakdown in animal and bacterial cells is to eliminate nonfunctional or denatured polypeptides whose accumulation *in vivo* could be quite toxic.[1–4] Proteins with highly abnormal structures may result from nonsense mutations, insertions or deletions, missense mutations, biosynthetic errors, incorporation of amino acid analogs into proteins, or postsynthetic damage to normal cell constituents. However, such polypeptides generally fail to accumulate *in vivo* to the levels of normal gene products because they are rapidly degraded to amino acids.

It is now clear that the ability of cells to hydrolyze selectively such abnormal proteins increases in a variety of stressful environmental conditions.[5,6] Important components of the degradative machinery are produced in higher amounts under a variety of potentially lethal conditions, such as upon exposure to increased environmental temperatures where

STEPHEN A. GOFF • Department of Physiology, Tufts Medical School, Boston, Massachusetts 02111.   RICHARD VOELLMY • Department of Biochemistry and Molecular Biology, University of Miami School of Medicine, Miami, Florida 33136.   ALFRED L. GOLDBERG • Department of Physiology and Biophysics, Harvard Medical School, Boston, Massachusetts 02115.   *Present address of S.A.G.:* Plant Gene Expression Center, U.S. Department of Agriculture, Albany, California 94710.

abnormal proteins may be generated in large amounts. Under such conditions, the pattern of gene expression changes, and cells overproduce a characteristic set of proteins, commonly referred to as heat-shock proteins ($hsps$).[7-10,160,161] Thus, the heat-shock response serves an adaptive function and increases the capacity of eukaryotic and prokaryotic cells to withstand such environmental stresses.[160] Although the heat-shock response has elicited much interest as an experimental system for studying the regulation of gene transcription and mRNA translation, the biological significance of the changes in gene expression and the function of the various heat-shock proteins remain poorly understood.[7,9]

Recent findings have indicated that a critical factor triggering this adaptive response in a variety of harsh environmental conditions is the appearance of large amounts of abnormal proteins within the cell.[6,10,11] Such observations suggest one important consequence of producing the $hsps$ is to increase the cell's capacity to eliminate these abnormal polypeptides and thus to help protect against the environmental conditions that generate them. In this chapter, we review our present understanding of the functions of the heat-shock proteins in bacterial and animal cells, and we focus on the interrelations between the heat-shock response and protein degradation.

2. THE UBIQUITIN-INDEPENDENT PROTEOLYTIC SYSTEM IN ESCHERICHIA COLI

A fundamental property of protein degradation in prokaryotic as in eukaryotic cells is its requirement for metabolic energy.[1,12] For example, in *E. coli*, the continuous production of ATP appears essential for the rapid degradation of highly abnormal proteins[3] as well as the slower hydrolysis of normal cell proteins.[13] An energy requirement for proteolysis would not be anticipated on thermodynamic grounds or from the properties of typical proteolytic enzymes. Studies of the biochemical basis of this energy requirement have led to important new insights about the responsible degradative systems in both bacterial and mammalian cells.

Knowledge in this area has developed rapidly in recent years owing to the development of cell-free preparations that carry out the selective degradation of abnormal proteins. Such extracts were first established from mammalian reticulocytes and led to the discovery of the nonlysosomal ATP-dependent pathway for proteolysis.[12,14] Subsequently, soluble proteolytic preparations were established from *E. coli*,[15] mitochondria,[16] and recently from nucleated cells[17,18] that also show the ATP requirement characteristic of proteolysis *in vivo*. The main properties of these various degradative systems are very similar: (1) they require ATP hydrolysis for proteolysis; (2) they are found in the $100,000g$ supernatant; (3) they show a slightly alkaline pH optimum; (4) proteins can be hydrolyzed all the way to amino acids; (5) they are comprised of multiple enzymes; and (6) they seem to perform selective degradation of proteins with abnormal conformations.

Despite these common features, very different mechanisms appear to be responsible for the energy requirement in the bacterial and eukaryotic cytosol.[12,14] The important studies in reticulocytes by Hershko, Ciechanover, and co-workers and more recent work from many other laboratories have shown that the stimulation by ATP involves the modification of protein substrates by conjugation to the heat-stable polypeptide, ubiquitin (Ub).[12] As discussed elsewhere in this volume, this covalent modification of substrates involves a multicomponent enzyme system that requires ATP and helps mark these proteins for rapid degradation by a multicomponent protease specific for Ub conjugates.[19,20] An initially attractive feature of the discovery of Ub conjugation was that it could account for the energy requirement for protein breakdown and made it unnecessary to hypothesize the existence of novel proteolytic enzymes that utilize ATP directly. However, studies in bacteria and mitochondria[21] have led to the discovery of such ATP-dependent proteases that function independently of Ub. Originally, Ub was so named because it was believed to be present in animals, plants, and bacteria, based largely on immunological findings (see Chapter 1). However, more recent work has shown that prokaryotes lack Ub (A. Hershko and A. Ciechanover; D. Finley and A. Varshavsky, personal communication), although they do have soluble ATP-dependent systems for degrading abnormal proteins. It is presently unclear how widely Ub-independent proteolytic systems are distributed and whether they play important roles in protein breakdown in cells other than bacteria.

In the proteolytic pathways in bacteria[22-24] and mitochondria[16,26] ATP is not necessary for the conjugation of Ub or another polypeptide to proteolytic substrates. Instead, bacteria and mitochondria contain a class of protease(s) that hydrolyzes ATP and protein in a coupled reaction mechanism.[22,26] Although the eukaryotic and prokaryotic proteolytic machineries thus differ in important ways, they may still utilize common enzymatic mechanisms. For example, recent studies indicate that in the Ub-dependent pathway, eukaryotic cells utilize ATP-hydrolyzing proteases[19,20] of unusually high molecular weight to digest the Ub–protein conjugates. In addition, certain mammalian cells (e.g., mouse erythroleukemia cells) contain Ub-independent ATP-requiring proteases that can hydrolyze unconjugated protein substrates[18,25] and thus appear similar to the *E. coli* enzymes. Moreover, components of the soluble systems that eliminate aberrant proteins in both eukaryotes and prokaryotes are regulated as part of the heat-shock response.

3. PROTEASE LA, THE *LON* GENE PRODUCT IN *ESCHERICHIA COLI*

Although *E. coli* has been a favorite organism for study by biochemists, proteases from these cells have only recently begun to be identified.[27] *E. coli* has been shown to contain at least nine soluble endoproteases, seven of which can degrade proteins *in vitro* to acid-soluble peptides. The physiological roles of most of these enzymes are uncertain at present, but it is clear that the energy-dependent proteases must play a critical role in intracellular proteolysis. In fact, the breakdown of most abnormal proteins appears to begin with cleavages by these proteases since depletion of intracellular ATP prevents the rapid degradation of such abnormal proteins and does not cause accumulation of partially digested molecules.[28] The best characterized of these ATP-dependent proteases is protease La, the product of the *lon* gene (also called *deg* and *capR*).[22–24]

A variety of observations indicate that the level of protease La in cells is a critical factor determining the rate of breakdown of some abnormal proteins. For example, mutations in the *lon* gene reduce by two- to fourfold degradation of incomplete polypeptides or of proteins containing amino acid analogs.[23] Mutations in *lon* also lead to a decreased breakdown of certain short-lived regulatory proteins,[29,30] and, as a result, these mutants show a variety of unusual phenotypic alterations. Furthermore when the cellular content of this enzyme is increased experimentally (e.g., in *E. coli* carrying the *lon* gene on plasmids or when the *lon* gene is expressed under control of an inducible promoter), the bacteria's capacity to degrade abnormal proteins is enhanced.[31] Such strains also display an increased rate of degradation of normal cell constituents and severely restricted growth.[31] Thus, it appears advantageous to the organism to regulate precisely the level of protease La (and perhaps other cellular proteases). These studies also demonstrate that the cellular content of this protease can limit the degradation of abnormal proteins; therefore, an increase in *lon* transcription, as occurs in heat shock (see Section 4), can enhance the cell's ability to dispose of potentially toxic molecules.

Protease La has been purified to homogeneity and its properties studied in depth.[32–37] In the presence of ATP, the purified enzyme degrades proteins to peptides that are about 1500 Da. Protease La is a tetramer containing a single 87,000 Da subunit, whose complete sequence has recently been elucidated.[162] La is unusually large for a proteolytic enzyme and bears no detectable homologies to other serine or thiol proteases. The enzyme shows both ATPase and proteolytic activities,[32,35] which are tightly coupled to each other. In fact, ATP hydrolysis is stimulated by the addition of protein substrates but not by the addition of native proteins, which are not degraded.[32] Two ATP molecules are hydrolyed for every peptide bond cleaved in proteins.[35]

Much has been learned about the mechanism of protease La using small fluorogenic peptides *in vitro* to determine the partial reactions involved in substrate hydrolysis.[33,34] Such studies suggest the following multistep ATP-dependent reaction cycle: (1) a protein substrate initially binds to a regulatory site of the protease; (2) ATP-binding then occurs and allows the proteolytic site to funtion; and (3) ATP hydrolysis occurs. This cyclic process occurs repetitively until the proteins are degraded to oligopeptides.

Perhaps the most intriguing feature of this complex mechanism is that activity of this protease can be carefully regulated throughout. Such a mechanism can prevent inappropriate degradation of normal cell polypeptides *in vivo*. For example, the pure enzyme has a limited capacity to hydrolyze peptide bonds, and this capacity increases when a proteiin substrate binds to the enzyme.[36] This activation results from an allosteric interaction of the protein substrate with a regulatory region that is distinct from the active site and that appears specific for unfolded polypeptides.[36] The enzyme does not funtion with ADP bound to it, but it has an affinity for ADP that is much higher than its affinity for ATP.[37,163] Consequently, protease La probably exists *in vivo* in an inhibited state with bound ADP. However, when it interacts with a protein substrate, the release of this bound ADP is stimulated as is the binding of new ATP moieties. By such a mechanism, proteolytic activity is specifically enhanced by substrates and falls again when the substrate is hydrolyzed.[37,163]

Even when the *lon* gene is deleted, the bacteria retain significant ability to degrade aberrant polypeptides or other short-lived proteins in an energy-dependent fashion,[38] and another ATP-dependent protease has been shown to exist in *E. coli*.[39,40] In addition, mutations in the locus that regulates the heat-shock response (*htpR* or *rpoH*, see Section 4) further reduce the cell's capacity for protein breakdown. In other words, these findings suggest that other ATP-dependent proteases function in these cells and that these other proteases may be regulated by the *htpR* locus.

4. THE HEAT-SHOCK RESPONSE IN *ESCHERICHIA COLI*

When *E. coli* growing exponentially at 30°C are shifted to 42°C, they respond by transiently decreasing protein synthesis[41] and stable RNA synthesis.[42] These "heat-shocked" cells also transiently increase the synthesis of a set of about 18–24 proteins that have been analyzed by two-dimensional gel electrophoresis.[43–46,160,161] Although the specific functions of most of these heat-shock proteins (hsps) remains unknown, the synthesis of hsps increases the organism's ability to withstand high temperatures[8,160] and a variety of other stressful conditions. Several of these hsps have been isolated and characterized biochemically. Two of the most abundant hsps from $E. coli$ are the products of the $dnaK$ gene[47,48] and the $groEL$ gene.[49,50] These genes were initially identified as loci required for bacteriphage replication or morphogenesis,[50,51] but they also have essential functions in uninfected cells.[52,53] Like these two polypeptides, the product of $grpE$ is involved in bacteriophage assembly.[54] The product of $lysU$ (a lysine tRNA synthetase) has also been identified as an hsp.[55] Among other hsps are the product of the $dnaJ$ gene[56] and the normal $E. coli$ σ factor of RNA polymerase, $σ^{70}$.[57] The exact roles of all these proteins in the heat-shock response and in enhancing thermotolerance remain unclear, but a recent report suggests a role for a number of these proteins in the pathway of protein degradation in $E. coli$[58] (see Section 6).

Prior to the discovery of the heat-shock response in $E. coli$, a temperature-sensitive nonsense mutant was isolated that displayed low rates of synthesis of the product of the $groE$ gene.[59,60] When its ability to synthesize heat-shock proteins following a thermal shift was tested,[61,62] these mutants failed to induce the hsps and were therefore designated $htpR$ mutants for regulator of the high-temperature proteins.[55,61] The hsp genes together with the regulatory locus controlling the heat-shock response ($htpR$) were collectively defined as a "high-temperature regulon" by Neidhardt and co-workers.[55] The product of the $htpR$ gene (also initially called hin[62]) has been cloned and shown to function $in vivo$ and $in vitro$ as a positive regulator of the heat-shock genes.[63–66]

Sequence analysis of the $htpR$ gene revealed extensive homology with the sigma subunit of RNA polymerase.[67] Subsequently, the $htpR$ protein was shown to function as an alternative sigma factor that redirects RNA polymerase to heat-shock genes under conditions of stress, such as exposure of cells to 42–45°C.[68] This factor has been designated $σ^{32}$ to distinguish it from the normal σ subunit of RNA polymerase, $σ^{70}$ (the product of the $rpoD$ locus),[68] and the $htpR$ gene has been renamed $rpoH$. An increase in the amount of the $rpoH(htpR)$ gene product has been shown to be sufficient to activate the transcription of the heat-shock genes,[58,164,170,171] and recent findings demonstrate that $σ^{32}$ complexed to core RNA polymerase can selectively promote transcription from heat-shock promoters $in vitro$.[66]

In addition to this positive control of the heat-shock genes by $σ^{32}$, there appears to exist negative regulation of the *hsp* genes, such that the increased expression of *hsp*s may help terminate the heat-shock response. The *dnaK* gene product appears important in this negative regulation.[69] For example, after a shift to 42°C, the *hsp*s are expressed at increased rates for only a limited period, typically 5–20 min. In a mutant carrying a defective *dnaK* gene product (*dnaK-756*), the heat-shock response is prolonged following a thermal shift,[69] and at low temperatures, the levels of *hsp*s are also increased. Thus, this protein serves to reduce its own expression and that of other *hsp*s and to terminate the heat-shock response.

Of particular physiological interest is the recent finding that protease La is a heat-shock protein.[5,70] This discovery provides an important clue to one of the functions of the heat-shock response. The purified protease was found to comigrate on two-dimensional gel electrophoresis with an *hsp* previously designated H94.0.[70] Furthermore, the activity of this enzyme was found to increase in wild-type cells shifted to 42°C but not in isogenic *htpR* strains.[5] Using a *lon-lacZ* operon fusion, the transcription of the *lon* gene was shown to increase at high temperatures or upon exposure of cells growing at low temperature (30°C) to 4% ethanol[5] (Figure 1) (S. A. Goff, L. P. Casson, and A. L. Goldberg, unpublished results). Ethanol, which is known to induce the heat-shock response,[71] can cause

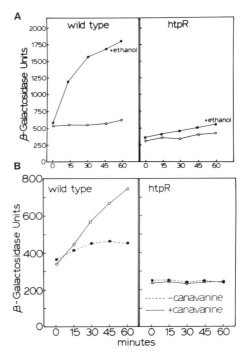

Figure 1. (A) Effects of the induction of the heat-shock response with ethanol on transcription of the *lon* gene. Wild-type cells and isogenic *htpR* strains were grown at 30°C and then exposed to 4% ethanol. This treatment is known to induce heat-shock genes in wild-type but not in the *htpR* cells. To follow transcription of this gene, we used strains carrying an operon fusion between the *lon* promoter and the *lacZ* gene.[5] (B) Effect of incorporation of the arginine analog canavanine on transcription of the *lon* gene in wild-type *E. coli* and in *htpR* mutants. Arginine auxotrophs were suspended in the presence of either canavanine or arginine for 30 min.[6] The cells were then washed and resuspended in the presence of arginine and the production of β-galactosidase was measured.

frequent translational errors,[72] and can also perturb the conformation of cellular proteins.[73] In contrast, no such induction of lon was seen in $htpR$ mutants exposed to 4% ethanol or shifted to 42°C. In addition, the basal transcription of the lon gene and the cell's content of protease La were significantly reduced in the $htpR$ strain.[5] Furthermore, these $htpR$ mutants were shown to degrade polypeptides containing amino acid analogs, protein fragments, and certain mutant proteins several-fold more slowly than the wild-type cell[5,74] (see Section 5).

5. ACTIVATION OF *LON* AND OTHER HEAT-SHOCK GENES BY ABNORMAL PROTEINS

These findings established a connection between the heat-shock response and the intracellular degradation of protein. Such a connection had also been suspected because cells growing at 40°C or higher exhibit significantly higher rates of protein degradation.[78,79] Since exposure to high temperature or to ethanol is likely to cause alterations in protein conformations, and since protease La plays a critical role in the degradation of various types of abnormal proteins,[4,5,75–77] we tested, in strains carrying a lon-$lacZ$ operon fusion, whether other conditions that generate large amounts of abnormal protein might also induce protease La and the other hsps.[5] The incorporation of amino acid analogs into proteins in place of the normal residues or the incorporation of puromycin to cause premature termination of polypeptides led to increased lon expression.[6] Following production of such "abnormal" or incomplete proteins, transcription of lon increased as shown by a rise in β-galactosidase activity (Figure 1). Such treatments obviously generate substrates for protease La[3] and lead to a large increase in the rates of protein breakdown. In addition, the expression of some other hsps also increased as shown by one-dimensional gel electrophoresis.[6] These treatments are likely to affect many cellular processes, since all newly synthesized proteins should be altered under these conditions. To determine if the presence of a single abnormal protein in high amounts increases lon expression, cells producing large amounts of a foreign protein, recombinant tissue plasminogen activator (TPA), were examined. This polypeptide contains many sulfhydral bridges and fails to fold properly in the bacterial cytosol. Induction of TPA resulted in an increased expression of the lon-$lacZ$ operon fusion (Figure 2) and increased synthesis of at least some other heat-shock proteins.[6] Thus, the production of a single abnormal protein is sufficient to stimulate transcription of the heat-shock genes in $E.$ $coli.$ The enhanced

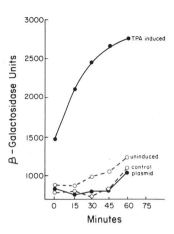

Figure 2. Increased transcription of the *lon* gene upon expression of a single-cloned protein, human tissue plasminogen activator. This cloned polypeptide fails to assume its native conformation in *E. coli* and is rapidly degraded.[6] Transcription of the *lon* gene was measured using a *lon-lacZ* fusion, as in Figure 1. Under these conditions, other heat-shock genes were also expressed at increased rates.[6] No such effects were seen in the *htpR* strain.

synthesis of protease La under these conditions should help the cell eliminate these misfolded or unfolded proteins.

It is noteworthy that in cells that have synthesized large amounts of abnormal proteins (e.g., following incorporation of amino acid analogs), the degradation of these proteins can be inhibited by chloramphenicol or rifampicin, which blocks protein or RNA syntheses, respectively. After analog incorporation, the expression of *lon* rises and apparently enhances the cell's degradative capacity. Accordingly, when similar experiments are done in *htpR* strains (A. L. Goldberg, unpublished results), the rates of protein breakdown are lower and are not reduced at all by inhibitors of protein or RNA synthesis. When wild-type cells produce only low amounts of abnormal proteins, protease La is not induced and the rate of protein degradation is not sensitive to inhibition of RNA or protein synthesis. In other words, when abnormal proteins are generated in wild-type cells in amounts that seem to exceed the cell's proteolytic capacity, the synthesis of protease La (and perhaps of other cell proteases) increases and the cell's capacity to eliminate such potentially damaging macromolecules rises.

To define further the mechanisms responsible for increased *hsp* synthesis in response to abnormal proteins, similar experiments were performed in cells containing a mutation in the regulatory locus *rpoH(htpR)*. The *rpoH(htpR)* mutants were found to exhibit lower basal expression of the *lon* gene[5] and did not respond to ethanol, production of protein fragments, or incorporation of amino acid analogs.[5,6] Since these cells have a lowered expressed of *lon*, they were tested for their ability to degrade abnormal cellular proteins. In wild-type cells,[3] proteins contain-

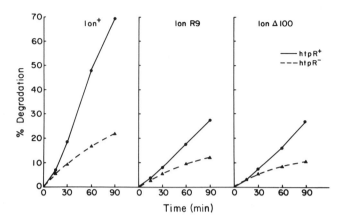

Figure 3. Effects of the *htpR* and different *lon* mutations on the degradation of proteins containing the arginine analog, canavanine. After incorporation of canavanine for 20 min and [³H]leucine for 5 min, the cells were resuspended at 30°C in the presence of a large excess of nonradioactive arginine and leucine. Rates of breakdown of the radioactive analog-containing proteins were then measured in isogenic wild-type *lon* and *htpR* strains or in the double mutants.

ing amino acid analogs and incomplete polypeptides containing puromycin are degraded very rapidly. However, in cells carrying the *rpoH(htpR)* mutation, the degradation of these proteins is greatly reduced[5] (Figure 3). Furthermore, specific abnormal proteins, such as the X-90 fragment of β-galactosidase, were also found to be more stable in *rpoH(htpR)* mutants.[74] This defect in proteolysis has been put to use in many laboratories to enhance the yield of foreign proteins. A number of cloned gene products are rapidly degraded in *E. coli*, including several of industrial interest (e.g., somatomedin C), but they are severalfold more stable in *htpR* mutants[80,81] and more so in *lon htpR* strains.

The defect in degradation of abnormal proteins observed in *rpoH(htpR)* cells was often greater than that observed in *lon* mutants, and the effects of these two mutations appeared to be additive.[5,74] The simplest explanation for these observations is that additional proteases are encoded by other heat-shock genes. We have thus far failed to obtain direct biochemical evidence that any of the other known *E. coli* proteases are under control of the *htpR* locus. However, Gross and co-workers have recently observed that mutations in several of the other known *hsp* genes, including *groE* and *dnaK*, also reduce the degradation of protein fragments relative to the wild-type cell.[58] It is uncertain whether these genes encode proteases or proteins that indirectly affect the degradative process.

It appears likely that denaturation of cell proteins occurs in most, if not all, of the environmental conditions that elicit increased synthesis of the $hsps$.[5,6] Proteins containing amino acid analogs, truncated proteins, many protein fusions, and many cloned foreign polypeptides are known to accumulate within eukaryotic and prokaryotic cells in large aggregates that are readily isolated by gel filtration or ultracentrifugation.[1,4,165–167] These structures appear to be amorphous clumps of denatured polypeptides. Aggregation of cellular protein has also been reported to occur in *E. coli* following a shift to growth at high temperature[82] as would be expected if high temperatures induced protein denaturation. It would appear advantageous to the cell to have evolved mechanisms for induction of protease(s) capable of removing such damaged proteins, and increased proteolytic capacity would appear to be of selective advantage in protecting the organism against an accumulation of nonfunctional and potentially toxic intracellular proteins.

6. PROTEOLYSIS AND REGULATION OF THE HEAT-SHOCK RESPONSE IN *ESCHERICHIA COLI*

It is unclear exactly how exposure to high temperatures or other stressful conditions are recognized by the cell and how transcription of the *hsp* genes is activated. One attractive possibility would involve cellular proteases in regulating induction of the heat-shock genes. The accumulation of abnormal proteins appears to be a common feature of the large variety of conditions that induce the heat-shock genes (Table I), and the level of abnormal protein is probably monitored by the cell as a "danger signal." When the amount of these abnormal proteins exceeds the cell's proteolytic capacity, the cell would respond by induction of heat-shock proteins.[5,6,10,83,84] The presence of large amounts of abnormal proteins may saturate the proteolytic system and thereby reduce the degradation of proteins otherwise rapidly eliminated by these proteolytic enzymes. It appears likely that the cell's maximal degradative capacity may be reached when they produce abnormal proteins in large amounts. A clear indication that protease La content can become rate limiting under such conditions is that a two- to fourfold increase in this protease enhances the rates at which both normal and aberrant proteins are degraded.[31]

To activate heat-shock genes, the positive regulatory protein produced by the *rpoH*(*htpR*) gene must be increased in either activity or cellular content.[160,164] Several findings indicate that this protein is short-lived within cells and is stabilized under conditions that generate large amounts of abnormal proteins. For example: (1) The *rpoH*(*htpR*) gene

Table I. Treatments Known to Induce the Heat-Shock Response

Prokaryotes	References
Temperature shifts	43, 44, 65
Ethanol	6, 71
Incorporation of amino acid analogs	6
Incorporation of puromycin	6
Bacteriophage infection	171, 172–174
Overproduction of certain cloned foreign proteins	6
Overproduction of certain fusion proteins	6
Increased translational errors	6
Treatment with protease inhibitors	6
DNA damage	175
Exposure to oxidants	176–177
Overproduction of certain bacteriophage proteins	171, 174
Mutations in $rpoD$	178
Starvation for a carbon source or amino acids	179

Eukaroytes	References
Temperature shifts	8
Ethanol	72
Incorporation of amino acid analogs	83, 84
Incorporation of puromycin	83, 84
Exposure to heavy metals, arsenite, iodoacetate, p-chloromercuri-benzoate	180–182
Exposure to oxidants	185
Ammonium chloride	183
Mitochondrial poisons	183
Hydroxylamine	183
Mutations reducing Ub conjugation	152, 153
Certain mutations in actin genes	155, 156
Viral infection	186, 187
Microinjection of abnormal proteins	11

product is present in low amounts although the transcript for this protein is fairly abundant[7]; (2) increased expression of the heat-shock genes appears to require an increase in the amount of the $rpoH(htpR)$ gene product[7,58]; and (3) the heat-shock response is very brief even in the absence of cell growth.[43] Therefore, the positive regulator seems to be rapidly removed with longer exposure to high temperatures, and the $rpoH(htpR)$ protein appeared to be very rapidly degraded in $vivo$.[58] For these reasons, we proposed that the activity of the degradative machinery may control transcription of hsps by regulating the half-time of the $rpoH(htpR)$ gene product.[5,6] In fact, Gross and co-workers recently showed that the half-life of the $rpoH(htpR)$ protein is only about 4 min in cells growing at 30°C, but becomes eightfold more stable immediately after a shift to 42°C.[164,170,171]

Such a model may help to explain a number of features of this adaptive response. For instance, since the generation of any of a variety of abnormal proteins or even single abnormal proteins[6] can cause an induction of $hsps$, the mechanism for sensing these proteins must be a very general one. A large variety of mutational events, biosynthesis errors, or postsynthetic damage are known to generate abnormal proteins that are rapidly eliminated.[1,2,4] Thus the cell's proteolytic systems have the capacity to detect the presence of a wide variety of abnormal proteins. If the $rpoH(htpR)$ gene product is a substrate for one of these same proteolytic enzyme(s), it is likely that a build up of abnormal proteins would compete with σ^{32} for proteolysis. Thus, the stability and cellular content of σ^{32} should increase and thereby elicit the heat-shock response.

Regulation of the levels of a critical cellular protein by altering its rates of degradation is not a novel phenomenon in bacteria. It is well established that bacteriophage λ prophage induction involves proteolytic cleavage of the λ repressor.[85] Gottesman and co-workers have demonstrated that filamentation of $E.\ coli\ lon$ mutants following DNA damage is due to a decreased degradation of the $sulA$ gene product.[29] In addition, lon mutations result in the overproduction of capsular polysaccharide owing to a failure to degrade a positive regulatory protein for genes involved in capsule synthesis.[86]

The temperature-sensitive mutants of $dnaK$ ($dnaK$-756) have been found to exhibit enhanced and prolonged synthesis of heat-shock genes following a temperature shift.[69] Attempts to explain these effects have postulated that the product of the wild type $dnaK$ allele directly antagonizes the role of the $rpoH(htpR)$ gene product.[69] However, the $dnaK$ gene has been shown to be evolutionarily conserved and very homologous to its eukaryotic counterpart hsp-70. The products of the hsp-70 gene and the closely related cognate genes form complexes with various proteins and have been proposed to reduce the rate of protein denaturation caused by thermal insults.[87–90] If the $dnaK$ protein has a similar role, the enhanced and prolonged heat-shock response in $dnaK$-756 mutants may result from greater accumulation of abnormal proteins.

In an attempt to obtain further evidence for a role of proteolysis in the regulation of the heat-shock response, we have added protease inhibitors to cultures of bacteria. Such cells display increased transcription from the lon promoter and increased synthesis of other heat-shock proteins following exposure to any of the three serine protease inhibitors tested[6] (S. A. Goff and A. L. Goldberg, unpublished results). These results also suggest that inhibition of proteolysis may directly activate heat-shock

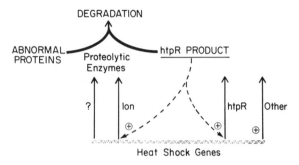

Figure 4. Proposed model to account for the induction of the heat-shock response by abnormal proteins. As summarized in the text, this model[6] assumes that a positive regulator of heat-shock gene transcription is subject to rapid degradation. The most likely candidate for this labile regulatory protein is the product of the *htpR(rpoH)* gene, the σ factor specific for *hsp* genes. The appearance of large amounts of abnormal proteins in the cell (e.g., at high temperatures) should saturate the cell's proteolytic machinery. Consequently, this regulatory factor should accumulate and promote the expression of the heat-shock genes. The increased production of *hsp*s would enhance the cell's capacity to degrade abnormal proteins. As these proteins are hydrolyzed, the degradation of the *htpR* protein is accelerated. Eventually, its level falls to that seen normally, and the heat-shock response is terminated.

genes. Our current working model for the regulation of *E. coli* heat-shock genes by protein degradation is illustrated in Figure 4. According to this model, the generation of abnormal proteins by various conditions that elicit a heat-shock response results in saturation of the proteolytic enzymes of the cell and thereby protects σ^{32} from proteolysis. The buildup of this positive regulator enhances transcription of the heat-shock genes. Since protein degradation is enhanced following increased synthesis of protease La and perhaps other protease(s) under control of the *rpoH(htpR)* locus, the system should return to its original level (or a new basal level) once the abnormal proteins are degraded. Under these conditions σ^{32} would regain its original short half-life, resulting in a low rate of transcription of the heat-shock genes. Accordingly, when cells have adapted to growth at high temperatures, the half-life of σ^{32} returns to that seen at 30°C.[164]

Conclusive evidence for this model will require further research. It will be necessary to demonstrate directly that the generation of abnormal proteins per se is capable of increasing the level of σ^{32} by saturating the cell's proteolytic machinery and that all the conditions that generate a heat-shock response do so by prolonging the half-life of σ^{32}. Strauss *et al.*[164] recently demonstrated that the amount of σ^{32} is also in part regulated at the translational level, and thus the model shown in Figure 4 cannot by itself account for the entire heat-shock response. However, if a positive regulator of translation of the $rpoH(htpR)$ message exists and is also rapidly degraded, then a very similar model may account for the increased translation of σ^{32} at high temperatures.

7. REGULATION OF *HSP* GENE EXPRESSION IN EUKARYOTIC CELLS

A shift to high temperatures and other stressful environmental conditions (e.g., exposure to amino acid analogs or heavy metals) cause the rapid transcriptional activation of heat-shock genes in eukaryotic cells[8-10] as in *E. coli* (Table I). It is likely that this response serves analogous functions in eukaryotic and prokaryotic cells for several reasons: (1) the conditions that induce the *hsps* are similar in both; (2) there is a marked evolutionary conservation of certain *hsps*; and (3) the resulting changes in gene expression lead to an increased ability of both types of cells to withstand these harsh conditons (reviewed in refs. 9, 91, 92, 160). Nevertheless, distinct differences exist in the nature of the response to stress in eukaryotes and prokaryotes. For example, in eukaryotic cells exposed to high temperatures both transcriptional and translational changes occur,[9,92-97,115] leading to increased expression of *hsp* genes and decreased synthesis of other proteins.

Of the major heat-shock proteins, the regulation of *hsp-70* has been investigated most thoroughly. The *hsp-70* is actually a member of a family of proteins, some of which are found in high amounts within cells under normal conditions. However, the rates of transcription of certain members of the *hsp-70* gene family may be 100- to 1000-fold higher in heat-treated cells than in unstressed controls.[98,99] The activation of these genes depends on upstream regulatory sequences, designated HSEs, which appear to function in a manner analogous to enhancers.[100-105] HSE sequences act as binding sites for factors that regulate transcription of the heat-shock genes[106,107]; such factors have been isolated from both yeast,[108,109] *Drosophila*,[108,169] and human cells.[168] Heat shock transcription factors function only in heat-shocked cells, but they must also be present in unstressed cells since inhibitors of protein synthesis do not block the increase in transcription of the heat-shock genes following a temperature shift.[9] Thus, the heat-shock transcription factor(s) appears to undergo some type of activation prior to the increase in transcription of the *hsp* genes.

Heat shock also affects gene expression post-transcriptionally. RNA splicing is inhibited, and this failure in RNA processing may account for the lack of introns in highly inducible *hsp* genes.[110] More importantly, *hsp* mRNAs are selectively translated during heat shock, while other mRNAs are excluded from translation[92,111-114] and overall protein synthesis falls. During recovery from heat shock, translation of other mRNAs is reinitiated. This selective translation of *hsp* mRNAs in the stressed cells involves a recognition of specific features present in the 5′ untranslated portion of the *hsp* mRNAs.[95–97,115] In addition to their preferential translation, *hsp* mRNAs are also more stable during heat shock than during recovery.[93]

The amount of *hsps* made in response to a particular stress is directly related to the severity of the stress applied.[94] Following a thermal shift, *hsp* gene transcription and translation appear to continue until a critical level of these proteins has been synthesized. Partial inhibition of either transcription or translation extends the period during which *hsp* synthesis occurs.[94] Thus, expression of *hsp* genes is apparently under feedback regulation, and a member of the *hsp-70* gene family appears to be involved in this feedback mechanism.[93,94] Interestingly, the *E. coli* equivalent of *hsp-70*, the *dnaK*-encoded protein, seems to play an analogous role in regulation of the heat-shock response in that organism[69] as discussed in Section 4.

8. HEAT-SHOCK PROTEINS IN EUKARYOTIC CELLS AND THEIR POSSIBLE FUNCTIONS

Most eukaryotic cells synthesize *hsps* of approximately 90,000, 70,000, 20,000–30,000, and 8000 Da molecular weight. The corresponding *hsps* from different organisms are closely related as shown by similarities in their peptide maps, cross-hybridization with antibodies, and nucleotide sequences (for reviews see refs. 9,10,116).

One of the *hsps* in yeast, mammalian, and chicken cells is Ub.[117,118] As discussed elsewhere in this volume, this protein is a critical factor in the ATP-dependent pathway for protein degradation in eukaryotic cells.[119,120] Ub conjugation to protein substrates leads to their rapid degradation,[12] although this process may also serve other important functions in the cell. The discovery that Ub is a heat-shock protein[117,118] suggests an important role for protein breakdown in the protection of cells from high temperatures and other types of stress. In fact, after a thermal shift, Ub conjugation increases markedly and the content of the free protein falls.[121,122] Under these conditions, there is also an increased rate of degradation of normally stable cell proteins,[121] and a surprising decrease in the breakdown of those proteins that are normally short-lived.[122] The latter result may reflect the temporary lack of sufficient Ub and competition between different proteins for proteolysis.

There exist several Ub-encoding genes in all the cells studied thus far, only one of which is induced as part of the heat-shock response.[123] The heat-inducible Ub gene from yeast has recently been demonstrated to code for five complete Ub proteins linked head to tail[123-124] (see also Chapter 2). Deletion of the heat-inducible, poly-Ub gene in yeast renders these cells temperature sensitive for growth and viability.[124] However, they can still grow at low temperatures where they contain a normal content of Ub owing to other Ub genes.[124] These additional Ub-encoding genes also have an unusual structure; the Ub-coding sequences are fused to sequences for other unidentified polypeptides whose function is currently unknown.[123] However, these studies indicate that an increased production of Ub is essential under stressful conditions such as heat shock[123] when the concentration of free Ub falls.[121,122] These results also strengthen the notion that enhanced proteolysis plays a critical role in protection of the organism from such harsh environmental conditions.

An interesting property shared by members of the *hsp-70* and *hsp-90* families is their tendency to bind to various classes of cellular proteins and multimeric protein complexes. Binding to 70,000 *hsps* may somehow protect cellular proteins against thermal denaturation. *Hsp-90* is one of the *hsps* that is also abundant in the cytoplasm of nonstressed cells.[128] It interacts transiently with a number of important regulatory proteins including several protein kinases and steroid receptors (for review see ref. 128). The binding properties of *hsp-90* suggest a role in either regulating the activity of critical proteins or protecting these macromolecules from damage. Another member of this family, *grp-94*, is not heat induced but is increased in cells starved for glucose or cells treated with inhibitors of N-linked glycosylation[129,130] or with the calcium ionophore A23187.[129] This protein is present in the lumen of the endoplasmic reticulum and is related to *hsp-90* in both biochemical properties[129,130] and DNA sequence.

Members of the *hsp-70* family, particularly the heat-inducible 72,000-Da protein from mammalian cells, concentrate in the nuclei and nucleoli of stressed cells.[131-138] Heat shock produces a drastic alteration in the integrity of nucleoli causing them to assume an unusually phase-dense and disrupted appearance.[131-138] In cells recovering from heat treatment, *hsp-72* was found only in nucleoli with visibly altered morphology.[138] In heat-shocked cells, *hsp-72* appears to reside in the granular region of nucleoli containing preribosomes, and during recovery this protein associates with ribosomes in the cytoplasm.[138] By contrast, in cells treated with arsenite, which induces *hsp* gene expression but does not cause obvious changes in nucleolar morphology, *hsp-72* is not concentrated in the nucleoli. Together, these observations suggest that members of the *hsp-70* family are involved in protecting preribosomes from heat-induced damage. One important finding consistent with this hypothesis is that overexpression of a *Drosophila hsp-70* gene in COS monkey cells speeds up nucleolar recovery following heat treatment.[139]

Among the other members of the *hsp-70* family is an ATPase that catalyzes the release of clathrin from coated vesicles.[141] This protein cross-reacts with antibodies raised against *hsp-72/73*.[89,90] The glucose-regulated protein *grp-78* is a member of this same protein family as shown by comparison of gene sequences and by biochemical characterization.[88] *grp-78* resides in the lumen of the endoplasmic reticulum and has been shown to be identical to the immunoglobulin heavy-chain-binding protein[88] that transiently associates with heavy chains in lymphoid cells.[144] The function of this intraluminal reaction is unclear. All the members of the *hsp-70* family have high affinity for ATP,[87,145] and upon incubation with ATP (but not with nonhydrolyzable ATP analogs), *hsp-72* dissociates from the complexes with isolated nuclei/nucleoli[87] or specific proteins and *grp-78* dissociates from the immunoglobulin heavy chains.[144] Presumably, *in vivo*, all these proteins must be functioning as ATPases, although it is uncertain exactly what ATP-requiring process they catalyze. One clue to the function of these reactions is provided by the uncoating ATPase, a specialized member of the *hsp-70* family that apparently has assumed a specific role in the uncoating of vesicles.[141]

Thus, members of the *hsp-70* family can interact reversibly with a variety of proteins that appear in some way abnormal and that might undergo further damage if not protected. These complexes include preribosomes and other specific proteins in heat-shocked cells, underglycosylated proteins in glucose-starved cells, free heavy-chain proteins in certain myelomas, or in normal lymphoid cells during the formation of immunoglobulin or mutant proteins, such as abnormal influenza hemagglutinins. It has been suggested that *hsp-70* associates with such proteins prior to their complete denaturation and prevents an irreversible unfolding or aggregation process.[140] Alternatively, it may associate with unfolded domains and promote their refolding. How *hsp-70* might recognize proteins in need of stabilization remains unclear. It has been suggested that members of the *hsp-70* family bind preferentially to hydrophobic surfaces.[140] However, this binding reaction has not yet been demonstrated *in vitro*, and it is unclear why ATP hydrolysis is required for the release of *hsp-70* proteins from such complexes.

Another major group of heat-shock proteins are the small *hsps* with molecular weights from 20,000 to 30,000 Da. In unstressed mammalian cells three out of the four isoforms of *hsp-28* are rapidly phosphorylated upon mitogenic stimulation.[146] The 28,000 proteins appear to be homologous to the *Drosophila* small *hsps*,[147] which in turn are structurally related to α-crystallin, a major lens protein[148,149] that forms large macromolecular complexes. By analogy, *hsp-28* and related *hsp*s may also participate in such macromolecular structures. The *Drosophila hsp*s of this class have been reported to associate with the cytoskeleton and to be present in distinct multiprotein complexes that resemble the ubiquitous particles originally called "prosomes."[150] The latter structures have recently been shown to be identical to the 700,000 Da multifunctional protease complex that is present in a latent form in the nucleus and cytosol of all eukaryotic cells.[151] These particles have therefore been renamed the "proteasome"; however, their function in protein degradation and their relation, if any, to the 28,000 *hsp* family are unknown at present.

9. LINKS BETWEEN THE EUKARYOTIC HEAT-SHOCK SYSTEM AND PROTEIN DEGRADATION

As has been mentioned in the previous sections, a wide variety of treatments induce *hsp* gene expression in both eukaryotes and prokaryotes. The only obvious common feature of these various inducers is that they can promote either the denaturation of preexisting cellular proteins or the synthesis of denatured proteins. Three types of observations indicate that the regulation of eukaryotic heat-shock gene expression is somehow linked to the accumulation of abnormal proteins. First, a mouse cell line, ts85, is temperature sensitive in the conjugation of Ub to proteins[152] and has a large defect in the degradation of short-lived proteins.[153] Because of this defect in proteolysis, the cells are likely to accumulate denatured proteins at the nonpermissive temperature. These mutants also synthesize elevated levels of *hsp*s under such conditions.[152] Second, as mentioned above, the connection between *hsp* gene expression and ATP-dependent proteolysis is further strengthened by the finding that Ub is itself a heat-shock protein whose induction is essential for viability.[118,124] Third, this model predicts that *hsp* synthesis should be induced in cells that produce certain mutant proteins in sufficiently high amounts to saturate the cell's capacity for protein degradation. Accordingly, in the flight muscles of mutant *Drosophila* strains that synthesize abnormal form of actin III, *hsp*s are produced constitutively.[154,155] Careful genetic analysis has shown that it is the synthesis of the altered actin protein that causes elevated *hsp* gene expression.[156]

We decided to test directly whether the accumulation of large amounts of denatured proteins triggers the activation of *hsp* genes in eukaryotic cells[11] as it does in *E. coli*.[6] To follow induction of the heat-shock response, a hybrid *hsp-70*–β-galactosidase gene fusion was constructed in which the heat-shock promoter directs the synthesis of the

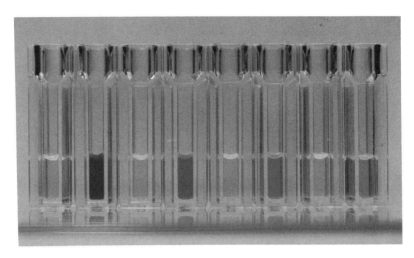

Figure 5. Induction of the heat-shock response in frog oocytes upon injection of denatured polypeptides. Transcription of a *Drosophila hsp-70* promoter was measured after injection of plasmids carrying the *lacZ* gene fused to the *hsp-70* promoter. Therefore, the appearance of β-galactosidase indicates increased transcription of *hsp-70* as occurs on a temperature shift. Similarly, injection of denatured albumin or denatured β-lactoglobulin or apohemoglobin causes induction of *hsp*s. The native polypeptides had no such effects.

easily measured *E. coli* β-galactosidase. When such *hsp* genes or gene fusions are microinjected into the nuclei of *Xenopus laevis* oocytes, they are transcribed from the correct start site in a heat-regulated fashion.[99,157] The heterologous proteins, bovine serum albumin and β-lactoglobulin, were coinjected into the oocytes either in their native forms or after denaturation by reduction and carboxymethylation (Figure 5). When the native forms of these proteins were microinjected, they did not increase the expression of the *hsp-70*–β-galactosidase fusion gene above background. In oocytes that had received the denatured proteins, a tenfold increase in enzyme activity was observed[11] (Figure 5). These experiments established a causal linkage between the accumulation of denatured proteins and induction of the heat-shock proteins.

Exactly how the presence of denatured proteins causes an activation of heat-shock genes will remain a matter of speculation until more is known about the properties of the heat-shock transcription factor(s). One possible mechanism is analogous to the model for regulation of heat-shock genes in bacteria described above (Figure 4). A critical assumption of these models[10,11] is that the large amount of denatured proteins generated at high temperature exceeds the degradative capacity of the cell's proteolytic machinery. As these abnormal proteins accumulate, they should compete with proteins that are otherwise rapidly degraded via Ub conjugation or direct proteolytic attack. Consequently, the rapidly degraded polypeptides are more stable and accumulate. According to this model, most of the heat-shock transcription factor is not functional in the unstressed cells because either it is rapidly degraded or another protein necessary for its activity or production is rapidly degraded. During heat shock the active factor builds up and promotes *hsp* transcription either because it is stabilized or because the activating enzyme is more stable. Two possible variations on this model would be that (1) the transcription factor assumes an active conformation upon interaction with abnormal proteins; or (2) a separate system senses the accumulation of abnormal proteins and somehow activates the transcription factor. One observation consistent with all these models is that the amounts of denatured proteins producing activation of *hsp* genes in *Xenopus* oocytes were sufficient to reduce the rate of degradation of another short-lived protein.[11] Furthermore, certain mutant proteins[158] and certain rapidly degraded analog-containing proteins were found to be more stable in heat-shocked cells.

A variant on this type of model[159] assumes that the heat-shock transcription factor is inactive when conjugated to Ub, but it becomes active after the loss of the Ub moiety. This model assumes that the transcription factor is an excellent substrate for Ub conjugation and that the resulting Ub conjugates are continually being dissociated by the cellular "isopeptidase" or are being degraded proteolytically. The accumulation of denatured proteins in heat-treated cells provides a large pool of substrates for Ub conjugation, and this process could lead to a severe shortage of Ub as suggested by the recent work of Kulka and co-workers[121] and Rechsteiner and co-workers.[122] Consequently, some of the proteins that are normally ubiquitinated would build up in an unconjugated form. In fact, heat shock is known to cause a wave of ubiquitination to cell proteins, which can deplete the pool of free Ub and reduce the level of ubiquitinated H2A.[121,122] Similarly, the heat-shock transcription factor or an activating polypeptide should accumulate in the nonconjugated state. Clearly, in the absence of information on whether the heat-shock transcription factor is a substrate for Ub conjugation or proteolysis, it is not yet possible to test these various models.

A complete explanation of the regulation of the heat-shock response must also account for the rapid termination of *hsp* transcription once these proteins have been synthesized and have performed some protective function. This apparent feedback regulation of heat-shock gene expression may involve members of the *hsp-70* family and may result from their tendency to associate reversibly with the damaged proteins or organelles. This association and the subsequent ATP-dependent dissociation of *hsp-*

70 may help refold partially damaged molecules and may prevent proteins from being recognized by the proteolytic machinery. The increased synthesis of *hsps* during heat shock or in other stressful conditions would result in a reduced rate of accumulation of the damaged proteins. Furthermore, the increased production of Ub during the heat-shock response may well lead to greater proteolytic capacity and consequently to a faster elimination of the abnormal polypeptides. It is also possible that one of the other eukaryotic heat-shock proteins is involved in Ub conjugation or is a proteolytic enzyme, as is the case with protease La from *E. coli*.

By lowering the amounts of substrate for Ub conjugation and proteolysis, the *hsps* should decrease the signal for their own production. Thus, in both eukaryotic and prokaryotic cells the heat-shock response represents an important form of autoregulation in which a perturbation of cellular composition is corrected through transcriptional and translational adaptations. According to the models described above, the *hsps* eventually help the cell eliminate denatured proteins which, if allowed to accumulate, could be lethal.

REFERENCES

1. Goldberg, A., and St. John, A. C., 1976, Intracellular protein degradation in mammalian and bacterial cells. Part II, *Annu. Rev. Biochem.* **45:** 747–803.
2. Pine, M. J., 1966, Metabolic control of intracellular proteolysis in growing and resting cells of *Escherichia coli, J. Bacteriol.* **92:** 847–850.
3. Goldberg, A. L., 1972, Degradation of abnormal proteins in *Escherichia coli, Proc. Natl. Acad. Sci. U.S.A.* **69:** 422–426.
4. Goldberg, A. L., and Goff, S. A., 1986, The selective degradation of abnormal proteins in bacteria, in: *Maximizing Gene Expression* (W. Reznikoff and L. Gold, eds.), Butterworth Press, Stoneham, MA, pp. 287–314.
5. Goff, S., Casson, L. P., and Goldberg, A. L., 1984, Heat shock regulatory gene *htpR* influences rates of protein degradation and expression of the *lon* gene in *Escherichia coli, Proc. Natl. Acad. Sci. U.S.A.* **81:** 6647–6651.
6. Goff, S. A., and Goldberg, A. L., 1985, Abnormal proteins stimulate expression of *lon* and other heat shock genes in *Escherichia coli, Cell* **41:** 587–595.
7. Neidhardt, F. C., VanBogelen, R. A., and Vaughn, V., 1984, The genetics and regulation of heat-shock proteins, *Annu. Rev. Genet.* **18:** 295–329.
8. *Heat-Shock: From Bacteria to Man* (M. Schlesinger, M. Ashburner, and A. Tissieres, eds.), Cold Spring Harbor Press, Cold Spring Harbor, NY.
9. Craig, E. A., 1985, The heat shock response, *Crit. Rev. Biochem.* **18:** 239–280.
10. Burdon, R. H., 1986, Heat shock and the heat shock proteins, *Biochem. J.* **240:** 313–324.
11. Ananthan, J., Goldberg, A. L., and Voellmy, R., 1986, Abnormal proteins serve as eukaryotic stress signal and trigger the activation of heat shock genes, *Science* **232:** 522–524.

12. Hershko, A., and Ciechanover, A., 1982, Mechanisms of intracellular protein breakdown, *Annu. Rev. Biochem.* **51:** 335–364.
13. Mandelstam, J., 1958, Turnover of protein in growing and non-growing populations of *Escherichia coli, Biochem. J.* **69:** 110–119.
14. Etlinger, J., and Goldberg, A. L., 1977, A soluble ATP-dependent proteolytic system responsible for the degradation of abnormal proteins in reticulocyte, *Proc. Natl. Acad. Sci. U.S.A.* **74:** 54–58.
15. Murakami, K., Voellmy, R., and Goldberg, A. L., 1979, Protein degradation is stimulated by ATP in extracts of *Escherichia coli, J. Biol. Chem.* **254:** 8194–8200.
16. Desautels, M., and Goldberg, A. L., 1982, Liver mitochondria contain an ATP-dependent, vanadate-sensitive pathway for the degradation of proteins, *Proc. Natl. Acad. Sci. U.S.A.* **79:** 1869–1873.
17. Fagan, J. M., Waxman, L., and Goldberg, A. L., 1987, Skeletal muscle and liver contain a soluble ATP + ubiquitin-dependent proteolytic system, *Biochem. J.* **243:** 313–323.
18. Waxman, L., Fagan, J. M., Tanaka, K., and Goldberg, A. L., 1985, A soluble ATP-dependent system for protein degradation from murine erythroleukemia cells, *J. Biol. Chem.* **260:** 11994–12000.
19. Waxman, L., Fagan, J., and Goldberg, A., 1987, Demonstration of two distinct high molecular weight proteases in rabbit reticulocytes, one of which degrades ubiquitin conjugates, *J. Biol. Chem.* **262:** 2451–2457.
20. Hough, R., Pratt, G., and Rechsteiner, M., 1986, Ubiquitin–lysozyme conjugates: Identification and characterization of an ATP-dependent protease from rabbit reticulocyte lysate, *J. Biol. Chem.* **261:** 2400–2408.
21. Goldberg, A. L., Voellmy, R., Chung, C. H., Menon, A. S., Desautels, M., Meixsell, T., and Waxman, L., 1985, The ATP dependent pathway for protein breakdown in bacteria and mitochondria, in: *Intracellular Protein Catabolism*, Alan R. Liss, New York, pp. 33–45.
22. Larimore, F., Waxman, L., and Goldberg, A. L., 1982, Studies of the ATP-dependent proteolytic enzyme, protease La, from *Escherichia coli, J. Biol. Chem.* **257:** 4187–4195.
23. Chung, C. H., and Goldberg, A. L., 1981, The product of the *lon* (*capR*) gene in *Escherichia coli* is the ATP-dependent protease, protease La, *Proc. Natl. Acad. Sci. U.S.A.* **78:** 4931–4935.
24. Charette, M. F., Henderson, G. W., and Markovitz, A., 1981, ATP-hydrolysis-dependent protease activity of the *lon* (*capR*) protein of *Escherichia coli* K12, *Proc. Natl. Acad. Sci. U.S.A.* **78:**4728–4732.
25. Tanaka, K., Waxman, L., and Goldberg, A. L., 1983, ATP serves two distinct roles in protein degradation in reticulocytes, one requiring and one independent of ubiquitin, *J. Cell. Biol.* **96:** 1580–1585.
26. Desautels, M., and Goldberg, A. L., 1982, Demonstration of an ATP-dependent, vanadate-sensitive endoprotease in the matrix of rat liver mitochondria, *J. Biol. Chem.* **257:** 11673–11679.
27. Goldberg, A. L., Swamy, K. H. S., Chung, C. H., and Larimore, F., 1982, Proteinases in *Escherichia coli,* in: *Methods in Enzymology* (L. Lorand, ed.), Vol. 80, Academic Press, New York, pp. 702–711.
28. Kowit, J., and Goldberg, A. L., 1977, Intermediate steps in the degradation of a specific abnormal protein in *Escherichia coli, J. Biol. Chem.* **252:** 8350–8357.
29. Mizusawa, S., and Gottesman, S., 1983, Protein degradation in *Escherichia coli:* The *lon* gene controls the stability of the *sulA* protein, *Proc. Natl. Acad. Sci. U.S.A.* **80:** 358–362.

30. Gottesman, S., Gottesman, M., Shaws, J. E., and Pearson, M. L. 1981, Protein degradation in *E. coli*: The *lon* mutation and bacteriophage lambda *N* and *cII* protein stability, *Cell* **24**: 225–233.

31. Goff, S. A., and Goldberg, A. L., 1987, An increased content of protease La, the *lon* gene product, increases protein degradation and blocks cell growth in *Escherichia coli*, *J. Biol. Chem.* **262**: 4508–4515.

32. Waxman, L., and Goldberg, A. L., 1982, Protease La from *Escherichia coli* hydrolyzes ATP and proteins in a linked fashion, *Proc. Natl. Acad. Sci. U.S.A.* **79**: 4883–4887.

33. Waxman, L., and Goldberg, A. L., 1985, Protease La, the *lon* gene product, cleaves specific fluorogenic peptides in an ATP-dependent reaction, *J. Biol. Chem.* **260**: 12022–12028.

34. Goldberg, A. L., and Waxman, L., 1985, The role of ATP hydrolysis in the breakdown of proteins and peptides by protease La from *Escherichia coli*, *J. Biol. Chem.* **260**: 12029–12034.

35. Menon, A., Waxman, L., and Goldberg, A. L., 1987, The energy utilized in protein breakdown by the ATP-dependent protease (La) from *Escherichia coli*, *J. Biol. Chem.* **262**: 722–726.

36. Waxman, L., and Goldberg, A., 1986, Selectivity of intracellular proteolysis: Protein substrates activate the ATP-dependent protease (La), *Science* **232**: 500–503.

37. Menon, A. S., and Goldberg, A. L., 1987, Binding of nucleotides to the ATP-dependent protease La in *Escherichia coli*, *J. Biol. Chem.* **262**: 14921–14928.

38. Maurizi, M. R., Trisler, P., and Gottesman, S., 1985, Insertional mutagenesis of the *lon* gene in *Escherichia coli*: *lon* is dispensable, *J. Bacteriol.* **164**: 1124–1135.

39. Katayama-Fujimura, Y., Gottesman, S., and Maurizi, M., 1987, A multiple-component, ATP-dependent protease from *Escherichia coli*, *J. Biol. Chem.* **262**: 4477–4485.

40. Hwang, B. J., Park, W. J., Chung, C. H., and Goldberg, A. L., 1987, *Escherichia coli* contains a soluble ATP-dependent protease (T1) distinct from protease La, *Proc. Natl. Acad. Sci. U.S.A.* **84**: 5550–5554.

41. Patterson, D., and Gillespie, D., 1972, Effect of elevated temperatures on protein synthesis in *Escherichia coli*, *J. Bacteriol.* **112**: 1177–1183.

42. Chaloner-Larsson, G., and Yamazaki, H., 1977, Adjustment of RNA content during temperature upshift in *Escherichia coli*, *Biochem. Biophys. Res. Commun.* **82**: 477–483.

43. Lemaux, P. G., Herendeen, S. L., Bloch, P. L., and Neidhardt, F., 1978, Transient rates of synthesis of individual polypeptides in *E. coli* following temperature shifts, *Cell* **13**: 427–434.

44. Yamamori, T., Ito, K., Nakamura, Y., and Yura, T., 1978, Transient regulation of protein synthesis in *Escherichia coli* upon shift-up of growth temperature, *J. Bacteriol.* **134**: 1133–1140.

45. Herendeen, S. L., VanBogelen, R. A., and Neidhardt, F., 1979, Levels of major proteins of *Escherichia coli* during growth at different temperatures, *J. Bacteriol.* **141**: 1409–1420.

46. Bloch, P. L., Phillips, T. A., and Neidhardt, F., 1980, Protein identifications on O'Farrell two dimensional gels: Locations of 81 *Escherichia coli* proteins, *J. Bacteriol.* **141**: 1409–1420.

47. Georgopoulos, G., Tilly, K., Drahos, D., and Hendrix, R., 1982; The B66.0 protein of *Escherichia coli* is the product of the *dnaK+* gene, *J. Bacteriol.* **149**: 1175–1177.

48. Georgopoulos, C. P., Lam, B., Lundquist-Heil, A., Rudolph, C. F., Yochem, J., and Feiss, M., 1979, Identification of the *E. coli dnaK(groPC756)* gene product, *Mol. Gen. Genet.* **172**: 143–149.

49. Neidhardt, F. C., Phillips, T. A., VanBogelen, R. A., and Smith, M. W., Georgalis, Y., and Subramanian, A. R., 1981, Identity of the B56.5 protein, the A-protein, and the *groE* gene product of *Escherichia coli*, *J. Bacteriol.* **145**: 513–520.

50. Hendrix, R. W., and Tsui, L., 1978, Role of the host in virus assembly: Cloning of the *Escherichia coli groE* gene and identification of its protein product, *Proc. Natl. Acad. Sci. U.S.A.* **75**: 759–763.

51. Georgopoulos, C. P., and Hohn, B., 1978, Identification of a host protein necessary for bacteriophage morphogenesis (the *groE* gene product), *Proc. Natl. Acad. Sci. U.S.A.* **75**: 131–135.

52. Wada, M., and Itikawa, H., 1984, Participation of *Escherichia coli* K-12 *groE* gene products in the synthesis of cellular DNA and RNA, *J. Bacteriol.* **157**: 694–696.

53. Zylicz, M., LeBowitz, J. H., McMacken, R., and Georgopoulos, C., 1983, The *dnaK* protein of *Escherichia coli* possesses an ATPase and autophosphorylating activity and is essential in an *in vitro* DNA replication system, *Proc. Natl. Acad. Sci. U.S.A.* **80**: 6431–6435.

54. Ang, D., Chandrasekhar, G. N., Zylicz, M., and Georgopoulos, C., 1986, *Escherichia coli grpE* gene codes for heat shock protein B25.3, essential for both λ DNA replication at all temperatures and host growth at high temperature, *J. Bacteriol.* **167**: 25–29.

55. Neidhardt, F. C., VanBogelen, R. A., and Lau, E. T., 1982, The high temperature regulon of *Escherichia coli*, in: *Heat Shock: From Bacteria to Man* (M. J. Schlesinger, M. Ashburner, and A. Tissieres, eds.) Cold Spring Harbor Press, Cold Spring Harbor, NY, pp. 139–145.

56. Bardwell, J. C. A., Tilly, K., Craig, E., King, J., Zylicz, M., and Georgopoulos, C., 1986, The nucleotide sequence of the *Escherichia coli* K12 *dnaJ*+ gene, *J. Biol. Chem.* **261**: 1782–1785.

57. Taylor, W. E., Straus, D. B., Grossman, A. D., Burton, Z. F., Gross, C. A., and Burgess, R. R., 1984, Transcription from a heat-inducible promoter causes heat shock regulation of the sigma subunit of *E. coli* RNA polymerase, *Cell* **38**: 371–381.

58. Gross, C., Cowing, D., Erickson, J., Grossman, A., Straus, D., Walter, W., and Zhou, Y.-N., 1987, Regulation of the heat-shock response in *Escherichia coli*, in: *New Frontiers in the Study of Gene Functions* (G. Poste and S. T. Crooke, eds.), Plenum Press, New York, pp. 21–32.

59. Cooper, S., and Ruettinger, T., 1975, A temperature-sensitive nonsense mutation affecting the synthesis of a major protein of *Escherichia coli*, *Mol. Gen. Genet.* **139**: 167–176.

60. Beckman, D., and Cooper, S., 1973, Temperature-sensitive nonsense mutations in essential genes of *Escherichia coli*, *J. Bacteriol.* **116**: 1336–1342.

61. Neidhardt, F. C., and VanBogelen, R. A., 1981, Positive regulatory gene for temperature-controlled proteins in *Escherichia coli*, *Biochem. Biophys. Res. Commun.* **100**: 894–900.

62. Yamamori, T., and Yura, T., 1982, Genetic control of heat-shock protein synthesis and its bearing on growth and thermal resistance in *Escherichia coli* K12, *Proc. Natl. Acad. Sci. U.S.A.* **79**: 860–864.

63. Neidhardt, F. C., VanBogelen, R. A., and Lau, E. T., 1983, Molecular cloning and expression of a gene that controls the high-temperature regulon of *Escherichia coli*, *J. Bacteriol.* **153**: 597–603.

64. Tobe, T., Ito, K., and Yura, T., 1984, Isolation and physical mapping of temperature-sensitive mutants defective in heat-shock induction of proteins in *Escherichia coli*, *Mol. Gen. Genet.* **195**: 10–16.

65. Yamamori, T., and Yura, T., 1980, Temperature-induced synthesis of specific proteins in *Escherichia coli*: Evidence for transcriptional control, *J. Bacteriol.* **142**: 843–851.

66. Bloom, M., Skelly, S., VanBogelen, R., Neidhardt, F., Brot, N., and Weissbach, H., 1986, *In vitro* effect of the *Escherichia coli* heat shock regulatory protein on expression of heat shock genes, *J. Bacteriol.* **166**:380–384.

67. Landick, R., Vaughn, V., Lau, E. T., VanBogelen, R. A., Erickson, J. W., and Neidhardt, F. C., 1984, Nucleotide sequence of the heat shock regulatory gene of *E. coli* suggests its protein product may be a transcription factor, *Cell* **38**:174–182.

68. Grossman, A. D., Erickson, J. W., and Gross, C. A., 1984, The htpR gene product of *E. coli* is a sigma factor for heat-shock promoters, *Cell* **38**: 383–390.

69. Tilly, K., McKittrick, N. M., Zylicz, M., and Georgopoulos, C., 1983, *The dnaK* protein modulates the heat-shock response of *Escherichia coli, Cell* **34**: 641–646.

70. Phillips, T. A., VanBogelen, R. A., and Neidhardt, F. C., 1984, The *lon(CapR)* gene product of *Escherichia coli* is a heat shock protein, *J. Bacteriol.* **159**: 283–287.

71. Travers, A. A., and Mace, H. A. F., 1982, The heat-shock phenomenon in bacteria—a protection against DNA relaxation? in: *Heat Shock: From Bacteria to Man* (M. J. Schlesinger, M. Ashburner, and A Tissieres, eds.), Cold Spring Harbor Press, Cold Spring Harbor, NY, pp. 127–130.

72. So, A. G., and Davie, E. W., 1964, The effects of organic solvents on protein biosynthesis and their influence on the amino acid code, *Biochemistry* **3**: 1165–1169.

73. Lapanje, S., 1978, *Physical Aspects of Protein Denaturation*, Wiley, New York, pp. 142–156.

74. Baker, T. A., Grossman, A. D., and Gross, C. A., 1984, A gene regulating the heat shock response in *Escherichia coli* also affects proteolysis, *Proc. Natl. Acad. Sci. U.S.A.* **81**: 6779–6783.

75. Bukhari, A. I., and Zipser, D., 1973, Mutants of *Escherichia coli* with a defect in the degradation of nonsense fragments, *Nature New Biol.* **243**: 238–241.

76. Shineberg, B., and Zipser, D., 1973, The *lon* gene and degradation of β galactosidase nonsense fragments, *J. Bacteriol* **116**: 1469–1471.

77. Gottesman, S., and Zipser, D., 1978, Deg phenotype of *Escherichia coli lon* mutants, *J. Bacteriol.* **133**: 844–851.

78. St. John, A., and Goldberg, A. L., 1978, Effects of reduced energy production on protein degradation, guanosine tetraphosphate, and RNA synthesis in *Escherichia coli, J. Biol. Chem.* **253**: 2705–2716.

79. Lin, S., and Zabin, I., 1972, β-galactosidase: Rates of synthesis and degradation of incomplete chains, *J. Biol. Chem.* **247**: 2205–2211.

80. Mandecki, W., Powell, B. S., Mollison, K. W., Carter, G. W., and Fox, J. L., 1986, High level expression of a gene encoding the human complement factor C5a in *Escherichia coli, Gene* **43**: 131–138.

81. Buell, G., Schultz, M.-F., and Selzer, G., Chollet, A., Movva, N. R., Semon, D., Escanez, S., and Kawashima, E., 1985, Optimizing the expression in *E. coli* of a synthetic gene encoding somatomedin-C (IGF-I), *Nucleic Acids Res.* **13**: 1923–1938.

82. Pellon, J. R., Gomez, R. F., and Sinskey, A., 1982, Association of the *Escherichia coli* nucleoid with protein synthesized during thermal treatments, in: *Heat Shock: From Bacteria to Man* (M. J. Schlesinger, M. Ashburner, and A. Tissieres, eds.), Cold Spring Harbor Press, Cold Spring Harbor, NY, pp. 121–126.

83. Hightower, L. E., 1980, Cultured animal cells exposed to amino acid analogs or puromycin rapidly synthesize several polypeptides, *J. Cell. Physiol.* **102**: 407–427.

84. Kelley, P. M., and Schlesinger, M. J., 1978, The effect of amino acid analogs and heat shock on gene expression in permissive cells, *J. Cell. Physiol.* **15**: 1277–1286.

85. Roberts, J. W., and Devoret, R., 1993, in: *Lambda II* (R. Hendrix, J. Roberts, F. Stahl, and R. Weisberg, eds.), Cold Spring Harbor Press, Cold Spring Harbor, NY, pp. 130–133.

86. Torres-Cabassa, A. S., and Gottesman, S., 1987, Capsule synthesis in *Escherichia coli* K-12 is regulated by proteolysis, *J. Bacteriol.* **169:** 981–989.
87. Lewis, M. J., and Pelham, H. R. B., 1985, Involvement of ATP in the nuclear and nucleolar functions of the 70-kd heat shock protein, *EMBO J.* **4:** 3137–3143.
88. Munro, S., and Pelham, H. R. B., 1986, An hsp-70-like protein in the endoplasmic reticulum: Identity with the 78-kilodalton glucose-regulated protein and immunoglobulin heavy chain binding protein, *Cell* **46:** 291–300.
89. Ungewickell, E., 1985, the 70-kilodalton mammalian heat shock proteins are structurally and functionally related to the uncoating protein that releases clathrin triskelia from coated vesicles, *EMBO J.* **4:** 3385–3392.
90. Chappell, T. G., Welch, B. J., Schlossman, D. M., Palter, K. B., Schlesinger, M.J., and Rothman, J. E., 1986, Uncoating ATPase is a member of the 70 kilodalton family of stress proteins, *Cell* **45:** 3–13.
91. Pelham, H. R. B., 1985, Activation of heat-shock genes in eukaryotes, *Trends Genet.* **1:** 31–35.
92. Lindquist, S., 1981, Regulation of protein synthesis during heat shock, *Nature* **293:** 311–314.
93. DiDomenico, B. J., Bugaisky, G. E., and Lindquist, S. S., 1982, Heat shock and recovery are mediated by different translational mechanisms, *Proc. Natl. Acad. Sci. U.S.A.* **79:** 6181–6185.
94. DiDomenico, B. J., Bugaisky, G. E., and Lindquist, S. S., 1982, The heat shock response is self-regulated at both the transcriptional and posttranscriptional levels, *Cell* **31:** 593–603.
95. DiNocera, P. P., and Dawid, I. B., 1983, Transient expression of genes introduced into cultured cells of *Drosophila*, *Proc. Natl. Acad. Sci. U.S.A.* **80:** 7095–7098.
96. McGarry, T. J., and Lindquist, S., 1985, The preferential translation of *Drosophila* hsp70 mRNA requires sequences in the untranslated leader, *Cell* **42:** 903–911.
97. Hultmark, D., Klemenz, R., and Gehring, W. J., 1986, Translational and transcriptional control elements in the untranslated leader of the heat-shock gene *hsp22, Cell* **44:** 429–438.
98. Dreano, M., Brochot, J., Myers, A., Cheng-Meyer, C., Rungger, D., Voellmy, R., and Bromley, P., 1987, High-level heat-regulated synthesis of proteins in eukaryotic cells, *Gene* **49:** 1–8.
99. Voellmy, R., Ahmed, A., Schiller, P., Bromley, P., and Rungger, D., 1985, Isolation and functional analysis of a human 70,000-dalton heat shock protein gene segment, *Proc. Natl. Acad. Sci. U.S.A.* **82:** 4949–4953.
100. Pelham, H. R. B., 1982, A regulatory upstream promoter element in the *Drosophila hsp 70* heat-shock gene, *Cell* **30:** 517–528.
101. Mirault, M.-E., Southgate, R., and Delwart, E., 1982, Regulation of heat-shock genes: A DNA sequence upstream of *Drosophila hsp 70* genes is essential for their induction in monkey cells, *EMBO J.* **1:** 1279–1285.
102. Dudler, R., and Travers, A. A., 1984, Upstream elements necessary for optimal function of the hsp 70 promoter in transformed flies, *Cell* **38:** 391–398.
103. Simon, J. A., Sutton, C. A., Lobell, R. B., Glaser, R., and Lis, J. T., 1985, Determinants of heat shock-induced chromosome puffing, *Cell* **40:** 805–817.
104. Bienz, M., and Pelham, H. R. B., 1986, Heat shock regulatory elements function as an inducible enhancer in the *Xenopus hsp70* gene and when linked to a heterologous promoter, *Cell* **45:** 753–760.
105. Amin, J., Mestril, R., Schiller, P., Dreano, M., and Voellmy, R., 1987, Organization of the *Drosophila melanogaster* hsp70 heat shock regulation unit, *Mol. Cell. Biol.* **7:** 1055–1062.

106. Parker, C. S., and Topol, J., 1984, A *Drosophila* RNA polymerase II transcription factor binds to the regulatory site of an *hsp70* gene, *Cell* **37:** 273–283.
107. Wu, C., 1984, Activating protein factor binds *in vitro* to upstream control sequences in heat shock gene chromatin, *Nature* **311:** 81–84.
108. Wiederrecht, G., Shuey, D. J., Kibbe, W. A., and Parker, C. S., 1987, The *Saccharomyces* and *Drosophila* heat shock transcription factors are identical in size and DNA binding properties, *Cell* **48:** 507–515.
109. Sorger, P. K., and Pelham, H. R. B., 1987, Purification and characterization of a heat shock element binding protein from yeast, *EMBO J.* **6:** 3035–3042.
110. Yost, H. J., and Lindquist, S., 1986, RNA splicing is interrupted by heat shock and is rescued by heat shock protein synthesis, *Cell* **45:** 184–193.
111. Mirault, M.-E., Goldschmidt-Clermont, M., Moran, L., Arrigo, A. P., and Tissières, A., 1977, The effect of heat shock on gene expression in *Drosophila melanogaster, Cold Spring Harbor Symp. Quant. Biol.* **42:** 819–827.
112. Storti, R. V., Scott, M. P., Rich, A., and Pardue, M. L., 1980, Translational control of protein synthesis in response to heat-shock in *D. melanogaster* cells, *Cell* **22:**825–834.
113. Krüger, C., and Benecke, B.-J., 1981, *In vitro* translation of *Drosophila* heat-shock and non-heat-shock mRNAs in heterologous and homologous cell-free systems, *Cell* **23:** 595–603.
114. Scott, M. P., and Pardue, M. L., 1981, Translational control in lysates of *Drosophila melanogaster* cells, *Proc. Natl. Acad. Sci. U.S.A.* **78:** 3353–3357.
115. Lawson, R., Mestril, R., Schiller, P., and Voellmy, R., 1984, Expression of heat shock–α-galactosidase hybrid genes, *Mol. Gen. Genet.* **198:** 116–124.
116. Subjeck, J. R., and Shyy, T.-T., 1986, Stress protein systems of mammalian cells, *Am. J. Physiol.* **250:** C1–C17.
117. Bond, U., and Schlesinger, M. J., 1985, Ubiquitin is a heat shock protein in chicken embryo fibroblasts, *Mol. Cell. Biol.* **5:** 949–956.
118. Bond, U., and Schlesinger, M. J., 1986, The chicken ubiquitin gene contains a heat shock promoter and expresses an unstable mRNA in heat-shocked cells, *Mol. Cell. Biol.* **6:** 4602–4610.
119. Hershko, A., Eytan, E., Ciechanover, A., and Haas, A. L., 1982, Immunochemical analysis of the turnover of ubiquitin–protein conjugates in intact cells, *J. Biol. Chem.* **257:** 13964–13970.
120. Ciechanover, A., Finley, D., and Varshavsky, A., 1984, The ubiquitin-mediated proteolytic pathway and mechanisms of energy-dependent intracellular protein degradation, *J. Cell. Biochem.* **24:** 27–53.
121. Parag, H. A., Raboy, B., and Kulka, R. G., 1987, Effect of heat shock on protein degradation in mammalian cells: Involvement of the ubiquitin system, *EMBO J.* **6:** 55–61.
122. Carlson, N., Rogers, S., and Rechsteiner, M., 1987, Microinjection of ubiquitin: Changes in protein degradation in HeLa cells subjected to heat-shock, *J. Cell Biol.* **104:** 547–555.
123. Ozkaynak, E., Finley, D., Solomon, M. J., and Varshavsky, A., 1987, The yeast ubiquitin genes: A family of natural gene fusions, *EMBO J.* **6:** 1429–1439.
124. Finley, D., Ozkaynak, E., and Varshavsky, A., 1987, The yeast polyubiquitin gene is essential for resistance to high temperatures, starvation and other stresses, *Cell* **48:** 1035–1046.
125. Mason, P. J., Hall, L. M. C., and Gausz, J., 1984, The expression of heat shock genes during normal development in *Drosophila melanogaster, Mol. Gen. Genet.* **194:** 73–78.

126. Gasc, J.-M., Renoir, J.-M., Radanyi, C., Joab, I., Tuohimaa, P., and Baulieu, E.-E., 1984, Progesterone receptor in the chick oviduct: An immunohistochemical study with antibodies to distinct receptor components, *J. Cell Biol.* **99:** 1193–1201.
127. Catelli, M. G., Binart, N., Jung-Testas, I., Renoir, J. M., Baulieu, E.-E., Feramisco, J. R., and Welch, W. J., 1985, The common 90-kd protein component of non-transformed "8S" steroid receptors is a heat-shock protein, *EMBO J.* **4:** 3131–3135.
128. Brugge, J. S., 1985, Interaction of the Rous sarcoma virus protein pp60^{src} with the cellular proteins pp50 and pp90, in: *Current Topics in Microbiology and Immunology* (P. D. Vogt, ed.), Vol. 123, Springer-Verlag, Berlin, pp. 1–22.
129. Welch, W. J., Garrels, J. I., Thomas, G. P., Lin, J. J. C., and Feramisco, J. R., 1983, Biochemical characterization of the mammalian stress proteins and identification of two stress proteins as glucose and calcium ionophore-regulated proteins, *J. Biol. Chem.* **258:** 7102–7111.
130. Sorger, P., and Pelham, H. R. B., 1987, The 94,000 dalton glucose-regulated protein is related to the 90,000 dalton heat shock protein, *J. Mol. Biol.* **194:** 341–344.
131. Welch, W. J., and Feramisco, J. R., 1984, Nuclear and nucleolar localization of the 72,000-dalton heat shock protein in heat-shock mammalian cells, *J. Biol. Chem.* **259:** 4501–4513.
132. Vincent, M., and Tanguay, R. M., 1979, Heat-shock induced proteins present in the cell nucleus of *Chironomus tentans* salivary gland, *Nature* **281:** 501–503.
133. Arrigo, A.-P., Fakan, St., and Tissieres, A., 1980, Localization of the heat-shock-induced proteins in *Drosophila melanogaster* tissue culture cells, *Dev. Biol.* **78:** 86–103.
134. Velazquez, J. M., DiDomenico, B. J., and Lindquist, S., 1980, Intracellular localization of heat shock proteins in *Drosophila*, *Cell* **20:** 679–689.
135. Levinger, L., and Varshavsky, A., 1981, Heat-shock proteins of *Drosophila* are associated with nuclease-resistant high-salt resistant nuclear structures, *J. Cell Biol.* **90:** 793–796.
136. Sinibaldi, R. M., and Morris, P. W., 1981, Putative function of *Drosophila melanogaster* heat shock proteins in the nucleoskeleton, *J. Biol. Chem.* **256:** 10735–10738.
137. Arrigo, A.-P., and Ahmad-Zadeh, C., 1981, Immunofluorescence localization of a small heat shock protein (hsp 23) in salivary gland cells of *Drosophila melanogaster, Mol. Gen. Genet.* **184:** 73–79.
138. Welch, W. J., and Suhan, J. P., 1986, Cellular and biochemical events in mammalian cells during and after recovery from physiological stress, *J. Cell Biol.* **103:** 2035–2052.
139. Pelham, H. R. B., 1984, Hsp70 accelerates the recovery of nucleolar morphology after heat shock, *EMBO J.* **3:** 3095–3100.
140. Pelham, H. R. B., 1986, Speculations on the functions of the major heat shock and glucose-regulated proteins, *Cell* **46:** 959–961.
141. Schlossman, D. M., Schmid, S. L., Braell, W. A., and Rothman, J. E., 1984, An enzyme that removes clathrin coats: Purification of an uncoating ATPase, *J. Cell Biol.* **99:** 723–733.
142. Morrison, S. L., and Scharff, M. D., 1975, Heavy chain-producing variants of a mouse myeloma cell line, *J. Immunol.* **114:** 655–659.
143. Haas, I. G., and Wabl, M., 1983, Immunoglobulin heavy chain binding protein, *Nature* **306:** 387–389.
144. Bole, D. G., Hendershot, L. M., and Kearney, J. F., 1986, Posttranslational association of immunoglobulin heavy chain binding protein with nascent heavy chains in nonsecreting and secreting hybridomas, *J. Cell Biol.* **102:** 1558–1566.
145. Welch, W. J., and Feramisco, J. R., 1985, Rapid purification of mammalian 79,000–

dalton stress proteins: Affinity of the proteins for nucleotides, *Mol. Cell. Biol.* **5:**1229–1237.

146. Welch, W. J., 1985, Phorbol ester, calcium ionophore, or serum added to quiescent rat embryo fibroblast cells are result in the elevated phosphorylation of two 28,000-dalton mammalian stress proteins, *J. Biol. Chem.* **260:** 3058–3062.

147. Hickey, E., Brandon, S. E., Potter, R., Stein, G., Stein, J., and Weber, L. A., 1986, Sequence and organization of genes encoding the human 27 kDa heat shock protein, *Nucleic Acids Res.* **14:** 4127–4145.

148. Ingolia, T. D., and Craig, E. A., 1982, Four small *Drosophila* heat shock proteins are related to each other and to mammalian α-crystallin, *Proc. Natl. Acad. Sci. U.S.A.* **79:** 2360–2364.

149. Southgate, R., Ayme, A., and Voellmy, R., 1983, Nucleotide sequence analysis of the *Drosophila* small heat shock gene cluster at locus 67B, *J. Mol. Biol.* **165:** 35–57.

150. Arrigo, A.-P., Darlix, J.-L., Khandjian, E. W., Simon, M., and Spahr, P.-F., 1985, Characterization of the prosome from *Drosophila* and its similarity to the cytoplasmic structures formed by the low molecular weight heat-shock proteins, *EMBO J.* **4:** 399–406.

151. Arrigo, A. P., Tanaka, K., Goldberg, A. L., and Welch, W. J., 1988, Identity of the 19S "prosome" particle with the large multifunctional protease complex of mammalian cells (the proteasome), *Nature* **331:** 192–194.

152. Finley, D., Ciechanover, A., and Varshavsky, A., 1984, Thermolability of ubiquitin-activating enzyme from the mammalian cell cycle mutant ts85, *Cell* **37:** 43–55.

153. Ciechanover, A., Finley, D., and Varshavsky, A., 1984, Ubiquitin dependence of selective protein degradation demonstrated in the mammalian cell cycle mutant ts85, *Cell* **37:** 57–66.

154. Hiromi, Y., and Hotta, Y., 1985, Actin gene mutations in *Drosophila*; heat shock activation in the indirect flight muscles, *EMBO J.* **4:** 1681–1687.

155. Okamoto, H., Hiromi, Y., Ishikawa, E., Yamada, T., Isoda, K., Maekawa, H., and Hotta, Y., 1986, Molecular characterization of mutant actin gene which induce heat-shock proteins in *Drosophila* flight muscles, *EMBO J.* **5:** 589–596.

156. Hiromi, Y., Okamoto, H., Gehring, W. J., and Hotta, Y., 1986, Germline transformation with *Drosophila melanogaster* mutant actin genes induces constitutive expression of heat shock genes, *Cell* **44:** 293–301.

157. Voellmy, R., and Rungger, D., 1982, Transcription of a *Drosophila* heat shock gene is heat-induced in *Xenopus* oocytes, *Proc. Natl. Acad. Sci. U.S.A.* **79:** 1776–1780.

158. Munro, S., and Pelham, H. R. B., 1984, Use of peptide tagging to detect proteins expressed from cloned genes: Deletion mapping functional domains of *Drosophila* hsp70, *EMBO J.* **3:** 3087–3093.

159. Munro, S., and Pelham, H. R. B., 1985, What turns on heat shock genes? *Nature* **317:** 477–478.

160. Neidhardt, F. C., and VanBogelen, R. A., 1987, Heat-shock response, in: *Escherichia coli and Salmonella typhimurium* (F. C. Neidhardt, ed.), American Society for Microbiology, Washington, D.C., pp. 1334–1345.

161. VanBogelen, R. A., Kelly, G. M., and Neidhardt, F. C., 1987, Differential induction of heat shock, SDS, and oxidative stress regulons and accumulation of nucleotides in *Escherichia coli, J. Bacteriol.* **169:** 26–32.

162. Chin, D. T., Goff, S. A., Webster, T., Smith, T., and Goldberg, A. L., 1988, DNA sequence of the *lon* gene in *Escherichia coli*: A Heat-shock gene which encodes the ATP-dependent protease La, *J. Biol. Chem.* (in press).

163. Menon, A. S., and Goldberg, A. L., 1987, Protein substrates activate the ATP-de-

pendent protease La by promoting nucleotide binding and release of bound ADP, *J. Biol. Chem.* **262:** 14929–14934.

164. Straus, D. B., Walter, W. A., and Gross, C. A., 1987, The heat shock response of *E. coli* is regulated by changes in the concentration of σ^{32}, *Nature* **329:** 348–351.
165. Prouty, W. F., and Goldberg, A. L., 1972, Fate of abnormal proteins in *E. coli*: Accumulation in intracellular granules before catabolism, *Nature New Biol.* **240:** 147–150.
166. Prouty, W. F., Karnovsky, M. J., and Goldberg, A. L., 1975, Degradation of abnormal proteins in *Escherichia coli*: Formation of protein inclusions in cells exposed to amino acid analogs, *J. Biol. Chem.* **250:** 1112–1122.
167. Klemes, Y., Etlinger, J. D., and Goldberg, A. L., 1981, Properties of proteins degraded rapidly in reticulocytes: Intracellular aggregation of the globin molecules prior to hydrolysis, *J. Biol. Chem.* **256:** 8436–8444.
168. Goldenberg, C., Luo, Y., Fenna, M., Baler, R., and Voellmy, R., 1987, Purified human factor, HHTF, activates a heat-shock promoter in a Hela cell-free transcription system (in press).
169. Wu, C., Wilson, S., Walker, B., Dawid, I., Paisley, T., Zimarino, V., and Ueda, H., 1987, Purification and properites of *Drosophila* heat shock activator protein, *Nature* **238:** 1247–1253.
170. Grossman, A. D., Straus, D. B., Walter, W. A., and Gross, C. A., 1987, σ^{32} synthesis can regulate the synthesis of heat shock proteins in *Escherichia coli*, *Genes Dev.* **1:** 179–184.
171. Bahl, H., Echols, H., Straus, D. B., Court, D., Crowl, R., and Georgopoulos, C. P., 1987, Induction of the heat shock response of *E. coli* through stabilization of σ^{32} by the phage cIII protein, *Genes Dev.* **1:** 57–64.
172. Drahos, D. J., and Hendrix, R. W., 1982, Effect of bacteriophage lambda infection on synthesis of *groE* protein and other *Escherichia coli* proteins, *J. Bacteriol.* **149:** 1050–1063.
173. Kochan, J., and Murialdo, H., 1982, Stimulation of *groE* synthesis in *Escherichia coli* by bacteriophage lambda infection, *J. Bacteriol.* **149:** 1166–1170.
174. Winter, R. B., and Gold, L., 1984, The maturation (A_2) protein from the RNA bacteriophage Qβ induces the synthesis of some *E. coli* heat-shock proteins, Cold Spring Harbor 1984 Bacteriophage Meeting, Cold Spring Harbor, NY (Abstr.).
175. Krueger, J. H., and Walker, G. C., 1984, *groEL* and *dnaK* genes of *Escherichia coli* are induced by UV irradiation and nalidixic acid in the $htpR^+$-dependent fashion, *Proc. Natl. Acad. Sci. U.S.A.* **81:** 1499–1503.
176. Christman, M. F., Morgan, R. W., Jacobson, F. S., and Ames, B. N., 1985, Positive control of a regulon for defenses against oxidation stress and some heat-shock proteins in *Salmonella typhimurium*, *Cell* **41:** 753–762.
177. VanBogelen, R. A., Kelley, P. M., and Neidhardt, F. C., 1987, Differential induction of heat shock, SOS, and oxidation stress regulons and accumulation of nucleotides in *Escherichia coli*, *J. Bacteriol.* **169:** 26–32.
178. Gross, C. A., Grossman, A. D., Liebke, H., Walter, W., and Burgess, R. R., 1984, Effects of the mutant $\sigma(rpoD800)$ on the synthesis of specific macromolecular components of the *Escherichia coli* K12 cell, *J. Mol. Biol.* **172:** 283–300.
179. Grossman, A. D., Taylor, W. E., Burton, Z. F., Burgess, R. R., and Gross, C. A., 1985, Stringent response in *Escherichia coli* induces expression of heat shock proteins, *J. Mol. Biol.* **186:** 357–365.
180. Levinson, W., Oppermann, H., and Jackson, J., 1980, Transition series metals and sulfhydryl reagents induce the synthesis of four proteins in eukaryotic cells, *Biochim. Biophys. Acta* **606:** 170–180.

181. Courgeon, A.-M., Maisonhaute, C., and Best-Belpomme, M., 1984, Heat shock proteins are induced by cadmium in *Drosophila* cells, *Exp. Cell Res.* **153:** 515–521.

182. Johnston, D., Oppermann, H., Jackson, J., and Levinson, W., 1980, Induction of four proteins in chick embryo cells by sodium arsenite, *J. Biol. Chem.* **255:** 6975–6980.

183. Ashburner, M., and Bonner, J. J., 1979, The induction of gene activity in *Drosophila* by heat shock, *Cell* **17:** 241–254.

184. Li, G. C., Shreve, D. C., and Werb, Z., 1982, Correlation between synthesis of heat shock proteins and development of tolerance to heat and to adriamysin in Chinese hamster fibroblasts: Heat shock and other inducers, in: *Heat Shock: From Bacteria to Man* (M. J. Schlesinger, M. Ashburner, and A. Tissieres, eds.), Cold Spring Harbor Press, Cold Spring Harbor, NY, pp. 395–404.

185. Rensing, L., 1973, Effects of 2,4-dinitrophenol and dinactin of heat-sensitive and eckysone-specific puffs of *Drosophila* salivary gland chromosomes *in vitro*, *Cell Diff.* **2:** 221–228.

186. Khandjian, E. W., and Turler, H., 1983, Simian virus 40 and polyoma virus induce synthesis of heat shock proteins in permissive cells, *Mol. Cell Biol.* **3:** 1–8.

187. Wu, B. J., Hirst, H. C., Jones, N. C., and Morimoto, R. I., 1986, The Ela 13S product of adenovirus type 5 activates transcription of the cellular human hsp70 gene, *Mol. Cell Biol.* **6:** 2994–2999.

Chapter 9

Lymphocyte Homing Receptors, Ubiquitin, and Cell Surface Proteins

Mark Siegelman and Irving L. Weissman

1. INTRODUCTION

The immune system, unlike most organ systems that are consolidated in one anatomic location, is dispersed over an entire organism. It exists as circulating elements in the blood, through which it gains access to nearly all body tissues, and as innumerable lymphoid aggregates throughout the body. Therefore, the immune system is placed under a special constraint, which is managed by substituting extensive cell–cell recognition and interactive events as a strategy for organization. Of course, these recognition events are not required for the imposed proximity of cells in architecturally localized organ systems. This layer of complexity must then be superimposed on the central function of the immune system, self–nonself-recognition and appropriate responses to nonself-encounters. These functions are largely executed by the major element of the immune system, the lymphocyte.

To fulfill its central function of self–nonself-recognition, each lymphocyte bears on its surface antigen receptors with identical binding sites. Therefore, the antigen-receptor specificity of antigen recognition on a

MARK SIEGELMAN and IRVING L. WEISSMAN • Laboratory of Experimental Oncology, Department of Pathology, Stanford University School of Medicine, Stanford, California 94305.

single lymphocyte is constant, and that property is transmitted to the progeny of these cells identically; that is, the antigen receptor is clonally expressed. The tremendous diversity of immune responsiveness derives from the fact that each lymphocyte expresses one of a large repertoire of antigen receptors. The diversity of antigen receptors also ensures that lymphocytes can recognize and respond to many different antigens with the requisite specificity.[1] B lymphocytes produce as recceptors immunoglobulin, and each B lymphocyte and its subsequent lineage of cells will produce immunoglobulin of identical binding sites and specificity.[2] T lymphocytes generally express receptors with a special recognition contingency; they require recognition of cell surface major histocompatibility antigenic determinants in conjunction with foreign antigen to trigger an appropriate immune response (reviewed in ref. 3). The molecular mechanism of the co-recognition of major histocompatibility molecules and antigen is as yet unclear (reviewed in refs. 4 and 5). Because the universe of antigenic determinants is enormous, the repertoire of antigen receptors must likewise be extensive enough to meet the challenge of the external environment, and to do so specifically. Because receptors are expressed on lymphocytes prior to antigen exposure, and since diversity of recognition is so rich, a lymphocyte with any given antigen specificity is expressed at a very low frequency in the total population of lymphocytes. The constraints imposed by a physically unmoored, blood-borne immune system containing a particular antigen-reactive lymphocyte at very low frequency demands additional organization to ensure appropriate interaction with antigen, regardless of the antigen's portal of entry. The dynamism of the circulating lymphoid system is therefore relieved by scattered solid collections of lymphoid elements such as thymus, lymph nodes, Peyer's patches, and spleen, which together constitute the lymphoid organs. These lymphoid organs drain defined tissues or vascular spaces and are architecturally organized so that proper inductive microenvironments ensure appropriate reception and processing of antigen and subsequent differentiation and maturation of the immune response. Lymph nodes collect antigen in lymphatic fluids from the blood and virtually all intercellular spaces in the body, excepting eye and brain, which are immunologically "privileged" sites, and the gastrointestinal tract, which possesses its own complement of lymphoid aggregates including adenoids, tonsils, Peyer's patches, and appendix.

An initial and fundamental event required for an appropriate progression of the immune response resides at the interface between a lymphocyte's mobile circulating phase and its relatively sessile phase within a partiicular lymphoid organ. Perpetual percolation of lymphocytes through lymphoid organs efficiently arms each with the entire repertoire

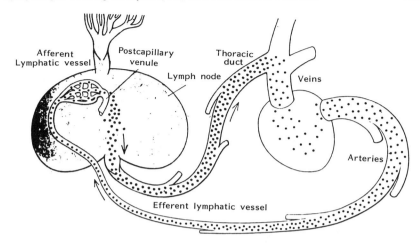

Figure 1. Lymphoid circulation. Small lymphocytes in the blood are pumped to the periphery by the heart, enter peripheral lymph nodes, pass through the capillary bed, and exit the blood circulation via specialized endothelia in postcapillary venules (see text). Maturation and differentiation of lymphocytes occur within the lymph node, and resultant lymphocytes are released into efferent lymphatics and returned to blood through the thoracic duct.

of antigen-receptive cells; lymphocytes recirculate from blood to lymphoid organs and back to blood, generally passing via efferent lymphatic vessels and their collecting ducts (Figure 1) (reviewed in ref. 6). In transit through lymphoid organs, lymphocytes pass regions of antigen-presenting cells to settle finally in specific B or T cell domains. The specific portal of entry of lymphocytes from bloodstream into peripheral lymphoid organ was identified as specialized postcapillary venules bearing unusually high-walled endothelia,[7,8] subsequently designated high endothelial venules (HEVs).[9] Recirculating lymphocytes, but not other blood-borne cells, specifically recognize, adhere to luminal walls, and migrate through this highly specialized endothelium into the lymphoid organ parenchyma proper (Figure 2). This migration of recirculating lymphocytes from bloodstream to particular lymphoid sites has been called *homing*, and the cell surface structures mediating recognition and adherence to lymphoid organ HEVs have been called *homing receptors*.[10] Therefore, lymphocyte homing appears to be regulated by the expression of complementary adhesion molecules on each of the two participants, the recirculating lymphocyte and the specialized lymphoid organ HEVs.[10,11] While the term homing has proved a useful convenience, which we shall apply in this chapter, it should be recognized that rigorously speaking the term is imprecise and occasionally misleading. The implication that a cell necessarily returns

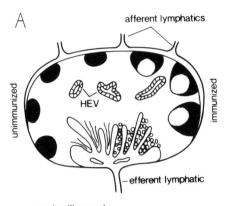

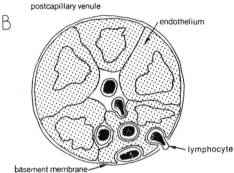

Figure 2. (A) Lymph node architecture. Germinal centers, in unimmunized or stimulated form, are shown at the periphery in the cortex. High endothelial venules (HEVs) are distributed around cortical medullary junctions. (B) Lymphocytes schematically shown in transit from the lumen of a postcapillary HEV into a lymph node.

to a site already encountered is not always true; homing from thymus to lymph node, for example, occurs without previous residence at the homing site. It should also not be inferred that in this instance homing implies a sensing of signals from a distance; no evidence exists that such is the case for lymphocyte homing.

2. *IN VITRO* MODELS OF LYMPHOCYTE HOMING

The first *in vitro* demonstration of a specific interaction between recirculating lymphocytes and lymphoid organ HEVs involved incubation of lymphocytes on thin tissue sections of lymphoid organs in the cold.[9] Lymphocytes were observed to adhere to HEVs, but not to non-HEV vessels in lymphoid organs and not to vessels in other organ tissues. Furthermore, an absolute correlation exists between lymphoid organ specificity of *in vitro* adhesion for a particular lymphoid cell line and its *in vivo* homing properties.[12] Development of the *in vitro* assay has allowed rapid and reproducible assessment of the adhesive component of the lymphocyte homing process. Such *in vitro* analysis with mouse tissues has shown genetic restrictions governing effective adhesiveness of lymphocytes to HEV; one polymorphism has been mapped to chromosome 7 in the mouse.[11]

2.1. Independent Homing Receptors

Migration of lymphocytes from blood to lymphoid organs has been shown to occur nonrandomly. Lymphocytes exiting peripheral lymph nodes, when collected and injected into the bloodstream, migrate preferentially to peripheral lymph node, whereas lymphocytes collected from gastrointestinal lymphatics light preferentially in Peyer's patches.[13-16] These predilections have been shown to be reflected in lymphocyte adherence to organ-specific HEVs *in vitro*.[17] The homing patterns are not derived by a lymphocyte passing through a particular lymphoid organ and microenvironment, thereby ensuring its return to that site. B lymphocytes of any origin exhibit preferential Peyer's patch homing when placed in the bloodstream; likewise, T lymphocytes retain their peripheral lymph node homing preferences regardless of derivation.[18] T cell subsets also have been shown to exhibit distinct homing patterns.[19] Both B and T lymphocytes enter lymphoid organs by common HEV.[20,21] These observations suggest that a homing phenotype is established independent of exposure to a homing site. More likely specificity of homing results from the particular complement of homing receptors already expressed by the lymphocyte. It is notable that B and T cell homing patterns are mirrored in the respective composition of lymphocyte subsets within lymphoid organs; Peyer's patches contain predominantly B cells, while the relatively few T cells are of the Lyt-2$^-$ (helper) phenotype subsets, and lymph nodes are heavily weighted toward T cell content, particularly of Lyt-2$^+$ (killer) phenotype. It is evident then that organ-specific HEVs and complementary receptors on lymphocytes that select a particular HEV provide a powerful force influencing distribution of lymphocytes within an organism. Presumably, such deployment of lymphocytes optimizes the effectiveness of an immune response in a particular lymphoid organ, site of infection, or microenvironment. For example, the preponderance of B cells and Peyer's patches may relate to the protective requirement of IgA responsiveness to organisms gaining entry through the gut.

Subsequent studies of lymphocyte populations and cell lines showed independent expression of homing to either lymph node HEV or Peyer's patch HEV, indicating that the two binding capacities are mediated by independent cell surface structures (Figure 3). A significant proportion

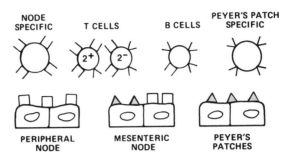

of murine lymphomas are unispecific with regard to Peyer's patch or peripheral lymph node HEV binding.[10,11] A population of lymphocytes embedded between gut epithelial cells appears to express exclusively Peyer's patch receptors.[22] There is also suggestive preliminary evidence that lymphomas with unispecific HEV binding activity disseminate as a solid tumor *in vivo* consistent with the distribution of that HEV type in an animal.[23] It is therefore conceivable that patterns of metastasis of lymphoid and possibly nonlymphoid tumors are influenced by such HEV homing receptors. In any case, the data together are compatible with a model involving complementary cell surface recognition structures mediating organ-specific lymphocyte HEV interactions.

2.2. Immunological Identification of Homing Receptors

The availability of clonal lymphoid cell lines with exclusive specificity for a single HEV type presented the opportunity to define directly the cell surface structures involved in the HEV recognition process. These cell lines can be used as immunogens to develop monoclonal antibodies specific for a particular homing receptor. One such monoclonal antibody, MEL-14, detects a cell surface determinant present only on those murine T and B cell lines that bind peripheral lymph node HEVs but not on those that bind either Peyer's patch HEV only or have no HEV binding activity at all.[10] Furthermore, presaturation of either normal or neoplastic lymphocyte populations, which normally bind peripheral lymph node HEV with MEL-14, specifically ablates binding to these HEVs. Finally, selection of clonal variants with altered expression of the MEL-14 determinant precisely parallels adherence activity to HEVs in the *in vitro* assay.[24] The concordance of these observations indicates that the actual receptor or a very closely coregulated structure is being examined. Studies in the rat have produced polyclonal as well as monoclonal reagents that specifically recognize both lymph node and Peyer's patch recognition structures.[25-29] Recently, monoclonal antibodies have been produced that detect cell surface glycoproteins likely representing the human homing receptor counterpart,[30] which appears to be weakly cross-reactive with MEL-14 (S. Jalkanen and W. M. Gallatin, unpublished results). The molecules thus far identified in mice and humans are glycoproteins of 90 ($\pm$ 10) kDa molecular weight. The development of antibodies identifying molecules associated with lymphocyte homing has been useful for characterizing the molecular species recognized and also for outlining the distribution of the putative receptor in various lymphoid organs and cell types during development.

2.3. Expression of the MEL-14 Epitope during Lymphocyte Activation

Recirculating naive immunocompetent lymphocytes are largely arrested in G_0 of the cell cycle.[31] Encounter of appropriate specific antigen or activation by nonspecific mitogenic stimuli leads to entry into the cell cycle, extensive proliferation, and differentiation along the pathway to either memory cells or to end-stage effector cells. In the immediate weeks following antigen exposure of a B cell, immunological memory to the inciting antigen is confined to the dividing activated cells,[32] while later this memory function reverts again to the recirculating sector of small G_0 lymphocytes.[33] In mice, upon activation of both B and T lymphocytes by antigen, *in vitro* or *in vivo* expression of the MEL-14 antigenic determinant is lost concomitantly with the ability to bind to lymphoid organ-specific HEVs.[34,35] Typically, *in vivo* antigen-activated lymphoblasts and T cell helpers agglomerate in specific architectural entities called germinal centers within lymphoid organs. These sites can be defined immunohistochemically by the absence of staining for MEL-14 epitopes and by the presence of bright staining with the galactose-specific lectin, peanut agglutinin (PNA).[34,36] It has been shown that the cells capable of adoptively transferring immunological memory soonest after antigenic activation are a PNA$^+$ MEL-14$^-$ subpopulation of lymphocytes, whereas later long-term memory is conferred exclusively by the PNA^{lo} MEL-14^{hi} recirculating lymphocytes.[32,37] The observation of an apparent inverse relation between glycosylation status of the cell surface (PNA staining) and expression of homing receptors is a provocative one; however, the relation is apparently more than simply inverse regulation since PNA^{hi} peripheral lymphocytes bind peripheral lymph node HEV quite well after neuraminidase treatment.[9] States of glycosylation or sialidation may rather be involved in anchoring cells in a particular microenvironment via lectins on stromal cells or elements appropriate to nurture a cell through a set of differentiative states. Indeed, we have proposed[38] that expression of molecules with specificties similar to the hepatic asialo-glycoprotein receptor[39] could easily be construed to explain retention of PNA^{hi} lymphocytes in germinal centers and thymic cortex.[34,40]

All murine antigen-specific helper and killer T cell clones thus far tested have not demonstrated binding to lymphoid organ HEVs, and they are devoid of detectable MEL-14 antigen.[35,41] These T cell clones, when placed in the bloodstream, fail to home specifically to lymphoid organs, but rather they appear to be trapped in spleen, liver, and lungs. The loss of homing properties in *in vitro*-derived cloned lines of known activity foils both *in vivo* analysis of their activity and immunotherapeutic avenues for these cells as well. For example, an efficient killer T cell clone, 1E4, with specificity for Abelson leukemia virus-induced lymphomas, is ineffective at clearing tumor in the peritoneum or in lymphatic metastases when administered intravenously. Yet, injection of the killer cells intraperitoneally clears both peritoneal cavity and regional lymph node metastases.[42] The nonexpression of homing receptors in cultured helper or killer lymphocyte lines may be species dependent, since at least some human T cell clones do appear to express the cell surface in a functional way.[43]

One might rationalize the disappearance of lymphocyte homing receptors on antigen-activated cells or in end-stage effector lymphocytes. Presumably, such lymphocytes are no longer subject to further influence in the lymphoid organs directing them through proliferative and differentiative states to their final end-stage state. Therefore, diversion of these mature cells back through lymphoid sites via lymphocyte homing receptors would be unnecessary and counterproductive. A more pertinent receptor, guiding localization for these lymphocytes, may be the antigen receptor, since it is the extralymphoidal sites harboring antigen where the activated lymphocyte must eventually light. Because entry of lymphocytes into extralymphoidal sites must occur via the blood vascular system, there may be separate homing pathways for end-stage or activated lymphocytes at inflammatory sites. These pathways could involve the entire panoply of inflammatory cells and mediators to ensure containment, neutralization, and eventual clearance of the inciting foreign presence. We can conceive of at least two classes of lymphocyte surface receptors involved in local endothelial cell recognition: antigen receptors and homing receptors specific for specialized endothelia. For the former to be operative, local endothelia must be capable of antigen presentation, perhaps following antigen processing and degradation elsewhere. For the latter mechanism of homing, local nonlymphoid endothelia must contain ligands allowing specific lymphocyte recognition; a case in point is the homing of lymphocytes to the skin in a variety of normal and pathological circumstances (reviewed in ref. 44). We have proposed[38] that these sites contain specialized endothelia that can be recognized by complementary homing receptors on a variety of cell types that require initial entry to the site, whether they are antigen-specific lymphocytes, nonspecific lymphocytes, or accessory cells. There may then be generic, inflammatory-site endothelial ligands recognized by common lymphocyte homing receptors on activated lymphocytes. Alternatively, there may be requirements for particular subsets of activated lymphocytes dictated by either the nature of the insult or the particular tissue affected. The latter contingency would require a large diversity of specific lymphocytes expressing homing receptors, and this situation could lead to specific diagnostic and therapeutic interventions where inappropriate (autoimmune) or undesired (transplant rejection) tissue-specific inflammatory or chronic immune responses persist. If such organism-wide tissue-specific lymphocyte homing receptors exist, it is conceivable that inappropriate expression in tumors, regardless of tissue origin, could contribute to metastatic patterns as well.

2.4. Regulation of Lymphocyte Homing Receptors in Early T and B Cell Development

Cells giving rise to hematolymphoid lineages initially reside in the yolk sac, and later they become invested in the fetal liver, spleen, and finally bone marrow in the adult.[45–47] In mammals, bone marrow appears to be the exclusive province of precursors that progress along the B lymphocyte pathway,[48] whereas the maturation of T lymphocytes occurs in the thymus (reviewed in ref. 49). Through a poorly understood process, lymphocytes emerge from their respective maturational microenvironments as virgin antigen-specific immunocompetent lymphocytes, and before they can perform their function, they must also acquire the ability to enter secondary lymphoid organs where antigen is collected and presented to them. It can then be gathered that the following properties have been acquired by the time lymphocytes are released to the bloodstream: (1) rearrangements of antigen receptor genes is completed to produce functional cell surface antigen ($\pm$ MHC) receptors; (2) expression of additional cell surface phenotypic markers correlated with functional cell subsets has occurred; (3) the ability to execute endowed function upon specific antigen stimulation exists; and (4) cell surface properties required for release from maturational microenvironments must be expressed while lymphocyte homing receptors come into play to allow localization in peripheral lymphoid organs. To characterize in detail the differentiative sequence of events, a description of the distribution and temporal course of expression of lymphocytic cell surface markers has been useful. In particular, MEL-14 antibody against the lymphocyte homing receptor has proved a useful marker in tracing the developmental fate of this antigen within the thymus.

Evidence suggests that a relatively infrequent (1–4%) population of thymic MEL-14^{hi} contains most, if not all, immunocompetent thymocyte cytotoxic T lymphocyte precursors.[50] The positive cells, which are restricted to the thymic cortex, constitute a major fraction of the glucocorticoid resistant cells.[51] The peak of diurnal glucocorticoid production is known to reach thymocytolytic concentrations *in vivo*.[50] Therefore, an argument can be made that the huge (more than 99%) attrition of thymocytes may result from glucocorticoid sensitivity and that the relative corticosteroid resistant population, which is highly enriched in MEL-14 expression, consists of successful maturants destined for survival. In addition to thymic emigrants, MEL-14^{hi} cells also include a population of dividing (apparently self-renewing) cells in the thymic subcapsular outer cortex.[53] It is likely that MEL-14^{hi} marks a differentiative state indicative of successful completion of a phase of maturation, including productive rearrangements and expression of antigen-specific T cell receptor genes.[54]

3. IS UBIQUITIN THE MEL-14 EPITOPE?

A cDNA library was constructed from mRNA isolated from the B cell lymphoma 38C13, the cell line used for immunization to produce the MEL-14 monoclonal antibody. The cDNA species existed as gene fusions to bacteriophage λ cDNA-lacZ such that expressed β-galactosidase–cDNA fusion proteins could be detected by antibody screening.[55] Screening with MEL-14 antibody resulted in detection of three independent cDNA clones, each of which encoded the well-characterized polypeptide ubiquitin (Ub).[56] Portions of two clones examined in detail encoded a protein identical to human Ub. Ub coding in one clone begins with the amino terminal methionine residue of Ub 4 amino acids from the in-frame fusion site with β-galactosidase and continues through the entire 76 amino acids of Ub, before diverging from the Ub sequence. In the other clone, an intact precise Ub monomer coding portion is preceded on the amino terminal side by 5 amino acids derived from the final carboxy-terminal residues of Ub and is flanked on the carboxy-terminal side with 25 amino acids of sequence, the first 15 of which are identical to the amino terminus of Ub; the remainder diverges from, but appears related to, that sequence. Thus, this clone represented a tandem head-to-tail Ub repeat cDNA with amino terminal Ub sequence adjacent to carboxy-terminal Ub sequences. Nonetheless, the striking feature common to these clones is the exact Ub coding unit, suggesting that the MEL-14 determinant resides within this protein sequence. These results suggested that Ub might in fact exist as all or a portion of this cell surface interactive molecule.

Ub is a small 8451 Da polypeptide first isolated from bovine thymus.[57] Primary structural analysis of various Ubs and their encoding genes have revealed remarkable evolutionary conservation, with virtually identical amino acid sequence existing between organisms as distant evolutionarily as insects and humans.[58-61] Such rigid conservation bespeaks a fundamental role for Ub in cellular function. Independent investigations of the protein structure of the nucleosomal protein A24 revealed Ub to be one moiety of a branched-chain complex with two amino termini and one carboxy terminus.[62] The linkage is a covalent bond between the carboxyl group of the carboxy-terminal glycine of Ub and the ε-amino group of lysine 119 of histone H2A. Subsequently, covalent conjugation of Ub to cytoplasmic proteins was demonstrated, further generalizing the principle of covalent conjugation of Ub to other proteins.[63,64] Functional roles attributed to Ub have included participation in regulation of intracellular protein degradation,[65-67] gene transcription and organization of chromatin structure,[68-70] and mitosis.[71] Whatever its requirement for participation in these processes, an absolute condition for the presence of an operational Ub conjugation system has been supported by genetic evidence.[69] The possibility that MEL-14 recognizes a Ub dependent determinant on the homing receptor protein would establish the presence of Ub or ubiquitinated proteins at the cell surface and would thereby extend the locus of this highly conserved peptide to this third major cellular compartment.

3.1. Purification of the Cell Surface Glycoprotein gp90^{MEL-14}

The cell line utilized for these studies was EL-4/MEL-14^{hi}, a variant of the continuous T cell lymphoma cell line EL-4, selected by fluorescence flow cytometry for high-level expression of the MEL-14 antigen, a property that cosegregates with the capacity to bind peripheral lymph node HEVs.[24] The molecular species specifically recognized by MEL-14 antibody was routinely isolated by immune complex formation between

MEL-14, an IgG_{2a} rat monoclonal antibody,[10] and affinity-purified goat anti-rat IgG antibody.

To isolate the MEL-14 reactive antigen directly from whole cell lysates labeled internally with tritiated amino acids, $[^{35}S]$methionine, or $[^{35}S]$cysteine, we interposed a lentil-lectin enrichment step before immunoprecipitation, thereby removing 90–95% of total lysate cpm.[72] Typical SDS–PAGE gel profiles of a MEL-14 antibody immunoprecipitation from a lectin-adherent pool of EL-4/MEL-14^{hi} labeled metabolically gave a single specific species of molecular weight 94,000, identical in size to that precipitated from cell surface iodinated lysates.[72] This species, gp90^{MEL-14}, was routinely eluted from appropriate gel fractions of electrophoresed immunoprecipitates from cells labeled with a variety of tritiated amino acids and $[^{35}S]$methionine.

3.2. Amino Acid Sequence Analysis of gp90^{MEL-14}

Approximately $2–3 \times 10^8$ EL-4/MEL-14^{hi} cells were metabolically labeled with single tritiated amino acids. In a typical experiment 20,000–50,000 cpm of affinity- and gel-purified material were subjected to automated, amino-terminal Edman degradation.[73] Anilino-thiazolinone derivates at each step were placed in scintillation fluid and counted directly. Positions containing radioactivity significantly above background indicated the presence of a particular tritiated amino acid residue at that position within a polypeptide chain. Amino acid sequence determinations of metabolically labeled gp90^{MEL-14} indicated the presence of two types of amino termini, easily distinguishable by their relative initial yields. A demonstration of this is given in Figure 4 for the purified MEL-14 antigen labeled with tritiated lysine. The predominant sequence identified is reflected in positions 8, 17, 20, and 32. A reproducible secondary sequence of lower specific activity is also present at positions 6, 11, 27, and 29. The secondary sequence corresponds precisely with the amino-terminal sequence of Ub.[58] The pattern of a major and minor amino-terminal sequence in absolute yields and relative proportions similar to that illustrated in Figure 4 was obtained with all amino acids labeled. A summary of the data for this sequence is given in Figure 5 in comparison to the known amino-terminal sequence of Ub.

It can be recalled that selection of cDNA clones in the λgt11 expression system suggested that MEL-14 recognizes Ub-associated determinants.[56] It was therefore important to demonstrate that the relatively minor Ub sequence was covalently associated with a specific core polypeptide and not representative of a general background of Ub or of ubiquitinated proteins precipitated by MEL-14 and distributed over the entire

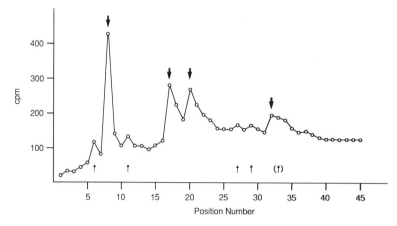

Figure 4. Amino-terminal automated Edman degradation of gp90^{MEL-14} intrinsically labeled with tritiated lysine.[72] A plot of position number against total number of counts per minute at that position is given. Downward arrows indicate the major high-specific-activity assignments. Upward errors indicate lower-specific-activity assignments. Upward arrow in parentheses indicates a tentative assignment.

SDS gel from which the specific MEL-14 species was isolated. This possibility was excluded by dividing the remainder of preparative gels into pools that were subjected to amino-terminal amino acid analysis. This analysis served to establish two points: (1) there were no detectable sequences corresponding to the Ub amino terminus in the other portions of the gels examined, and (2) for the pools just neighboring MEL-14, this implied that the relatively minor sequence of Ub in the MEL-14 specific species could not be explained by contamination with neighboring molecules containing Ub as a major sequence. Thus, by two such divergent approaches as amino acid sequence analysis and filter paper immunoassay to select the cDNA clones,[56] evidence indicates that Ub is indeed a component of a cell surface protein associated with lymphocyte homing.

```
          5         10        15        20        25        30        35
M Q I  F V K T L T G K T I T L E V E P S D T I E N V K A K I Q D K E G
M - I  F V K T L T - K T I T L - V - - - - (T) I - - - K - K - - (K) - -
```

Figure 5. Compilation of amino-terminal amino acid sequence assignments for the low-specific-activity component in gp90^{MEL-14}.[72] The sequence is compared to the previously reported sequence[58] of the amino terminus of Ub given in the top line. The second line represents positions determined from independent single amino acid labels. A dash indicates that no assignment at that position has been made. The parentheses indicate tentative assignments.

```
        5         10        15        20        25        30
 - T - H - - - K - M - - - - - - K F - K - - - - - - V (V) I - L K - -
```

Figure 6. Compilation of amino-terminal amino acid sequence assignments for the major high-yield sequence in gp90^{MEL-14}.[72] Determinations were made from independent single amino acid labels. Dashes indicate that no assignment at the position has been made. The parentheses at position 28 indicate a tentative assignment.

A summary of the second (major) sequence obtained is given in Figure 6. The repetitive yield of lysine calculated from the two peak heights at position 8 and 17 (92%) compares well with the repetitive yield calculated, subtracting background, for the lysines in the Ub chain at positions 6 and 11. This observation is consistent with the interpretation that the major sequence like the Ub sequence is derived from a single polypeptide chain. In thorough computer searches of data files, the sequence appears to be unique, although limited information prohibits adequate searches for homologies. It should be noted that the relatively lysine-rich amino terminus should be regarded as a minimal estimate of the number of lysine positions. Ub is known to conjugate to proteins in an isopeptide bond between its carboxy-terminal glycine and ε-amino groups of lysines.[58] Should such a conjugated lysine exist in the amino-terminal 35 residues of the core polypeptide, it is likely that after Edman degradation a 77 residue Ub polypeptide (or Ub multimer) attached to this lysine would not be extracted from the filter and would remain undetected by this analysis. The finding that Ub may be synthesized as a poly-Ub transcriptional unit in head-to-tail array[56,74,77] raises the issue of whether Ub exists as a concatamer in association with proteins as well. This has been suggested as one explanation for high molecular weight species of Ub–lysozyme conjugates.[78,79]

This issue was addressed by cyanogen bromide (CNBr) cleavage of filter-bound gp90^{MEL-14}.[72] Glass fiber filters to which protein samples are applied for amino acid sequence analysis in gas phase sequencers can be subjected to CNBr cleavage reactions even after amino-terminal sequence analysis of the protein on the filter has been performed.[80] This allows one to assess the presence of internal methionine residues by the generation of additional amino termini. After CNBr cleavage, Edman degradation is again performed. This CNBr maneuver was conducted on purified MEL-14 antigen labeled with single tritiated amino acids following 30 Edman degradation cycles. For example, a shift by one in the valine from position 5 in the native form to position 4 in digested material was noted.[72] Ub contains a single methionine in its sequence, located at position 1. Amino-terminal analysis of CNBr digested poly-Ub would result in a shift in a sequence by one position as was seen for this tritiated valine run. Single

Ub subunits would generate no additional methionines following this maneuver. A similar shift by one on CNBr digestion was seen for the first Ub lysine and threonine positions on labeling separately with these tritiated amino acids as well. Since the amino-terminal Ub sequence through the first 30 positions is presumably absent, a second, shifted sequence after CNBr digestion suggests that a Ub sequence exists internally within this complex, consistent with a head-to-tail arrangement. Alternatively, it is possible that one or more Ub monomers are present and that a portion of the Ub amino termini, blocked to Edman degradation, are released for Edman degradation after CNBr digestion. Isolation of purified CNBr-derived fragments will be required to resolve the question.

3.3. gp90^{MEL-14}—A Ubiquitinated Cell Surface Protein

A precedent for Ub–protein conjugates was set by the nuclear protein A24, which was sequenced in its entirety, including a detailed analysis of the Ub–H2A junction.[62] Since isolation of gp90^{MEL-14} is performed after complete reduction and since Ub contains no cysteine residues, a disulfide protein linkage between the chains is not possible. We therefore expect that some other covalent bond exists linking Ub to the core polypeptide of this murine lymphocyte homing receptor.

Other explanations for the data should be considered. The two amino-terminal sequences could derive from completely separate and unassociated chains, a Ub sequence from a Ub protein polymer of approximately 10–12 sequential Ub subunits giving the appropriate molecular weight species. The other sequence would derive from a completely unrelated glycoprotein, possibly representing the "true" homing receptor, which fortuitously bears a determinant cross-reactive with, but unrelated to, Ub sequences. Both chains would be recognized and precipitated by the MEL-14 antibody. This alternative can be regarded as highly unlikely for several reasons: (1) The intrinsically labeled receptor migrates as a single major species on two-dimensional gel analysis, which is identical to the cell surface iodinated molecule. (2) Since Ub is not known to be glycosylated and contains no canonical sequences that would serve as an acceptor site for N-linked glycosylation, free or unmodified poly-Ub should be selected against in the lectin adherence step used to enrich for glycoproteins. In addition, if a Ub polymer were glycosylated in O-linked form, it would be expected to be insensitive to Endo F digestion, which is contrary to our findings.[72] (3) Ub contains a single methionine at position 1, so that complete CNBr digestion of a hypothetical 10–12 unit concatamer should result in a 10–12-fold increase in a Ub-associated sequence, and though such a sequence was identified,[72] this extent of pos-

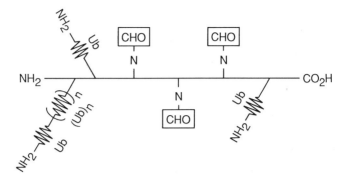

Figure 7. Hypothetical model of gp90$^{\text{MEL-14}}$, a putative lymphocyte homing receptor for peripheral lymph node high endothelial venules.[72] CHO, sites of capital N-linked glycosylation; Ub, ubiquitin subunits. The latter are represented in both monomeric (Ub), and in head-to-tail arrangement (Ub)$_n$. The disposition and number of Ub moieties is arbitrary, as is the placement of carbohydrates. Both carbohydrates and Ub chains are attached to a central core polypeptide.

sible concatamerization was not seen in our experiments. (4) The gp90$^{\text{MEL-14}}$ migrates in nonreducing SDS–PAGE more rapidly than under reducing conditions,[10] a property of proteins rich in internal disulfide bonds; Ub contains no cysteines. Therefore, we favor the Ub–core polypeptide relation indicated in Figure 7.

Other details regarding the relation of Ub to the core polypeptide remain to be ascertained. From our available data, we cannot determine the number of Ub units associated with the receptor, since the stoichiometry in the yield in counts per minute in the two polypeptide chains is unlikely to reflect their molar ratios (see below). Experiments using a purified appropriate isopeptidase, when available, to strip Ub from the core polypeptide should allow us to determine the molecular weight of the core peptide and the number of Ub subunits present. Another unresolved issue is whether Ub exists in linear arrangements of more than 1 Ub unit in association with this receptor (Figure 7). Evidence, discussed above, from sequence analysis of CNBr digests of the receptor molecule on filters that have already been subjected to 30 cycles of amino-terminal amino acid degradation,[72] suggests that such head-to-tail organization might exist in this system.

It has been noted that, by amino acid sequence analysis, the yield in counts per minute for the major polypeptide is quite different from the counts in the Ub sequence, which generally represents 10–20% of the yield found in the putative core polypeptide. Since we have reason to believe from λgt11 expressed cDNA clones[57] that the MEL-14 determinant is Ub dependent, we expect that MEL-14 may only recognize a ubiquitinated form of the receptor. Therefore, one expects at least a 1:1 molar ratio of Ub/core protein, assuming at least 1 Ub per receptor complex. As already outlined, we have ruled out the trivial explanation that MEL-14 immunoprecipitates contain a general background of ubiquitinated proteins. We also excluded the alternative possibility that after immunoprecipitation Ub–core polypeptide conjugates spontaneously dissociate. This follows from amino acid sequence analysis of pools containing the smallest species on SDS polyacrylamide gels expected to contain Ub; no trace of such a Ub sequence was found. We cannot formally exclude the possibility that the extracellular domain of gp90^{MEL-14} bears an antigenic determinant only cross-reactive with Ub and that this region is functionally active. The Ub sequence would then be obtained from Ub molecules conjugated to intracellular domains of a portion of the homing receptors to give the yields found in these studies.

Another explanation for the relatively low yield of Ub sequence is that the amino terminus of Ub chains are preferentially blocked to Edman degradation, either as a consequence of cell culture or during protein isolation. We regard as more likely an alternative that takes into consideration the intracellular synthetic properties that might be surmised about Ub. Ub is likely to comprise a large intracellular pool for conjugation to a large number of proteins destined for various cellular compartments. Based on evidence of its refractoriness to a battery of proteolytic digestions in its native state,[57] and requirement for maneuvers such as maleation before yielding to enzymatic digestion for peptide isolation, in addition to NMR studies[81] and recent x-ray crystallography[82] indicating that Ub assumes a highly globular compact stable conformation, one might conclude that Ub is relatively resistant to degradation. Since Ub appears destined for all cellular compartments, different pathways and sequestrations of Ub, depending on its ultimate destination, may also modify its longevity. It follows that, by intrinsic labeling, it might be particularly difficult to pulse efficiently a given tritiated amino acid through a large pool of relatively long-lived Ub over the 4–5 hr period of labeling, compared to a molecule of faster turnover, as we would suggest the core polypeptide may be. Therefore, the differential yields in the two polypeptide amino termini may simply reflect differential effective turnover rates.

3.4. N-Linked Glycosylation of the MEL-14 Antigen

Characterization by SDS gels of peripheral lymph node lymphocyte homing receptor isolated from cell surface iodinated or intrinsically labeled preparations reveals a molecular weight range for the species of

85–95kDa. An additional isoelectric dimension reveals extensive micro-heterogeneity in a highly acidic molecule, pI 4-4.5,[72,83] suggesting a highly glycosylated molecule.

The previous discussion has demonstrated one potential source of heterogeneity in this molecule to be the degree of ubiquitination. Whether this form of heterogeneity exists, however, requires a determination of the molecular weight of the core polypeptide, which may be obscured by the presence of numerous oligosaccharide side chains. To ascertain the relative contribution of oligosaccharide side chains, Endoglycosidase F (Endo F) digestions of isolated receptor were performed. Endo F digestions of isolated $gp90^{MEL-14}$ analyzed by SDS–PAGE revealed at least three digestion products, which all resolved to the major 55kDa species by 22 hr of digestion.[72] The stepwise reduction in molecular weight suggested a minimal estimate of three N-linked glycosylation sites. Digestion over shorter time intervals may also reveal additional intermediates. The sequence of the rigidly conserved Ub polypeptide contains no typical glycosylation triplets (Asn-X-Ser/Thr). Therefore, independent of sequence analysis, the homing receptor must contain another polypeptide to which both Ub and oligosaccharide side groups are linked. The evidence implies that the peptide representing the major amino acid sequence is this polypeptide core. Since O-linked groups are unaffected by Endo F under these conditions and since analysis was performed on immune complexes where access to some Endo F susceptible groups might be sterically hindered, it should be emphasized that the 55 kDa estimate represents an upper limit to the size of the core polypeptide. Subtraction of one Ub leaves about 46.5 kDa for the upper size of the core polypeptide. Since exhaustive digestion resulted in a single species, and not in two or more products separated in size by 8500 Da (the unit size of Ub), it is unlikely that this core polypeptide is ubiquitinated with a variable number of Ub subunits.

3.5. Ubiquitin as a Unique Conformational Determinant of $gp90^{MEL-14}$

The convergence of observations from two such disparate approaches as amino acid sequence analysis of a cell surface protein purified with a monoclonal antibody to a lymphocyte homing receptor[72] and the isolation of antibody-reactive expressed cDNA clones[56] provides compelling evidence that this antibody recognizes a Ub dependent determinant. However, the specificity of this antibody by cell and tissue section staining and by functional assays[12,18,34,41] raises a possible paradox. How can a determinant on a polypeptide as prevalent as Ub be so selectively expressed and what might that determinant be? Crystallographic studies of Ub reveal a tight globular conformation,[82] and native Ub monomers may therefore present relatively few epitopes at the solvent surface, which would serve as immunogens. Furthermore, since the amino acid sequence of Ub is highly conserved, unless immunological tolerance is broken, we expect only a very restricted number of Ub-associated determinants to be immunogenic. The most logical site for a relatively unique Ub determinant to be generated would be at its carboxy-terminal end, where conjugation to other proteins is effected. Evidence attesting to the relative availability of the carboxy terminus derives from x-ray crystallography of Ub, demonstrating that the carboxy-terminal amino acids extend away from the otherwise globular structure and that this portion of the molecule appears particularly free of constraints on its motion.[82] This deduction was tested directly using a radioimmunoassay detecting binding to intact Ub and several of its isolated enzymatically derived peptides.[84] Two of the independently derived anti-human Ub monoclonal antibodies 42D8 and 12H11 bind to intact Ub, but not to amino- and carboxy-terminal peptides.[84] MEL-14 binds specifically to the carboxy-terminal 13 residue peptide, but not to intact Ub, unless previously denatured in SDS.[56] Additionally, the avidity of binding to the carboxy-terminal peptides appeared relatively low compared to antibodies made to intact Ub and recognizing specific peptides. This further suggests that MEL-14 recognizes the carboxy terminus of Ub only as a portion of the entire MEL-14 determinant.

MEL-14 therefore likely recognizes a Ub–core polypeptide conjugate-specific determinant, or a conformational alteration induced by Ub conjugation to the core polypeptide distant from the Ub moiety. It should be noted that the MEL-14 determinant was presented only rarely in the set of all β-galactosidase–Ub fusion proteins.[56] In these experiments only three cDNAs were detected by MEL-14, a frequency of approximately 10^{-5}. Screening the library with a nucleic acid probe encoding Ub indicated the frequency of Ub-containing clones to be 1:200, one-sixth of which can be expected statistically to be productive in-frame fusions with β-galactosidase.[56] The expression of the MEL-14 determinant is indeed rare among Ub cDNA clones. We have postulated that the particular Ub-containing cDNA clones selected by MEL-14 produce proteins with amino acid sequences carboxy terminal to the Ub sequence that mimics the Ub–receptor conjugate determinant, or that these sequences induce conformational changes in Ub similar to those in the receptor complex. An alternate explanation is that the stringency and thermodynamics of antibody binding under the conditions of nitrocellulose filter immunoassays are quite different from complex formation in solution or in cell surface staining. The stringency may be low enough for the former that the contribution of Ub to the antigenic determinant is sufficient to select these clones, while the other assays require intact Ub–receptor conjugate. In any case, these observations are provocative and suggest that the system may prove informative at a fundamental level about the nature of what constitutes an antigenic determinant.

4. A MODEL FOR THE HEV HOMING RECEPTOR

We have reviewed initial amino-terminal sequence analysis and structural characterization of $gp90^{MEL-14}$, a putative lymphocyte homing receptor, which we conclude undergoes a complex series of post-translation modifications.[72] Our evidence suggests the receptor exists as a branched-chain polypeptide consisting of a core polypeptide in covalent linkage to Ub and N-linked carbohydrate as well. This represented, to our knowledge, the first description of a ubiquitinated cell surface protein. A schematic model for the overall structure of the murine lymphocyte homing receptor for peripheral lymph nodes is given in Figure 7.[72] A core protein of maximum size 46.5 kDa is modified by a minimum of three N-linked glycosylations, constituting approximately 45% of the mass of this molecule. This is a lower estimate of the extent of glycosylation, since the presence of O-linked oligosaccharides is not addressed by these studies. The second major and rather unexpected structural feature is the covalent association of Ub with the core polypeptide, inferred from the identification of a clear Ub amino-terminal amino acid sequence present in addition to that of the putative core polypeptide. We have only begun to characterize the polypeptide that we suggest is the core polypeptide of this lymphocyte homing receptor. Although the size of this core polypeptide is estimated at a maximum of 46.5 kDa, a single *di*ubiquitin, if present, would lower that estmate to 38 kDa.

A notable feature about the amino-terminal sequence of the core protein is its enrichment in lysine, with four such residues in the first 35 amino acids. This is of interest for several reasons. Isopeptide bonds between Ub and other proteins classically occur at the ε-amino group of lysines, so it is curious that, if representative, abundant opportunity for such bonds might exist at the amino terminus of the presumed core polypeptide, where interactiion with the cell surface of HEV may be likely to occur. Moreover, it has been shown in rats[9] that HEV binding lymphocytes treated with mild trypsinization lose their capacity to bind to HEV and are believed to lose an amino-terminal fragment of the homing receptor molecule. Since MEL-14 completely inhibits peripheral lymph node HEV binding and appears to recognize a Ub-dependent cell surface determinant, the evidence together suggests that functional HEV binding may require an amino-terminal trypsin-sensitive ubiquitinated peptide. Finally, a relatively positively charged amino terminus would be expected to help stabilize interaction with a generally negatively charged cell surface of an HEV. Complete sequence information of the core polypeptide will likely require a full-length cDNA clone whose sequence encodes the known amino acid positions.

5. ADDITIONAL UBIQUITINATED CELL SURFACE PRODUCTS

When it became clear that one cell surface molecule, $gp90^{MEL-14}$, was associated with Ub, it was of interest to test whether other cell surface species are conjugated to Ub. Anti-human Ub mouse monoclonal antibodies have been developed and characterized.[84] The specificities of these monoclonal antibodies have been thoroughly assessed by solid phase immunoassay; they generally show relatively poor binding to free Ub monomers in solution. Immunofluorescent staining and FACS analysis of EL-4/MEL-14^{hi}, the cell line with high-level expression of $gp90^{MEL-14}$ and the line used for the protein studies described above, revealed unequivocal staining above background with two of the ten anti-Ub monoclonal antibodies.[56] However, the antibodies stained MEL-14 negative cell lines as well, indicating species other than $gp90^{MEL-14}$ bear Ub epitopes on cell surfaces. The inability to demonstrate staining with the other Ub antibodies was likely due to low affinity, the inaccessibility of the epitope, or possibly more significantly their inability to recognize particular conformations of Ub in association with other proteins. Precipitations of lymphoid cell lysates after cell surface iodination were also performed with these antibodies.[72] The results revealed a group of specifically precipitated cell surface species, one of which corresponds to the molecular weight of MEL-14. Two larger species have mobilities of 150 and 125 kDa. These experiments did not formally exclude the possibility that recognition of species by anti-Ub monoclonals is a result of fortuitous cross-reactions with determinants unrelated to Ub. However, since at least two of the specific bands were precipitated by two monoclonals recognizing independent Ub determinants,[72] fortuitous cross-reactions are less likely. By this analysis, it therefore appears that ubiquitination of cell surface molecules may be not an exclusive feature of a particular lymphocyte homing receptor but a more general phenomenon. It should be pointed out that cell surface transmembrane proteins could be ubiquitinated either on extracellular or cytoplasmic domains; staining of viable cells will detect only the former, while immunoprecipitation of solubilized molecules will isolate either form.

6. POSSIBLE MECHANISMS FOR UBIQUITINATION OF CELL SURFACE PROTEINS

The studies reviewed here establish a precedent for the existence of a cell surface ubiquitinated protein, a putative lymphocyte homing receptor. We presented additional evidence supporting that other cell surface proteins may also be ubiquitinated.[72] Recently, Yarden et $al.$[85] have concluded that another cell surface molecule, the platelet-derived growth factor receptor, is also covalently bound to Ub.

Assignments of a particular role for Ub on cell surfaces can only be speculated on at this point. The remarkable degree of conservation of Ub throughout evolution clearly suggests a fundamental cellular function. One role, for which a body of evidence already exists, proposes that it is required in pathways of intracellular protein degradation.[63–66] It is conceivable that ubiquitination of cell surface proteins is incidental and simply serves as a tag for surface proteins that are destined for degradation. Indeed, it may be that a surface molecule such as a lymphocyte homing receptor has a special requirement for rapid internalization and degradation by either high endothelial cells or the lymphocytes bearing it so that entry into the secondary lymphoid organ is directional and not easily reversible. Moreover, it has been shown that in the rat[9] the adhesive properties have particular proteolytic sensitivities. Ubiquitination may then ensure rapid turnover or heighten proteolytic sensitivities. This could be a general property of developmental cell surface molecules that target cells to particular sites.

It has been shown that some sugars appear specifically to inhibit binding of lymphocytes to peripheral lymph node HEV, in particular free mannose-6-phosphate or the polymer 6-phosphomannosyl-conjugated mannan.[86] By inference, then, the murine lymphocyte homing receptor for peripheral lymph nodes might be considered to be a lectin. This binding has also been shown to be influenced by the extent of sialydation.[87] Since the Ub moiety of this complex also appears vital to HEV binding, the possibility suggests itself that Ub or the microenvironment around the ubiquitination site dictates the lectin activity. There is as yet no evidence addressing this point. It should be recalled, however, that Ub is predominantly a cytoplasmic protein. It is then of interest that several intracellular acid hydrolases also contain 6-phosphomannosyl residues and that a specific class of cytoplasmic proteins target these hydrolases to lysosomal sites in the cytoplasm.[88-92] We have suggested that perhaps pathways of ubiquitination provide specificity for directing intracellular protein trafficking as well.[38]

Consideration should also be given to a possible independent role for Ub at the cell surface. It can be recalled that MEL-14 antibody specifically and completely inhibits binding of lymphocytes to peripheral lymph node HEV and that, from our evidence, it appears to recognize an extracellular Ub-dependent determinant. Given the size of this receptor complex, it is likely that this class of cell surface molecules can be bound by antibody without interfering with functional HEV attachment, and it implies that the ubiquitinated region of this receptor, and possibly Ub itself, may be critical for binding. We may extrapolate this principle to suggest that Ub may subserve a vital role in cell–cell interaction and adhesion. It is possible that along with the degree of motion allowed the carboxy-terminal portion of Ub,[82] the number and size of Ub side chains, in combination with the potential for conjugation to a number of cell surface proteins, provide a sufficiently varied conformational repertoire at the site of conjugation to determine, to a great degree, receptor specificity. In general, a strong evolutionary pressure for the conservation of Ub primary structure may be its requirement to interact with large numbers of proteins around such sites of conjugation. Alternatively, the homing receptor backbone could provide specificity for a particular cell type, in this instance high endothelium, while Ub might provide additional stability to the adhesive interactions. X-ray crystallography indicates that β-pleated sheet comprises 37% of the Ub chain and appears concentrated along one outer aspect of the molecule.[82] We can speculate that external β-pleated regions may present a broad hydrophilic surface with which a complementary surface can interact. Such avid associations between two surfaces may be important for a lymphocyte to migrate through a vessel wall to a lymphoid organ. It would be of interest to examine whether specific cell–cell junctions are involved in the adhesion and/or in transport of lymphocytes through blood vessel endothelium. This mechanism could be a general requirement in many biological systems in which transmigration through vessel walls or other tissues is essential. Such processes could be imagined to occur with some frequency during development when vectorial mass migration of cells must occur to form distinct tissues or organs; for example, neural crest cells or germ cells have extended and specific migration pathways.

Whereas ubiquitination as a post-translational modification poses no special problems for nuclear and cytoplasmic proteins, ubiquitination of extracellular domains of cell surface proteins is more difficult to envision. If ubiquitination occurs cotranslationally on the cytoplasmic face of the rough endoplasmic reticulum (ER), transport of a branched-chain polypeptide to the lumen of the ER may present unique problems. Alternatively, it is more difficult to imagine the ubiquitination event to occur within the lumen of the ER or golgi, as Ub and so far all Ub transcripts do not encode a typical amino-terminal signal sequence, which we would expect as a requirement for independent entry to these compartments. The process of translocation of bulky charged macromolecules across a lipid bilayer is a complex one whose molecular details and requirements remain largely unresolved.[93] In particular, while the initiating events in signal sequence recognition are well characterized and explain initial anchoring to a membrane site, it is unclear whether the remaining nascent chain translocates spontaneously, is extruded to the luminal side by the ribosome, or whether a proteinaceous "tunnel" is formed to guide the chain across in an energy-dependent fashion.[94,95] If ubiquitination occurs on nascent chains at the cytoplasmic face of the rough ER, it may be necessary to invoke unique translocation mechanisms for the accommodation of the additional bulkiness contributed by the Ub domain. Yet, some evidence to the contrary exists. For example, it appears that a partially folded globin domain is likely presented to microsomal membranes during synthesis of a nascent fusion protein between globin and prolactin,[94] suggesting that presentation and translocation of three-dimensionally complex conformations do occur. If, however, the standard mechanism for translocation *is* overburdened by branched proteins, perhaps a new form of transport in a unique vesicle is involved.

A sequence of events might also be constructed that does not require a translocation of the receptor in branched-chain form at all. It has been suggested, though not yet demonstrated *in vivo*, that ligation of Ub to the amino-terminal α-amino groups of target proteins, to form linear Ub conjugates, is chemically feasible.[96] Such proteins have been artificially constructed as chimeric genes and expressed *in vivo*. One such gene encoded yeast Ub linked to β-galactosidase of *E. coli*.[97] Studies using derivatives of this gene varying at the amino-terminal residue of the β-galactosidase component resulted in the proposal of the "N-end rule." The astonishing finding indicated that the particular amino-terminal residue in large measure determines *in vivo* half-lives in a dramatic range of 20 hr to less than 3 min.[97] We now know that the amino terminus of the presumed core polypeptide is a tryptophan (M. Siegelman and I. L. Weissman, unpublished data), a relatively infrequent amino acid and one not addressed directly in these proteolytic studies. However, the destabilizing amino acids were noted to share the property of having relatively larger radii of gyration than stabilizing amino acids. Tryptophan would therefore likely be included in the destabilizing category. This would be consistent with observations of other compartmentalized proteins, for example, those destined for secretion, which also generally have destabilizing amino-terminal amino acids. Since amino-terminal residues can be critical for directing proteolytic events, perhaps the amino-terminal tryptophan of the core polypeptide is involved in diverting molecules away from digestive pathways and toward other compartmentalizations. If amino-terminal ligation of Ub does occur, we might hypothesize that a cotranslational event in the synthesis of gp90^{MEL-14} is the ligation of Ub or even poly-Ub to the α-amino group of the nascent core polypeptide, which would presumably include an amino-terminal signal sequence. It has been shown in *in vitro* models that signal sequences engineered to internal positions within a protein result in the translocation of the synthesized protein on both sides of the signal sequence into microsomal membrane vesicles.[94] Provided amino-terminal Ub domains assume conformations compatible with translocation, the entire receptor complex could then enter the ER in linear form. This would circumvent the need for passage of branched protein structures through a lipid bilayer, the energy requirement for which might be prohibitive. Once on the luminal side of the ER, the machinery for the remainder of ubiquitination events would have to be extant in that compartment. Ub or poly-Ub would then be cleaved and attached to appropriate branch sites. Alternatively, other specialized proteins could transport Ub to this compartment by a similar mechanism, thereby supplying a continual pool of Ub to this compartment.

It shall also be of great interest to determine whether isopeptide bond formation occurs only with lysines surrounded by a "canonical" sequence, and if so, with what frequency such sequences are ubiquitinated. Why only a subset of cell surface molecules is ubiquitinated is unclear. Resolution of this should reveal new insights into functions of ubiquitination and protein targeting mechanisms.

ACKNOWLEDGMENTS.   We are grateful to Janice Mason for preparation of the manuscript. This work was supported by USPHS grant AI 19512. Mark Siegelman was supported by NIH fellowship award IF32 A107304 01.

REFERENCES

1. Hood, L. E., Weissman, I. L., Wood, W. B., and Wilson, J. H., 1985, *Immunology*, Benjamin Cummings, Menlo Park, CA.
2. Joho, R., Nottenburg, C., Coffman, R. L., and Weissman, I. L., 1983, Immunoglobulin gene rearrangement and expression during lymphocyte development, in: *Current Topics*

in Developmental Biology, Vol. 18 (A. A. Moscona and A. Monroy, eds.), Academic Press, New York, pp. 16–58.

3. Doherty, P. C., Blanden, R. V., and Zinkernagel, R. M., 1976, Specificity of virus-immune effector T-cells for H-2K or H-2D compatible interactions: Implications or H-antigen diversity, *Transplant. Rev.* 29: 89–124.

4. Pernis, B., and Axel, R., 1985, A one and a half receptor model for MHC-restricted antigen recognition by T lymphocytes, *Cell* 41: 13–16.

5. Hood, L. E., Kronenberg, M., and Hunkapiller, T., 1985, T cell antigen receptors and the immunoglobulin supergene family, *Cell* 40: 225:229.

6. Ford, W. L., 1975, Lymphocyte migration and immune responses, *Prog. Allergy* 19: 1–59.

7. Pirro, A. F., 1954, Le vene postcapillari del parenchima nella tonsilla palatina umana, *Quaderni di Anatomia Pratica* X(1–2): 86–125.

8. Gowans, J. L., and Knight, E. J., 1964, The route of recirculation of lymphocytes in the rat, *Philos. Trans. R. Soc. (Lond.) B* 159: 257–282.

9. Stamper, H. B., Jr., and Woodruff, J. J., 1976, Lymphocyte homing into lymph nodes: *in vitro* demonstration of the selective affinity of recirculating lymphocytes for high-endothelial venules, *J. Exp. Med.* 144: 828–833.

10. Gallatin, W. M., Weissman, I. L., and Butcher, E. C., 1983, A cell-surface molecule involved in organ-specific homing of lymphocytes, *Nature* 304: 30–34.

11. Butcher, E. C., and Weissman, I. L., 1980, Cellular, genetic, and evolutionary aspects of lymphocyte interactions with high endothelial venules, *Ciba Foundation Symp.* 71: 265–286.

12. Butcher, E., Scollay, R. G., and Weissman, I. L., 1979, Lymphocyte adherence to high endothelial venules: Characterization of a modified *in vitro* assay, and examination of the binding of syngeneic and allogeneic lymphocyte populations, *J. Immunol.* 123: 1996–2003.

13. Griscelli, C., Vassali, P., and McCluskey, R. T., 1969, Distribution of large dividing lymph node cells in syngeneic recipient rats after intravenous injection, *J. Exp. Med.* 130: 1427–1451.

14. Guy-Grand, D., Griscelli, C., and Vassali, P., 1974, The gut-associated lymphoid system: Nature and properties of the large dividing cells, *Eur. J. Immunol.* 4: 435–443.

15. Scollay, R., Hopkins, J., and Hall, J., 1976, Possible role of surface Ig in non-random recirculation of small lymphocytes, *Nature* 260: 528–529.

16. Smith, M. E., Martin, A. F., and Ford, W. L., 1980, Migration of lymphoblasts in the rat. Preferential localization of DNA-synthesizing lymphocytes in particular lymph nodes and other sites, *Mongr. Allergy* 16: 203–232.

17. Butcher, E. C., Scollay, R. G., and Weissman, I. L., 1980, Organ specificity of lymphocyte migration: Mediation by highly selective lymphocyte interaction with organ-specific determinants on high endothelial venules, *Eur. J. Immunol.* 10: 556–561.

18. Stevens, S. K., Weissman, I. L., and Butcher, E. C., 1982, Differences in the migration of B and T lymphocytes: Organ-selective localization *in vivo* and the role of lymphocyte–endothelial cell recognition, *J. Immunol.* 128: 844–851.

19. Kraal, G., Weissman, I. L., and Butcher, E. C., 1983, Differences in *in vivo* distribution and homing of T cell subsets to mucosal vs. non-mucosal lymphoid organs, *J. Immunol.* 130:145–151.

20. Howard, J. C., Hunt, S. V., and Gowans, J. L., 1972, Identification of marrow-derived and thymus-derived small lymphocytes in the lymphoid tissue and thoracic duct lymph of normal rats, *J. Exp. Med.* 135: 200–219.

21. Gutman, G. A., and Weissman, I. L., 1975, Evidence that uridine incorporation is not a selective marker for mouse lymphocyte subclasses, *J. Immunol.* 115: 939–940.

22. Schmitz, M., Nunez, D., and Butcher, E. C., Selective recognition of mucosal endothelium by gut intraepithelial leukocytes, *Gastroenterology* (in press).
23. Jalkanen, S., Reichert, R. A., Gallatin, W. M., Bargatze, R. F., Weissman, I. L., and Butcher, E. C., 1986, Homing receptors and the control of lymphocyte migration, *Immunol. Rev.* **91:** 39–60.
24. Gallatin, W. M., and Weissman, I. W., in preparation.
25. Chin, Y.-H., Rasmusen, R., Cakiroglu, A. G., and Woodruff, J. J., 1984, Lymphocyte recognition of lymph node high endothelium. VI. Evidence of distinct structures mediating binding to high endothelial cells of lymph nodes and Peyer's patches. *J. Immunol.* **133:** 2961–2965.
26. Carey, G. D., Chin, Y.-H., and Woodruff, J. J., 1981, Lymphocyte recognition of lymph node high endothelium. III. Enhancement of a component of thoracic duct lymph, *J. Immunol.* **127:** 976–979.
27. Chin, Y.-H., Carey, G. D., and Woodruff, J. J., 1982, Lymphocyte recognition of lymph node high endothelium. IV. Cell surface structures mediating entry into lymph nodes, *J. Immunol.* **129:**1911–1915.
28. Chin, Y.-H., Carey, G. D., and Woodruff, J. J., 1983, Lymphocyte recognition of lymph node high endothelium. V. Isolation of adhesion molecules from lysates of rat lymphocytes, *J. Immunol.* **131:** 1368–1374.
29. Rasmussen, R. A., Chin, Y.-H., Woodruff, J. J., and Easton, T. G., 1985, Lymphocyte recognition of lymph node high endothelium. VII. Cell surface proteins involved in adhesion defined by monoclonal anti-$HEBF_{LN}$ (A.11) antibody, *J. Immunol.* **135:** 19–24.
30. Jalkanen, S. T., Bargatze, R. F., Herron, L. R., and Butcher, E. C., 1986, A lymphoid cell surface glycoprotein involved in endothelial cell recognition and lymphocyte homing in man, *Eur. J. Immunol.* **16:** 1195–1202.
31. Gowans, J. L., McGregor, D. D., and Cowen, D. M., 1962, Initiation of immune responses by small lymphocytes, *Nature* **196:** 651–655.
32. Coico, R. F., Bhogsal, B. S., and Thorbecke, G. J., 1983, Relationship of germinal centers in lymphoid tissue to immunologic memory. VI. Transfer of B cell memory with lymph node cells fractionated according to their receptors for peanut agglutinin, *J. Immunol.* **131:** 2254–2257.
33. Gowans, J. L., and Uhr, J. W., 1966, The carriage of immunological memory by small lymphocytes in the rat, *J. Exp. Med.* **124:** 1017–1030.
34. Reichert, R. A., Gallatin, W. M., Weissman, I. L., and Butcher, E. C., 1983, Germinal center B cells lack homing receptors necessary for normal lymphocyte recirculation, *J. Exp. Med.* **157:** 813–827.
35. Dailey, M. O., Gallatin, W. M., and Weissman, I. L., 1985, The *in vivo* behavior of T cell clones altered migration due to loss of the lymphocyte surface homing receptor, *J. Mol. Cell. Immunol.* **2:** 27–36.
36. Rose, M. L., Birbeck, M. S. C., Wallis, V., Forrester, J. A., and Davies, A. J. S., 1980, Peanut lectin binding properties of germinal centres in mouse lymphoid tissue, *Nature* **284:** 364–366.
37. Kraal, G., Weissman, I. L., and Butcher, E. C., in preparation.
38. Gallatin, M., St. John, T. P., Siegelman, M., Reichert, R., Butcher, E. C., and Weissman, I. L., 1986, Lymphocyte homing receptors, *Cell* **44:** 673–680.
39. Schwartz, A. L., Marshak-Rothstein, A., Rup, D., and Lodish, H. F., 1981, Identification and quantification of the rat hepatocyte asialoglycoprotein receptor, *Proc. Natl. Acad. Sci. U.S.A.* **78:** 3348–3352.
40. Weissman, I. L., Rouse, R. V., Kyewski, B. A., Lepault, F., Butcher, E. C., Kaplan,

H. S., and Scollay, R. G., 1982, Thymic lymphocyte maturation in the thymic microenvironment, in: *Behring Institute Mitteilungen, The Influence of the Thymus on the Generation of the T Cell Repertoire* (F. R. Seiler and H. G. Schwick, eds.), Die Medizinische Verlagsgesellschaft, Marburg, Germany, pp. 242–251.

41. Dailey, M. O., Fathman, C. G., Butcher, E., Pillemer, E., and Weissman, I. L., 1982, Abnormal migration of T lymphocyte clones, *J. Immunol.* **128:** 2134–2136.

42. Dailey, M. O., Pillemer, E., and Weissman, I. L., 1982, Protection against syngeneic lymphoma by a long-lived cytotoxic T-cell clone, *Proc. Natl. Acad. Sci. U.S.A.* **79:** 5384–5387.

43. Navarro, R. F., Jalkanen, S. T., Hsu, M., Soenderstrup-Hansen, O., Goronzy, J., Weyand, C., Fathman, C. G., Clayberger, C., Krensky, A. M., and Butcher, E. C., 1985, Human T cell clones express functional homing receptors required for normal lymphocyte trafficking, *J. Exp. Med.* **162:** 1075–1080.

44. Spangrude, G. J., Samlowski, W. E., and Daynes, R. A., 1986, Physical consideration in experimental photoimmunology, in: *Experimental and Clinical Photoimmunology* (R. A. Daynes and G. K. Krueger, eds.), CRC Press, Boca Raton, FL, pp. 3–29.

45. Ford, C. E., Micklem, H. S., Evans, E. P., Gray, J. G., and Ogden, D. A., 1966, The inflow of bone-marrow cells to the thymus. Studies with part body-irradiated mice injected with chromosome-marked bone marrow and subjected to antigenic stimulation, *Ann. N.Y. Acad. Sci.* **129:** 283–295.

46. Metcalf, D., and Moore, M., 1971, *Frontiers of Biology,* Vol. 24, North-Holland, Amsterdam.

47. Weissman, I., Papaioannou, V., and Gardner, R., 1978, Fetal hematopoietic origins of the adult hematolymphoid system, in: *Differentiation of Normal and Neoplastic Hematopoietic Cells* (B. Clarkson, P. A. Marks, and J. E. Tills, eds.), Cold Spring Harbor Laboratory, Cold Spring Harbor, NY, pp. 33–47.

48. Osmond, D. G., and Nossal, G. J. V., 1974, Differentiation of lymphocytes in mouse bone marrow. II. Kinetics of maturation and renewal of antiglobulin-binding cells studied by double labeling, *Cell Immunol.* **13:** 132–145.

49. Miller, J. F. A. P., 1979, Experimental thymology has come of age, *Thymus* **1:** 3–25.

50. Fink, P. J., Gallatin, W. M., Reichert, R. A., Butcher, E. C., and Weissman, I. L., 1985, Homing receptor-bearing thymocytes: An immunocompetent cortical subpopulation, *Nature* **313:** 233–235.

51. Reichert, R. A., Gallatin, W. M., Butcher, E. C., and Weissman, I. L., 1984, A homing receptor-bearing cortical thymocyte subset: Implications for thymus cell migration and the nature of cortisone-resistant thymocytes, *Cell* **38:** 89–99.

52. Weissman, I. L., and Levy, R., 1975, *In vitro* cortisone sensitivity of *in vivo* cortisone resistant thymocytes, *Israel J. Med. Sci.* **11:** 884–888.

53. Reichert, R. A., Jerabek, L., Gallatin, W. M., Butcher, E. C., and Weissman, I. L., 1986, Ontogeny of lymphocyte homing receptor expression in the mouse thymus, *J. Immunol.* **136:** 3535–3542.

54. Haars, R., Kronenberg, M., Gallatin, W. M., Weissman, I. L., Owen, R. L., and Hood, L., 1986, Rearrangement and expression of T cell antigen receptor and gamma genes during thymic development, *J. Exp. Med.* **164:** 1–24.

55. Young, R. A., and Davis, R. W., 1983, Efficient isolation of genes by using antibody probes, *Proc. Natl. Acad. Sci. U.S.A.* **80:** 1194–1198.

56. St. John, T., Gallatin, W. M., Siegelman, M., Smith, H. T., Fried, V. A., and Weissman, I. L., 1986, Expression cloning of a lymphocyte homing receptor cDNA: Ubiquitin is the reactive species, *Science* **231:** 845–850.

57. Goldstein, G., Scheid, M., Hammerling, U., Schlesinger, D. H., Niall, H. D., and Boyse,

E. A., 1975, Isolation of a polypeptide that has lymphocyte-differentiating properties and is probably represented universally in living cells, *Proc. Natl. Acad. Sci. U.S.A.* **72:** 11–15.

58. Schlesinger, D. H., Goldstein, G., and Niall, H. D., 1975, The complete amino acid sequence of ubiquitin, and adenylate cyclase stimulating polypeptide probably universal in living cells, *Biochemistry* **14:** 2214–2218.
59. Schlesinger, D. H., and Goldstein, G., 1975, Molecular conservation of 74 amino acid sequence of ubiquitin between cattle and man, *Nature* **25:** 423–424.
60. Watson, D. C., Levy, W. B., and Dixon, G. H., 1978, Free ubiquitin is a non-histone protein of trout testis chromatin, *Nature* **276:** 196–198.
61. Gavilanes, J. G., Gonzalez de Buitrago, G., Perez-Castells, R., and Rodriguez, R., 1982, Isolation, characterization, and amino acid sequence of a ubiquitin-like protein from insect eggs, *J. Biol. Chem.* **257:** 10267–10270.
62. Goldknopf, I. L., and Busch, H., 1977, Isopeptide linkage between nonhistone and histone 2A polypeptides of chromosomal conjugate-protein A24, *Proc. Natl. Acad. Sci. U.S.A.* **74:** 864–868.
63. Hershko, A., and Ciechanover, A., 1982, Mechanisms of intracellular protein breakdown, *Annu. Rev. Biochem.* **51:** 335–364.
64. Ciechanover, A., Finley, D., and Varshavsky, A., 1984, The ubiquitin-mediated proteolytic pathway and mechanisms of energy-dependent intracellular protein degradation, *J. Cell. Biochem.* **24:** 37–53.
65. Hershko, A., 1983, Ubiquitin: Roles in protein modification and breakdown, *Cell* **34:** 11–12.
66. Hough, R., and Rechsteiner, M., 1984, Effects of temperature on the degradation of proteins in rabbit reticulocyte lysates and after injection into HeLa cells, *Proc. Natl. Acad. Sci. U.S.A.* **81:** 90–94.
67. Finley, D., Ciechanover, A., and Varshavsky, A., 1984, Thermolability of ubiquitin-activating enzyme from the mammalian cell cycle mutant ts85, *Cell* **37:** 43–55.
68. Goldknopf, I. L., Wilson, G., Ballal, N. R., and Busch, H., 1980, Chromatin conjugate A24 is cleaved and ubiquitin is lost during chicken erythropoiesis, *J. Biol. Chem.* **255:** 10555–10558.
69. Finley, D., Ciechanover, A., and Varshavsky, A., 1984, Thermolability of ubiquitin-activating enzyme from the mammalian cell cycle mutant ts85, *Cell* **37:** 43–45.
70. Levinger, L., and Varshavsky, A., 1982, Selective arrangement of ubiquitinated and D1 protein-containing nucleosomes within the *Drosophila* genome, *Cell* **28:** 375–385.
71. Matsui, S. I., Seon, B. K., and Sandburg, A. A., 1979, Disappearance of a structural chromatin protein A24 in mitosis: Implications for molecular basis of chromatin condensation, *Proc. Natl. Acad. Sci. U.S.A.* **76:** 6386–6390.
72. Siegelman, M., Bond, M. W., Gallatin, W. M., St. John, T., Smith, H. T., Fried, V. A., and Weissman, I. L., 1986, Cell surface molecule associated with lymphocyte homing is a ubiquitinated branched-chain glycoprotein, *Science* **231:** 823–829.
73. Hewick, R. M., Hunkapiller, M. W., Hood, L. E., and Dreyer, W. J., 1981, A gas–liquid solid phase peptide and protein sequenator, *J. Biol. Chem.* **256:** 7990–7997.
74. Dworkin-Rastl, E., Shrutkowski, A., Dworkin, M. B., 1984, Multiple ubiquitin mRNAs during *Xenopus laevis* development contain tandem repeats of the 76 amino acid coding sequence, *Cell* **39:** 321–325.
75. Ozkaynak, E., Finley, D., and Varshavsky, A., 1984, The yeast ubiquitin gene: Head-to-tail repeats encoding a polyubiquitin precursor protein, *Nature* **312:** 663–666.
76. Bond, U., and Schlesinger, M. J., 1985, Ubiquitin is a heat shock protein in chicken embryo fibroblasts, *Mol. Cell Biol.* **5:**949–956.

77. Wibog, O., Pedersen, M. S., Wind, A., Berglund, L. E., Marcker, K. A., and Vuust, J., 1985, The human ubiquitin multigene family: Some genes contain multiple directly repeated ubiquitin coding sequences, *EMBO J.* **4:** 755–759.

78. Hershko, A., and Heller, H., 1985, Occurrence of a polyubiquitin structure in ubiquitin-protein conjugates, *Biochem. Biophys. Res. Commun.* **128:** 1079–1086.

79. Hough, R., and Rechsteiner, M., 1986, Ubiquitin–lysozyme conjugates. Purification and susceptibility to proteolysis, *J. Biol. Chem.* **261:** 2391–2399.

80. Urdal, D. L., Mochizuki, D., Conlon, P. J., March, C. J., Remerowski, M. L., Eisenman, J., Ramthun, C., and Gillis, S., 1984, Lymphokine purification by reversed-phase high-performance liquid chromatography, *J. Chromatogr.* **296:** 171–179.

81. Lenkinski, R. E., Chen, D. M., Glickson, J. D., and Goldstein, G., 1977, Nuclear magnetic resonance studies of the denaturation of ubiquitin, *Biochim. Biophys. Acta* **494:** 126–130.

82. Vijay-Kumar, S., Bugg, C. E., Wilkinson, K. D., and Cook, W. J., 1985, Three-dimensional structure of ubiquitin at 2.8 Å resolution, *Proc. Natl. Acad. Sci. U.S.A.* **82:** 3582–3585.

83. Jalkanen, S. T., and Butcher, E. C., in preparation.

84. Smith, H. T., Morrison, M., and Fried, V. A., in preparation.

85. Yarden, Y., Escobeda, J. A., Kuang, W.-J., Yang-Feng, T. L., Daniel, T. O., Tremble, P. M., Chen. E. Y., Ando, M. E., Harkins, R. N., Francke, U., Fried, V. A., Ullrich, A., and Williams, L. T., 1986, Structure of the receptor for platelet-derived growth factor helps define a family of closely related growth factor receptors, *Nature* **323:** 226–237.

86. Stoolman, L. M., Tenforde, T. S., and Rosen, S. D., 1984, Phosphomannosyl receptors may precipitate in the adhesive interaction between lymphocytes and high endothelial venules, *J. Cell. Biol.* **99:** 1535–1540.

87. Rosen, S. D., Singer, M. S., Yednock, T. A., and Stoolman, L. M., 1985, Involvement of sialic acid on endothelial cells in organ-specific lymphocyte recirculation, *Science* **228:** 1005–1007.

88. Sly, W. S., and Fischer, H. D., 1982, The phosphomannosyl recognition system for intracellular and intercellular transport of lysosomal enzymes, *J. Cell. Biochem.* **18:** 67–85.

89. Sahagian, G. G., Distler, J., and Jourdian, G. W., 1981, Characterization of a membrane-associated receptor from bovine liver that binds phosphomannosyl residues of bovine testicular β-galactosidase, *Proc. Natl. Acad. Sci. U.S.A.* **78:** 4289–4293.

90. Steiner, A. W., and Rome, L. H., 1982, Assay and purification of a solubilized membrane receptor that binds the lysosomal enzyme α-L-iduronidase, *Arch. Biochim. Biophys. Acta* **214:** 681–687.

91. Goldberg, D. E., Gabel, C. A., and Kornfeld, S., 1983, Studies of the biosynthesis of the mannose-6-phosphate receptor in receptor-positive and -deficient cell lines, *J. Cell Biol.* **97:** 1700–1706.

92. Mitchell, C., Maler, T., and Jourdian, G. W., 1984, Detergent dissociation of bovine liver phosphomannosyl binding protein, *J. Cell. Biochem.* **24:** 319–330.

93. Wickner, W. T., and Lodish, H. F., 1985, Multiple mechanisms of protein insertion into and across membranes, *Science* **230:** 400–407.

94. Perara, E., and Lingappa, V. R., 1985, A former amino terminal signal sequence engineered to an internal location directs translocation of both flanking protein domains, *J. Cell Biol.* **101:** 2292–2301.

95. Gilmore, R., and Blobel, G., 1985, Translocation of secretory proteins across the mi-

crosomal membrane occurs through an environment accessible to aqueous perturbants, *Cell* **42:** 497–505.

96. Hershko, A., Heller, H., Eytan, E., and Reiss, Y., 1986, The protein substrate binding site of the ubiquitin–protein ligase system, *J. Biol. Chem.* **261:**11992–11999.

97. Bachmair, A., Finley, D., and Varshavsky, A., 1986, *In vivo* half-life of a protein is a function of its amino-terminal residue, *Science* **234:** 179–186.

Chapter 10

Role of Transfer RNA in the Degradation of Selective Substrates of the Ubiquitin- and ATP-Dependent Proteolytic System

Aaron Ciechanover

1. INTRODUCTION

Degradation of intracellular proteins via the ubiquitin (Ub) pathway involves several steps. Initially, Ub is covalently linked to the protein substrate in an ATP-dependent reaction. Following Ub conjugation, the protein is selectively degraded with the release of free and reusable Ub. Ub modification of a variety of protein targets in the cell appears important in a number of basic cellular functions. For example, modification of core nucleosomal histones may regulate gene expression at the level of chromatin structure. Ub attachment to cell surface proteins can play a role in the process of cell–cell interaction and adhesion. Conjugation of Ub to other, yet to be identified, protein(s) is probably involved in the progression of cells from one stage to another in the cell cycle.

Despite the considerable progress in elucidating the mode of action and roles of the Ub pathway, major problems remain unsolved. A problem of central importance is what determines the specificity of the Ub ligation

AARON CIECHANOVER • Unit of Biochemistry, Faculty of Medicine, Technion–Israel Institute of Technology, Haifa 31096, Israel.

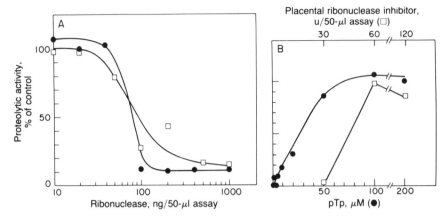

Figure 1. (A) Inhibition of the Ub- and ATP-dependent proteolytic system by RNase A (●) or micrococcal nuclease (□). Crude reticulocyte fraction II was preincubated with the indicated concentrations of the ribonucleases, followed by initiation of the proteolytic reaction with the addition of [^{125}I]BSA, Ub, and ATP. (B) Effect of inhibitors of RNases on the inhibitory effect of the enzymes on the Ub- and ATP-dependent proteolytic system. RNase A was preincubated with the indicated amounts of human placental ribonuclease inhibitor (□), and micrococcal nuclease was preincubated with the indicated amounts of pTp (●) prior to their addition to a complete proteolytic system containing [^{125}I]BSA, crude reticulocyte fraction II, Ub, and ATP. 100% proteolytic activity was measured in a system to which nucleases were not added (adapted from ref. 1).

system for commitment of certain proteins for degradation. It was recently found that a free α-NH$_2$ group on the substrate is an important feature recognized by the Ub ligation pathway. It was also found that tRNA is necessary for the conjugation of Ub to certain proteolytic substrates and for their subsequent degradation. The latter findings, described in detail in this chapter, provide insight on the specificity of the Ub system.

2. TRANSFER RNA AND THE DEGRADATION OF BOVINE SERUM ALBUMIN

We recently observed that Ub- and ATP-dependent degradation of labeled bovine serum albumin (BSA) is strongly and specifically inhibited by ribonucleases.[1] Ribonuclease A from bovine pancreas and micrococcal nuclease (MNase) at concentrations exceeding 2 μg/ml inhibited the degradation of [^{125}I]BSA by 80–90% (see Figure 1A). The inhibition was specific to ribonucleases since snake venom endonuclease and RNases T$_1$ and T$_2$ all showed strong inhibition at low concentrations, whereas DNase I at high concentration did not inhibit the proteolytic system.[1]

By inhibiting the activity of the RNases before adding them to the proteolytic system, we confirmed that the inhibitory effect of the RNases was indeed due to their enzymatic activities rather than to some other features of the protein molecules. Preincubation of RNase A with human placental ribonuclease inhibitor completely abolished its inhibitory effect. Preincubation of MNase with thymidine-3′,5′-diphosphate (pTp), a specific inhibitor of the enzyme,[2] also relieved the inhibition. Likewise, omission of Ca^{2+}, which is essential for MNase activity,[2] resulted in full proteolytic activity[1] (see Figure 1B).

To test the notion that RNA is required for the activity of the Ub- and ATP-dependent proteolytic system, phenol-extracted total RNA from crude reticulocyte fraction II was added to an RNA-depleted proteolytic system. As shown in Figure 2B, the added RNA completely restored proteolytic activity. Thus, we concluded that reduced proteolysis in the cell-free system is due to the destruction of an essential endogenous RNA component.

The obvious question then became: How many RNA species were necessary to restore proteolytic activity? When total fraction II RNA was separated by gel electrophoresis, cytoplasmic RNAs such as 7SL, 5.8S, 5S, and transfer RNAs were most abundant, whereas the nuclear RNAs U1, U2, U4, U5, and U6 were not detected (see Figure 2A). Individual RNAs were extracted from each band and added separately to a nuclease-treated proteolytic system in which the MNase had been previously inhibited by EGTA (or pTp). Only the tRNA-sized molecules restored the activity of the nuclease-treated system; equivalent amounts of 7SL, 5.8S, and 5S RNAs had no stimulatory activity[1] (see Figure 2B).

Since the active component comigrated with tRNA, it was desirable to determine whether any individual tRNA species would suffice to reconstitute the proteolytic activity. Certain patients with autoimmune diseases, such as systemic lupus erythematosus and polymyositis, produce autoantibodies directed against subsets of tRNAs.[3] We used sera from three such patients to isolate pure tRNA species for addition to the nuclease-treated proteolytic system. RNAs precipitated from ^{32}P-labeled NIH/3T3 cell extracts by patients sera were analyzed by two-dimensional polyacrylamide gel electrophoresis. Serum MN, which is of the anti-Jo-1 specificity,[4] precipitated a single RNA species, previously identified by RNA sequence analysis as $tRNA^{His}$ (see Figure 3A).[4] The antigenic protein is histidyl-tRNA synthetase.[5] The other two sera, LL and SU, precipitate several uncharacterized tRNA-sized molecules (see Figure 3A). To characterize these RNAs, the precipitated species were subjected to fingerprint analysis. Serum SU was found to precipitate five major species, as well as several minor ones. One of the most prominent species

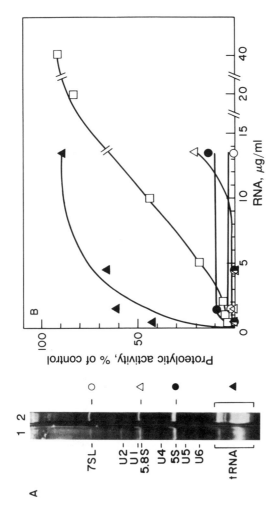

Figure 2. (A) Polyacrylamide gel electrophoresis of RNA extracted from NIH/3T3 cells (lane 1) or from crude reticulocyte fraction II (lane 2). (B) Ability of purified RNA subfractions to restore activity to a RNase-inhibited Ub- and ATP-dependent proteolytic system. RNA was separated on a polyacrylamide gel (see Figure 2A, lane 2), and the bands were visualized, excised, extracted, and ethanol precipitated. The indicated amounts of RNA fractions were added to a micrococcal nuclease-treated complete proteolytic system (following inhibition of the ribonuclease), and the degradation of $[^{125}I]BSA$ was monitored. □, total RNA; ○, 7SL RNA; △, 5.8S RNA; ●, 5S RNA; ▲, tRNA (adapted from ref. 1).

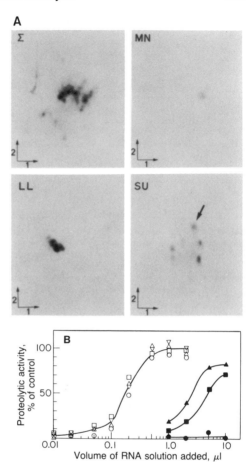

Figure 3. (A) Subsets of tRNAs present in immunoprecipitates of NIH/3T3 cell extracts. Cells were labeled with ^{32}p prior to RNA extraction and preparation of tRNA. Σ, total tRNA; MN, SU, LL, tRNAs precipitated using MN, Su, and LL autoimmune sera, respectively. (B) Effect of purified immunoprecipitated tRNAs on the Ub- and ATP-dependent proteolytic system. RNAs extracted from the immunoprecipitates ($\blacktriangle$, MN; $\blacksquare$, SU; $\bullet$, LL) or from the supernatants after immunoprecipitation ($\triangle$, Mn; $\square$, Su; $\bigcirc$, LL) were added to a micrococcal nuclease-inhibited Ub system (following inhibition of the enzyme), and the degradation of [^{125}I]BSA was monitored. Serum MN precipitated about 10% of the total cellular tRNA^{His}, and serum SU precipitated only <5%. The activity of the tRNA in the supernatant was therefore 9–23-fold higher than that of the tRNA extracted from the precipitate (in different experiments) and hence very similar to that of total tRNA prior to precipitation (∇) (adapted from ref. 1).

was tRNA^{His}; the other tRNAs were not identified. Protein is required for immunoprecipitation, suggesting that serum SU might recognize a protein synthesis elongation factor or a tRNA-modifying enzyme. Of the three major RNAs precipitated by serum LL, none could be identified as tRNA^{His} (see Figure 3A).[1]

When tRNAs isolated from NIH/3T3 cell immunoprecipitates were added to the inactivated proteolytic extract, tRNA^{His} (precipitated by serum MN) was sufficient to restore >80% of the proteolytic activity. In addition, tRNAs precipitated by serum SU (which include tRNA^{His}) restored the protein degradation activity, but the tRNAs precipitated by serum LL had no effect[1] (see Figure 3B).

The effect of tRNA was very specific. tRNA from another mammalian source, mouse NIH/3T3 cells, could restore the proteolytic activity to an RNase-treated system, whereas yeast tRNA and non-tRNA small RNAs isolated from the same cells using antibodies directed against small ribonucleoproteins (Sm, Ro, and La[3]) did not. Human DNA and poly(A)$^+$ mRNA (from hepatoma HepG2 cells) likewise had no effect, as was the case for the polyanions poly(I), poly(IC), poly(U), poly(C), heparin, and protamine sulfate.[1]

3. REQUIREMENT OF tRNA FOR UBIQUITIN CONJUGATION

Further studies on the tRNA requirement for Ub-mediated proteolysis revealed that while the degradation of BSA was sensitive to ribonucleases and required tRNA, the degradation of lysozyme was not affected by RNA destruction.[1,6] This finding indicated that the degradation of proteins in a Ub-dependent mode probably occurs via two distinct pathways. Accordingly, we postulated that the two pathways should have some distinct enzymatic component(s), though other components might be shared by both systems. Subsequent experiments supported this hypothesis.

We identified additional proteolytic substrates, the degradation of which was either sensitive or insensitive to ribonucleases. Classification of additional substrates allowed us to rule out the possibility that the ribonuclease sensitivity of BSA degradation was unique to this substrate. The degradation of both bovine α-lactalbumin (bovine α-LA) and soybean trypsin inhibitor (STI) was found to be sensitive to treatment with ribonuclease; the degradation of reduced and carboxymethylated BSA (rcmBSA) was sensitive as well (see Figure 4). Although native BSA is a very good substrate for the cell-free, Ub-dependent proteolytic pathway, Ub–BSA conjugates could not be demonstrated using the native molecule (A. Ciechanover and A. Hershko, unpublished results). Therefore, we chose rcmBSA because this protein is not only degraded in a Ub- and ATP-dependent mode, but also Ub–rcmBSA conjugates could be demonstrated.[7] The kinetics of formation and degradation of Ub–rcmBSA conjugates permits visualization of these conjugates, thus making rcmBSA a better substrate for analysis of the ribonuclease-sensitive step in the Ub pathway. It should be noted that RNase A was slightly more active than MNase in inhibiting the degradation of the three substrates, α-LA, STI, and rcmBSA[1,6] (see Figures 1 and 4). This may reflect the fact that RNase A can more efficiently digest the reticulocyte tRNA component necessary for the degradation of these substrates.

As in an earlier study,[1] we demonstrated that the inhibition of deg-

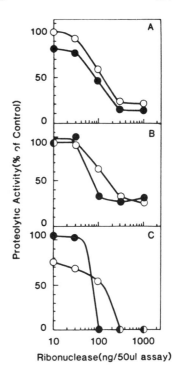

Figure 4. Effect of ribonucleases on the degradation of (A) rcmBSA, (B) bovine α-LA, and (C) STI. The degradation of the ^{125}I-labeled substrates was determined in the presence of the indicated concentrations of RNase A ($\bullet$) or micrococcal nuclease ($\circ$) as described in the legend to Figure 1A (adapted from ref. 6).

radation of rcmBSA, bovine α-LA, and STI was due to destruction of tRNA. Fraction II was treated with MNase followed by the addition of pTp or the Ca^{2+} chelator, EGTA. The subsequent addition of tRNA was found to restore almost completely the inhibited proteolytic activity.[6] Thus, it is clear that the tRNA requirement is a general feature of the system.

In principle, tRNA might participate either in conjugation of Ub to the proteolytic substrate or in the degradation of the Ub–protein conjugates. To distinguish between these possibilities, crude reticulocyte fraction II was incubated with ^{125}I-labeled Ub and either rcmBSA, bovine α-LA, or STI. Distinct conjugates were formed between Ub and the added exogenous substrates[6] (see Figure 5, compare lanes 1 and 2). Preincubation of crude reticulocyte fraction II with MNase prior to addition of the substrate and labeled Ub almost completely abolished the formation of conjugates between Ub and the exogenous substrates (see Figure 5, lane 3).[6] Pretreatment of crude reticulocyte fraction II with RNase A resulted in a similar inhibition of specific conjugate formation (not shown). The formation of conjugates between Ub and most endogenous protein

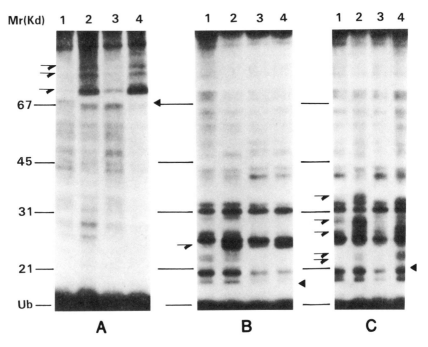

Figure 5. Effect of micrococcal nuclease and tRNA on the formation of conjugates of [^{125}I]Ub with (A) rcmBSA, (B) bovine α-LA, and (C) STI. Crude reticulocyte fraction II was preincubated with or without micrococcal nuclease, followed by inhibition of the nuclease and initiation of the reaction with the addition [^{125}I]Ub, ATP, and the appropriate substrate. The conjugates were resolved by SDS–PAGE followed by autoradiography. Lanes 1, with no addition of substrate; lanes 2, with added exogenous substrate; lanes 3, treated with micrococcal nuclease; lanes 4, same as lanes 3, but in addition tRNA was added. Arrows on the left side of the panels indicate specific conjugates. Arrows on the right side of the panels indicate the molecular weights of the added exogenous substrates (adapted from ref. 6).

acceptors was not inhibited by the nuclease treatment (see Figure 5, compare lanes 1 and 3). However, some endogenous protein–Ub conjugates disappeared after nuclease treatment. (See bottom of Figure 5, lanes 1 and 3 for representative results.)

When tRNA was added to the MNase-treated system following inhibition of the nuclease, the specific conjugates were formed (see Figure 5, lane 4).[6] For nuclease-sensitive substrates, a strong correlation therefore exists between inhibition of conjugate formation and inhibition of degradation; added tRNA restores both conjugation and degradation. Similar results were obtained using unlabeled Ub and ^{125}I-labeled STI or bovine α-LA (not shown). These results, which indicate that the formed adducts are indeed conjugates between Ub and the added exogenous substrate, rule out the possibility that the added substrates exert their effect by stimulating formation of conjugates between Ub and some endogenous acceptors.

The exact structure of these Ub–protein conjugates is not known. Previously, from the molecular weight of the conjugates, we tried to determine their structure and the number of the Ub molecules attached to a single substrate molecule.[8] With better understanding of the system, it has been recognized that this is no longer possible, as the system contains many enzymes that can rapidly cleave Ub–protein conjugates.[9] The existence of isopeptidase(s) that cleaves the isopeptide bond between Ub and proteins or peptides[9,10] and of poly-Ub structures,[11] as well as an ATP-dependent protease,[12] makes a simple analysis of the structure of Ub–protein conjugates difficult.

We have already noted that the degradation of lysozyme is not affected by nucleases.[1,6] Since it was also clear from conjugation assays that the effect of MNase is highly selective and that most endogenous substrates are not affected by the nuclease treatment, we sought to identify more ribonuclease-resistant proteolytic substrates. The degradation of ^{125}I-labeled oxidized RNase A, α-casein, β-lactoglobulin, and lysozyme (see Figure 6, panels A, B, C, and D, respectively) was not inhibited by ribonucleases. On the contrary, the degradation of oxidized RNase A, α-casein, and lysozyme was significantly stimulated when fraction II was preincubated with the nucleases (see Figure 6, panels A, B, and D). It is possible that ribonuclease treatment inhibits the flow of endogenous substrates into the system in an early step of the pathway, thus preventing competition between exogenously added substrates and endogenous substrates on commonly shared enzymatic components. Conceivably, the two distinct pathways converge at one point and thereafter share common components. To corroborate the relation between Ub conjugation to the substrate and its subsequent degradation, we incubated crude reticulocyte fraction II with ^{125}I-labeled Ub and unlabeled lysozyme, oxidized RNase A, and β-lactoglobulin. Distinct conjugates between Ub and the proteolytic substrates were formed. When the enzyme preparation was preincubated with MNase, no change in the conjugate pattern could be demonstrated (not shown). Thus, for both the nuclease-sensitive and -insensitive substrates, a strong correlation exists between Ub conjugation to the substrate and its subsequent degradation.

4. ROLE OF tRNA IN THE UBIQUITIN PATHWAY

What can be the role of tRNA in the Ub- and ATP-dependent proteolytic pathway? tRNA probably participates in covalent modification

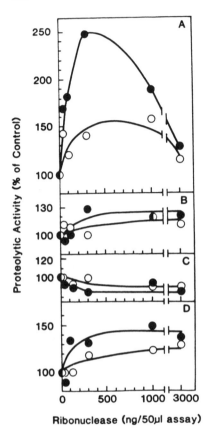

Figure 6. Effect of RNase A (●) and micrococcal nuclease (○) on the degradation of ^{125}I-radiolabeled oxidized RNase A (panel A), α-casein (panel B), β-lactoglobulin (panel C), and lysozyme (panel D). The degradation of the ^{125}I-radiolabeled substrates was determined as described in the legend to Figure 1A (adapted from ref. 6).

of selective proteolytic substrates, which renders them more susceptible to the Ub proteolytic system. It should be noted that all three ribonuclease-sensitive substrates examined have an acidic α-NH$_2$ terminus (aspartic acid in STI[13] and BSA,[14] and glutamic acid in bovine α-LA[15]). Recent data from our laboratory indicate that an acidic α-NH$_2$ terminus can indeed serve as a recognition marker for the tRNA-dependent reaction. Other proteins with an acidic α-NH$_2$ terminus, such as the kappa light chain of the human immunoglobulin (Bence–Jones protein)[16] and thyroglobulin,[17] are degraded via the Ub pathway in a tRNA-dependent mode.[18] The degradation of human α-LA, which although 75% homologous to the bovine molecule has a lysine residue at the α-NH$_2$ terminus, is independent of tRNA.[18]

Both bacteria and eukaryotes contain an unusual class of enzymes, aminoacyl tRNA-protein transferases, which catalyze post-translational conjugation of specific amino acid residues to the mature amino termini of acceptor proteins. Among the eukaryotic enzymes, the best studied is arginyl tRNA-protein transferase.[17] We tested the hypothesis that tRNA-dependent post-translational addition of an arginine moiety to an acidic terminus of a substrate is required for its subsequent degradation. First, we showed that crude reticulocyte fraction II does indeed contain an arginyl tRNA-protein transferase activity specific to substrates with acidic amino termini. The arginylation was dependent on enzyme and substrate concentrations, metabolic energy, and tRNA.[18] The incorporation of arginine was almost completely dependent on the addition of the appropriate exogenous substrate with very little incorporation into fraction II endogenous proteins. Next, we examined whether tRNA-dependent arginylation of substrates with acidic amino termini is required for their Ub-mediated degradation and whether ribonucleases inhibit the degradation of these substrates by destroying the tRNA component of the system, thus preventing this initial modification. As can be seen in Table I, the degradation of ^{125}I-labeled BSA, STI, and bovine α-LA was sensitive to ribonuclease and could be restored by addition of exogenous tRNA. In striking contrast, the degradation of the same substrates in which the amino termini were modified by the addition of [^{3}H]arginine was insensitive to ribonuclease (Table I). It is not clear yet whether other amino acids can participate in post-translational modification of substrates of the Ub system. In preliminary studies, we have found a tRNA- and energy-dependent histidylation of substrates with acidic amino termini (S. Arfin and A. Ciechanover, unpublished results). The relevance of this modification to Ub-dependent proteolysis is being investigated.

We recently purified and characterized the arginyl tRNA-protein transferase from rabbit reticulocytes. It is a high molecular weight complex (360,000 Da) between arginyl tRNA synthetase and arginyl tRNA-protein transferase. This complex is required in addition to E1, E2, and E3 in order to conjugate Ub to substrates with acidic N termini (A. Ciechanover, S. Ferber, S. Arfin, S. Elias, A. Hershko, and D. Ganoth, unpublished results). It is possible that the addition of a basic residue to the acidic N terminal of the substrate is necessary for recognition and binding of the substrate to Ub–protein ligase (E3) before conjugation to Ub. Perhaps the site on E3 to which the amino terminus of a protein substrate binds is negatively charged, so that only proteins with positively charged amino termini form complexes with E3. Proteins with neutral or basic amino-terminal residues fulfill this requirement. Proteins with acidic amino-terminal residues have no net charge at this site and therefore cannot bind to E3. Indeed, a selective binding site of E3 for proteins with

Table I. Effect of Micrococcal Nuclease on the Degradation of ^{125}I-Labeled and [^{3}H]Arginine-Labeled Substrates of the Ubiquitin Pathway[a]

Substrate reaction conditions	[^{125}I]α-LA	[^{3}H]Arg α-LA	[^{125}I]STI	[^{3}H]Arg-STI	[^{125}I]BSA	[^{3}H]Arg-BSA
Complete system	68	74	26	50	22	40
− Ub	19	16	7	23	3	9
− ATP	16	13	5	20	1	4
+ MNase + EGTA (or pTp)	22	79	8	53	2	44
+ MNase + EGTA (or pTp) + tRNA	67	—	20	—	15	—

[a] Values given are degradation in percent. The [^{3}H]arginylated substrates were prepared by incubating crude reticulocyte fraction II, [^{3}H]arginine, the appropriate exogenous substrate, ATP, and an ATP-regenerating system. Following trichloroacetic acid precipitation, the protein pellet was dissolved and dialyzed (to remove free labeled arginine). Since greater than 90% of the protein label is in the exogenously added substrate (not shown),[18] this protein mixture was used as a source of the [^{3}H]arginine-labeled protein (adapted from ref. 18).

basic and neutral amino termini has recently been identified (Y. Reiss, D. Kaim, and A. Hershko, in press).

Recently, it has been reported that the *in vivo* half-life of a protein is a function of its amino-terminal residue. β-Galactosidase with 16 different genetically engineered N termini demonstrated half-lives that ranged from 2 min to more than 30 hr.[19] An "N-end rule" was proposed, according to which long-lived proteins have "stabilizing" amino termini, whereas short-lived proteins have "destabilizing" amino-terminal residues. β-Galactosidase with glutamic or aspartic acid in its α-NH_2 position has a very short half-life (10–30 min). It is possible that these residues are not destabilizing as such but become so only through their ability to be conjugated (e.g., in a tRNA-dependent mode) to other destabilizing residues (see Chapter 11 for more details on the N-end rule).

While the N terminus of a protein is certainly an important structural determinant for its recognition by the Ub system, other findings suggest it might not be the only recognition signal.[20] For example, RNase A, which has lysine in its amino terminus (a strongly destabilizing residue[19]), does not bind to Ub–protein ligase (E3) and therefore is not degraded by the Ub system.[20] Oxidation of methionine residues causes the protein to bind tightly to E3 and to become an efficient proteolytic substrate. So it seems that in addition to its N-terminal residue, other signals or alterations in the three-dimensional structure of a protein can also serve as important recognition markers for Ub ligation.

It is interesting to note that tRNA-dependent post-translational addition of amino acids to proteins *in vivo* is increased under conditions of stress, for example, after physical injury to axons of nerve cells.[21] Although the relevance of this modification to proteolysis by the Ub system is not known, it seems relevant that the Ub system is also activated under conditions of stress, such as heat shock.[22-24] It is possible that after thermal insult, protein turnover is stimulated to remove heat-denatured proteins. One can therefore imagine that protein conformations that increase their recognition by the Ub system are generated by heat or other denaturing conditions.

5. CONCLUSION

In the past few years considerable progress has been made in elucidating the mode of action and cellular roles of the Ub pathway. Powerful tools are now available to study the biological functions of the Ub system, including cloned genes, specific antibodies, a temperature-sensitive mutant, and microinjection techniques (see Chapters 2, 4, and 7). Still, many major problems remain unsolved, and the unknown greatly exceeds what we presently know about the Ub system. What determines the specificity of the Ub conjugation system for commitment of a certain protein for degradation? It is reasonable to assume that a free α-NH$_2$ group is only one of several features on the protein structure recognized by the Ub ligation system. How does the system distinguish between Ub ligation leading to protein breakdown and that involved in protein modification? What determines whether a particular protein is conjugated with a single Ub or with multiple Ub molecules? While we have no answers to these questions at present, available information suggests that there are several different Ub–protein ligation systems and that these may act on different types of cellular proteins. Examples are the tRNA-dependent conjugation reaction, which is specific to certain proteins, and the ligation of Ub to lysine 115 or *Dictyostelium* calmodulin, which contains a blocked amino terminus[25] (see Chapter 11 for discussions on free α-NH$_2$ terminus recognition). It also remains to be seen whether the different substrate-specific conjugation systems are carried out by different species of E2 (see Chapter 3) or E3 or by additional factors.

ACKNOWLEDGMENTS. The author is indebted to Dr. Alan L. Schwartz for his very generous help. The author would like to thank Ireta Ashby for her skillful help and patience in typing the manuscript. The research was supported by grant No. 85-00059/2 from the Deutche Forschungsgemeinschaft, the United States–Israel Binational Science Foundation, and by grants from the Vice President of the Technion for Research (Lovengart Fund), and the Foundation for Promotion of Research at Technion. The author is a recipient of an Israel Cancer Research Fund Research Career Development Award.

REFERENCES

1. Ciechanover, A., Wolin, S. L., Steitz, J. A., and Lodish, H. F., 1985, Transfer RNA is an essential component of the ubiquitin and ATP-proteolytic system, *Proc. Natl. Acad. Sci. U.S.A.* **82:** 1341–1345.
2. Cuatrecasas, P., Fuchs, S., and Anfinsen, C. B., 1967, Catalytic properties and specificity of the extracellular nuclease of *Staphylococcus aureus, J. Biol. Chem.* **242:** 1541–1547.
3. Hardin, J. A., Rahn, D. R., Shen, C., Lerner, M. R., Wolin, S. L., Rosa, M. D., and Steitz, J. A., 1982, Antibodies from patients with connective tissue diseases bind specific subsets of cellular RNA-protein particles, *J. Clin. Invest.* **70:** 141–147.
4. Rosa, M. D., Hendrick, J. P., Lerner, M. R., Steitz, J. A., and Reichlin, M., 1983, A

mammalian $tRNA^{His}$ containing antigen is recognized by the polymyositis-specific antibody anti-Jo-1, *Nucleic Acids Res.* **11**: 853–870.

5. Mathews, M. B., and Bernstein, R. M., 1983, Myositis autoantibody inhibits histidyl-tRNA synthetase: A model for autoimmunity, *Nature* **304**: 177–179.

6. Ferber, S., and Ciechanover, A., 1986, Transfer RNA is required for conjugation of ubiquitin to selective substrates of the ubiquitin- and ATP-dependent proteolytic system, *J. Biol. Chem.* **261**: 3128–3134.

7. Evans, A. C., and Wilkinson, K. D., 1985, Ubiquitin-dependent proteolysis of native and alkylated bovine serum albumin: Effects of protein structure and ATP concentration on selectivity, *Biochemistry* **24**: 2915–2923.

8. Hershko, A., Ciechanover, A., Heller, H., Haas, A. L., and Rose, I. A., 1980, Proposed role of ATP in protein breakdown: Conjugation of proteins with multiple chains of the polypeptide of ATP-dependent proteolysis, *Proc. Natl. Acad. Sci. U.S.A.* **77**: 1783–1786.

9. Hershko, A., Leshinsky, E., Ganoth, D., and Heller, H., 1984, ATP-dependent degradation of ubiquitin–protein conjugates, *Proc. Natl. Acad. Sci. U.S.A.* **81**: 1619–1623.

10. Matsui, S. I., Sandberg, A. A., Negoro, S., Seon, B. K., and Goldstein, G., 1982, Isopeptidase: A novel eukaryotic enzyme that cleaves isopeptide bonds, *Proc. Natl. Acad. Sci. U.S.A.* **79**: 1535–1539.

11. Hershko, A., and Heller, H., 1985, Occurrence of a polyubiquitin structure in ubiquitin–protein conjugates, *Biochem. Biophys. Res. Commun.* **128**: 1079–1086.

12. Hough, R., Pratt, G., and Rechsteiner, M., 1987, Purification of two high molecular weight proteases from rabbit reticulocyte lysate, *J. Biol. Chem.* **262**: 8303–8313.

13. Ikenaka, T., Shimada, K., and Matsushima, Y., 1963, The N-terminal amino acid of soybean trypsins inhibitor, *J. Biochem.* **54**: 193–195.

14. Shearer, W. T., Bradshaw, R. A., Gurd, F. R. N., and Peters, T., Jr., 1967, The amino acid sequence and copper(II)-binding properties of peptide (1–24) of bovine serum albumin, *J. Biol. Chem.* **242**: 5451–5459.

15. Brew, K., Castellino, F. J., Vanaman, T. C., and Hill, R. L., 1970, The complete amino acid sequence of bovine α-lactalbumin, *J. Biol. Chem.* **245**: 4570–4582.

16. Kabat, E. A., Wu, T. T., Bilofsky, H., and Reid-Miller, M., 1983, *Sequences of Proteins of Immunological Interest,* U.S. Department of Health and Human Services, Public Health Service, National Institutes of Health, Washington, DC, pp. 14–29.

17. Soffer, R. L., 1973, Post-translational modification of proteins catalyzed by aminoacyl-tRNA-protein transferases, *Mol. Cell. Biochem.* **2**: 3–14.

18. Ferber, S., and Ciechanover, A., 1987, Role of arginine-tRNA in protein degradation by the ubiquitin pathway, *Nature* **326**: 808–811.

19. Bachmair, A., Finley, D., and Varshavsky, A., 1986, *In vivo* half-life of a protein is a function of its amino terminal residue, *Science* **234**: 179–186.

20. Hershko, A., Heller, H., Eytan, E., and Reiss, Y., 1986, The protein substrate binding site of the ubiquitin–protein ligase system, *J. Biol. Chem.* **261**: 11992–11999.

21. Shyne-Athwal, S., Riccio, R. V., Chakraborty, G., and Ingoglia, N. A., 1986, Protein modification by amino acid addition is increased in crushed sciatic but not optic nerves, *Science* **231**: 603–605.

22. Finley, D., Ciechanover, A., and Varshavsky, A., 1984, Thermolability of ubiquitin-activating enzyme from mammalian cell cycle mutant ts85, *Cell* **37**: 43–55.

23. Ciechanover, A., Finley, D., and Varshavsky, A., 1984, Ubiquitin dependence of selective protein degradation demonstrated in the mammalian cell cycle mutant ts85, *Cell* **37**: 57–66.

24. Bond, U., and Schlesinger, M. J., 1985, Ubiquitin is a heat-shock protein in chicken embryo fibroblasts, *Mol. Cell. Biol.* **5:** 949–956.
25. Gregori, L., Marriott, D., West, C. M., and Chau, V., 1985, Specific recognition of calmodulin from *Dictyostelium discoideum* by the ATP/ubiquitin-dependent degradative pathway, *J. Biol. Chem.* **260:** 5232–5235.

The N-End Rule of Selective Protein Turnover
Mechanistic Aspects and Functional Implications

Alexander Varshavsky, Andreas Bachmair, Daniel Finley, David Gonda, and Ingrid Wünning

1. INTRODUCTION

In both bacterial and eukaryotic cells, relatively long-lived proteins, whose half-lives are close to or exceed the cell generation time, coexist with proteins whose half-lives can be less than 1% of the cell generation time. Rates of intracellular protein degradation are a function of the cell's physiological state and appear to be controlled differentially for individual proteins.[1-7] In particular, damaged and some otherwise abnormal proteins are metabolically unstable.[1-10] It is also clear that many otherwise undamaged regulatory proteins are extremely short-lived *in vivo*.[1-23] Metabolic instability of such proteins allows for efficient temporal control of their intracellular concentrations through regulated changes in rates of their synthesis or degradation. Instances in which the metabolic instability of an intracellular protein is known to be directly relevant to its function

ALEXANDER VARSHAVSKY, ANDREAS BACHMAIR, DANIEL FINLEY, DAVID GONDA, and INGRID WÜNNING • Department of Biology, Massachusetts Institute of Technology, Cambridge, Massachusetts 02139.

include the cII protein of bacteriophage λ (cII is the essential component of a molecular switch that determines whether λ grows lytically or lysogenizes an infected cell),[16–18] the σ^{32} factor of the *Escherichia coli* RNA polymerase (σ^{32} confers on RNA polymerase the specificity for promoters of heat-shock genes),[15,19–21] and the HO endonuclease of the yeast *Saccharomyces cerevisiae* (HO is a site-specific endodeoxyribonuclease that initiates the process of mating-type interconversion in yeast).[22,23]

Most of the selective degradation of intracellular proteins under normal metabolic conditions is ATP dependent and (in eukaryotes) nonlysosomal.[1–8,24–27] Recent biochemical and genetic evidence indicates that, in eukaryotes, covalent conjugation of ubiquitin (Ub) to short-lived intracellular proteins is essential for their selective degradation.[1,2,4,26–28] Ub, a 76 residue protein, is found in eukaryotes either free or covalently joined via its carboxyl-terminal Gly residue to a variety of cytoplasmic, nuclear, and integral membrane proteins (for reviews, see refs. 1, 2, 4, and 7). Although *E. coli* almost certainly lacks highly conserved counterparts of either eukaryotic Ub or Ub-specific enzymes,[29–31] the possibility remains that a functional counterpart of Ub exists in bacteria but has so far escaped detection owing to its insufficiently strong sequence homology to eukaryotic Ub.

A few years ago, we initiated a systematic study of the Ub system in yeast, where biochemical approaches can be combined with the powerful methods of molecular genetic analysis. The yeast Ub system was found to be closely related to that of mammals.[1,28–32] Results of molecular genetic analysis of specific components of the Ub system are discussed in Chapter 2 of this book. The use of plasmids encoding well-defined proteolytic substrates forms a part of another genetically based approach to the problem of selective protein turnover[31] and is considered below.

The understanding that ubiquitinated intracellular proteins include obligatory intermediates in selective protein turnover leaves untouched the problem of targeting. How are intracellular proteins initially recognized as substrates for selective degradation? By analogy with the signal sequences that confer on a protein the ability to enter distinct cellular compartments,[33–35] it could be expected that proteins also contain sets of specific amino acid sequences that alone or in combination would act to determine the half-life of each protein *in vivo*. *A priori*, the sequence complexity of such a "half-life rule" might be at least comparable to that of other known signal sequence systems in proteins. Using a new approach, which makes it possible to expose *in vivo* different amino acid residues at the amino termini of otherwise identical test proteins, we have recently found that an important component of the half-life rule (the "N-end rule") is remarkably simple and involves the amino-terminal residue of a target protein.[31] The initial recognition step that involves the amino-terminal residue of the target appears to be an essential component of a pathway which depends on at least one more determinant. The results that have led to the concept of the N-end rule,[31] some of the implications of these results, and the more recent findings are considered in this chapter.

2. UBIQUITIN FUSION PROTEINS AS *IN VIVO* PROTEOLYTIC SUBSTRATES

2.1. Rapid *in Vivo* Deubiquitination of a Nascent Ubiquitin–βgal Fusion Protein

Branched conjugates in which the carboxyl-terminal Gly of Ub moieties is joined via an isopeptide bond to the ε-amino groups of internal Lys residues in proteins apparently constitute the bulk of Ub conjugates in eukaryotic cells.[1,2,26,27,36-38] Joining of Ub to the amino-terminal α-amino groups of target proteins, to yield linear Ub conjugates, should also be chemically feasible.[39] Whether or not linear Ub–protein fusions are actually synthesized *in vivo* through post-translational enzymatic conjugation of Ub to protein amino termini, such proteins can also be produced by constructing appropriate chimeric genes and expressing them *in vivo*. Construction of one such gene, which encodes yeast Ub (refs. 28–30) linked to βgal of *E. coli* is shown in Figure 1. When this gene is expressed in *E. coli*, the resulting βgal-containing protein has an apparent molecular mass approximately 6 kDa greater than that of the control βgal, a value consistent with the presence of Ub in the protein encoded by the chimeric gene (Figure 2, lanes a and c). In contrast, when the same gene is expressed in yeast, the corresponding βgal protein is electrophoretically indistinguishable from the control βgal (Figure 2, lanes a and c). This result is independent of the length of the [^{35}S]methionine labeling period (between 1 and 30 min) (see the legend to Figure 2). Determination of the amino-terminal residue in the putative (deubiquitinated) Met–βgal (half-life, $t_{1/2} > 20$ hr) by Edman degradation of the *in vivo* labeled, gel-purified βgal (Figure 2, lane d) directly confirmed the presence of the expected

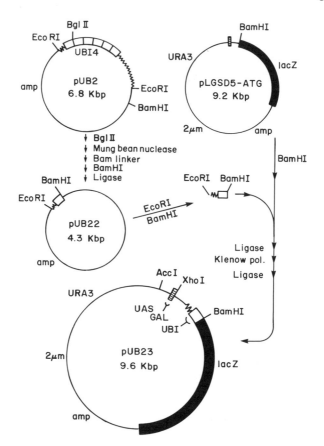

Figure 1. Construction of a Ub–*lacZ* fusion. pUB2, a pBR322-based genomic DNA clone,[29] contains six repeats of the yeast Ub-coding sequence (open boxes) together with the flanking regions (jagged lines). pUB2 was modified as shown in the diagram and described in ref. 31. This allowed the construction of an in-frame fusion (confirmed by nucleotide sequencing) between a single Ub repeat and the *lacZ* gene of the expression vector pLGSD5-ATG (called G2 in ref. 72). The term "2 μm" denotes a region of pLGS5D-ATG that contains the replication origin and flanking sequences of the yeast 2 μm plasmid.[72] See Figure 3B for the amino acid sequence of the fusion protein in the vicinity of the Ub–βgal junction. (From ref. 31, with permission. Copyright 1986 by AAAS.)

Met residue (Figure 3A and Table I) at its amino terminus.[31] Further evidence that cleavage of the Ub fusion proteins occurs immediately after the last Gly residue of Ub is presented in Section 2.3. We conclude that, in yeast, Ub is efficiently cleaved off the nascent Ub–βgal fusion protein, yielding a deubiquitinated βgal. More recent experiments have shown that

Table I. The N-End Rule[a]

Residue X in Ub–X–βgal	Deubiquitination of Ub–X–βgal	$t_{1/2}$ of X–βgal
Met	+	
Ser	+	
Ala	+	
Thr	+	>20 hr
Val	+	
Gly	+	
Cys	+	
Ile	+	~30 min
Glu	+	
Tyr	+	
Gln	+	~10 min
His	+	
Phe	+	
Leu	+	
Trp	+	~3 min
Asp	+	
Asn	+	
Lys	+	
Arg	+	~2 min
Pro	– [b]	~7 min

[a] The original N-end rule of ref. 31 is updated here by inclusion of four more amino-terminal amino acids—Cys, His, Trp, and Asn. These more recent data complete the N-end rule as defined with βgal in the yeast *S. cerevisiae*. See ref. 31 for technical details.

[b] The rate of *in vivo* deubiquitination of Ub–Pro–βgal is extremely low in both yeast and mammalian cells (see Section 2.5). The $t_{1/2}$ shown is that of the initial Ub–Pro–βgal fusion protein.[31] Pro–βgal, the product of slow deubiquitination of Ub–Pro–βgal is a long-lived protein *in vivo*[31] (see also Section 2.5).

a protease with a similar if not identical substrate specificity exists in mammalian cells as well (see Section 7).

The Ub–βgal junction encoded by the chimeric gene, Gly–Met (Figures 1 and 3B), is identical to the junctions between adjacent repeats in the polyubiquitin precursor protein, which is efficiently processed into mature Ub.[28–30] Thus, it is likely that the same protease is responsible both for the conversion of polyubiquitin into mature Ub and for the deubiquitination of the nascent Ub–βgal protein. If so, one possible way to inhibit the *in vivo* deubiquitination of the Ub–βgal (and thereby to allow analysis of metabolic consequences of a stable Ub attachment to βgal) would be to convert the Met residue of βgal at the Ub–βgal junction (Figure 3B) into other amino acid residues (Figure 3A). The unexpected results of such an approach[31] are described in Section 2.2.

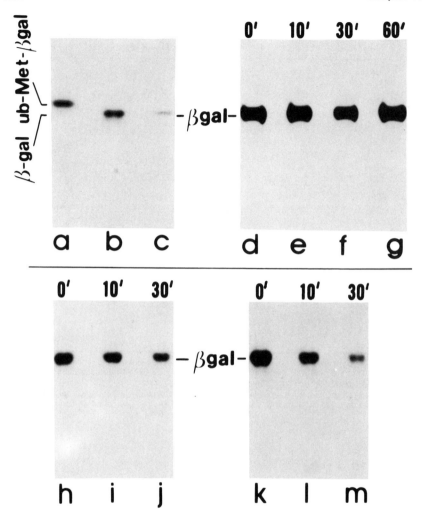

Figure 2. The *in vivo* half-life of βgal is a function of its amino-terminal residue. (Lane a) Minicells isolated from an *E. coli* strain carrying pUB23, the initial Ub–*lacZ* fusion (Figures 1 and 3B), were labeled with [^{35}S]methionine for 60 min at 36°C, with subsequent extraction, immunoprecipitation, and electrophoretic analysis[31] of βgal. The same result was obtained when the labeled minicell SDS extract was combined with an unlabeled yeast SDS extract before immunoprecipitation of βgal. (Lane b) *Saccharomyces cerevisiae* cells carrying pUB23 (Figure 1), which encodes Ub–Met–βgal (Figure 3B), were labeled with [^{35}S]methionine for 5 min at 30°C, with subsequent analysis of βgal. The same result was obtained with the lengths of the [^{35}S]methionine labeling periods from 1 to 30 min, and with yeast extracts produced either by mechanical disruption of cells in the presence of protease inhibitors or by boiling the cells directly in an SDS-containing buffer.[31] (Lane c) Same as lane a, but with *E. coli* cells carrying the control plasmid pLGSD5 (called G1 in ref. 72), which encodes βgal. (Lanes d–g) *S. cerevisiae* cells carrying pUB23 (Fig-

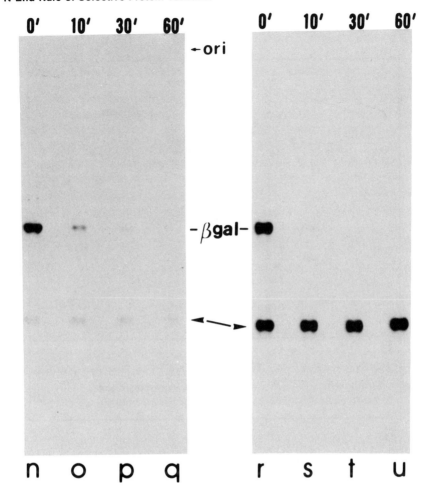

ure 1), which encodes Ub–Met–βgal (Figure 3A), were labeled with [^{35}S]methionine for 5 min at 30°C (lane d) followed by a chase in the presence of cycloheximide for 10, 30, and 60 min (lanes e–g), extraction, immunoprecipitation, and electrophoretic analysis of βgal. (Lanes h–j) Same as lanes d–f but with Ub–Ile–βgal (see Figure 3A). (Lanes k–m) Same as lanes d–f, but with Ub–Gln–βgal. (Lanes n–q) Same as lanes d–g, but with Ub–Leu–βgal. (Lanes r–u) Same as lanes d–g, but with Ub–Arg–βgal. Designations: ori, origin of the separating gel; Ub, ubiquitin; βgal, an electrophoretic band of the βgal protein containing a specific amino-terminal residue; in this terminology, the Met–βgal portion of Ub–Met–βgal is designated as βgal. Arrowheads denote a metabolically stable, ~90 kDa degradation product of βgal, which is formed apparently as the result of an *in vivo* endoproteolytic cleavage of a proportion of short-lived βgal proteins such as Leu–βgal and Arg–βgal (lanes n–u). (From ref. 31, with permission. Copyright 1986 by AAAS.)

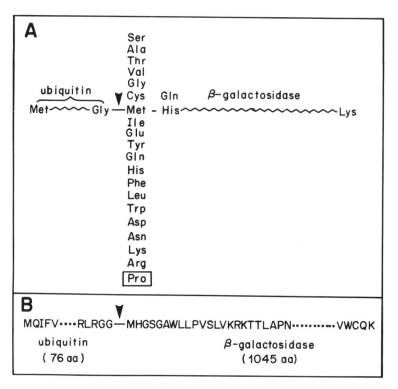

Figure 3. Changing amino acid residues of βgal at the Ub–βgal junction. (A) The initial plasmid, pUB23 (Figure 1), which encodes Ub–Met–βgal, was mutagenized[40,41] to convert the original Met codon ATG at the Ub–βgal junction into codons specifying 19 amino acids other than Met. (The original set of changes encompassed 15 amino acids[31] and was more recently expanded to include the remaining four amino acids—Cys, His, Trp, and Asn.) The arrowhead indicates the site of the deubiquitinating *in vivo* cleavage in the nascent fusion protein that occurs with all the fusion proteins except Ub–Pro–βgal (see text). All the constructions shown encode His as the second βgal residue. In addition, in some of the constructions (Ub–Met–His–Gly–βgal, Ub–Met–Gln–Gly–βgal, and Ub–Met–Gln–His–Gly–βgal, the last one produced by an insertion mutation; see Table II), either His or Gln followed Met at the Ub–βgal junction, with indistinguishable consequences for the metabolic stabilities of the corresponding βgal proteins. (B) The amino acid (aa) sequence (in single-letter abbreviations) of Ub–Met–βgal, the initial fusion protein (Figure 1), in the vicinity of the Ub–βgal junction. Single-letter amino acid abbreviations: A, Ala; C, Cys; D, Asp; E, Glu; F, Phe; G, Gly; H, His; I, Ile; K, Lys; L, Leu; M, Met; N, Asn; P, Pro; Q, Gln; R, Arg; S, Ser; T, Thr; V, Val; W, Trp; Y, Tyr. (Modified from ref. 31, with permission. Copyright 1986 by AAAS.)

2.2. The *in Vivo* Half-Life of βgal Is a Function of Its Amino-Terminal Residue

The ATG codon that specifies the original Met residue of βgal at the Ub–βgal junction (Figure 3B) was converted by site-directed mutagenesis[40,41] into codons specifying 19 other amino acids; the resulting constructions differ exclusively in the first codon of βgal at the Ub–βgal junction (Figure 3A). After each of the 20 plasmids thus designed was introduced into yeast, analysis of the corresponding βgal proteins pulse labeled *in vivo* led to the following results (Figures 2 and 4 and Table I):

1. With one exception (see below), the efficient deubiquitination of the nascent Ub–βgal occurs regardless of the nature of the amino acid residue of βgal at the Ub–βgal junction. Thus, the apparently Ub-specific protease that cleaves the original Ub–βgal protein at the Gly–Met junction is generally insensitive to the nature of the first residue of βgal at the junction (Figure 3A and Table I). This result, in effect, makes it possible to expose different amino acid residues at the amino termini of the otherwise identical βgal proteins produced *in vivo*.

2. The *in vivo* half-lives of the βgal proteins thus designed vary from more than 20 hr to less than 3 min, depending on the nature of the amino acid residue exposed at the amino terminus of βgal (Figures 2 and 4 and Table I). Specifically, deubiquitinated βgal proteins with either Met, Ser, Ala, Thr, Val, Cys, or Gly at the amino terminus have relatively long *in vivo* half-lives of ~20 hr or more (Figure 2, lanes d–g, and Table I), similar to the half-life of a control βgal whose gene had not been fused to that of Ub. In contrast, the βgal proteins with either Arg, Lys, Phe, Trp, Leu, Asn, or Asp at the amino terminus have extremely short half-lives, between approximately 2 min for Arg–βgal and approximately 3 min for Lys–βgal, Phe–βgal, Trp–βgal, Leu–βgal, Asn–βgal, and Asp–βgal (Figure 2, lanes n–u, and Table I). The half-life of βgal proteins with amino-terminal residues of either Gln, Tyr, or His is approximately 10 min (Figure 2, lanes k–m, and Table I), while an amino-terminal Ile or Glu confers on βgal a half-life of approximately 30 min (Figure 2, lanes h–j, and Table I). Both pulse-chase and continuous labeling techniques were used in these experiments and yielded similar results.

We conclude[31] that the set of 20 individual amino acids can be ordered with respect to the half-lives that they confer on βgal when exposed at

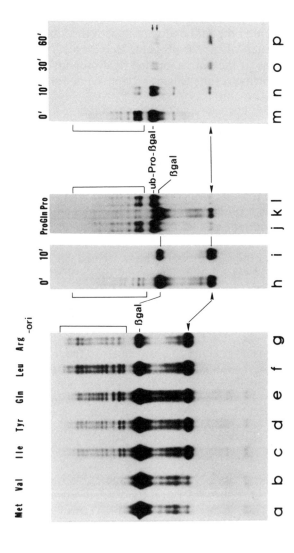

Figure 4. Ub–βgal is short-lived if not deubiquitinated. (Lanes a–g) *S. cerevisiae* cells carrying plasmids encoding Ub–X–βgal fusion proteins, in which X is the residue indicated at the top of each lane, were labeled for 5 min at 30°C with [^{35}S]methionine, with subsequent extraction, immunoprecipitation, and analysis of βgal. Fluorographic exposures for these lanes were several times longer than those for similar patterns in Figure 2 to reveal the multiple ubiquitination of short-lived βgal proteins. (Lanes h and i) Fluorographic over-exposure of lanes n and o in Figure 2 reveals the ladder of multiply ubiquitinated Leu–βgal proteins in a pulse-chase experiment (zero and 10 min chase, respectively). (Lane j) Same as lanes a–g, but with Ub–Pro–βgal. (Lane k) Same as lane j but with Ub–Gln–βgal. (Lane l) Same as lane j. (Lanes m–p) *S. cerevisiae* cells carrying a plasmid encoding Ub–Pro–βgal were labeled for 5 min at 30°C with [^{35}S]methionine (lane m) followed by a chase in the presence of cycloheximide for 10, 30, and 60 min (lanes n–p).[31] The upper small arrow to the right of lane p denotes Ub–Pro–βgal, a small proportion of which is still present after a 1 hr chase. The lower small arrow indicates an apparently deubiquitinated Pro–βgal that slowly accumulates during chase and is metabolically stable. The dot to the left of lane m denotes an endogenous yeast protein that is precipitated in some experiments by the antibody used. Square brackets denote the multiply ubiquitinated βgal species.[31] Other designations are as in Figure 2. (From ref. 31, with permission. Copyright 1986 by AAAS.)

its amino terminus. The resulting rule (Table I) is referred to as the "N-end rule."

2.3. Amino-Terminal Location of an Amino Acid Is Essential for Its Effect on βgal Half-Life

We used site-directed mutagenesis to insert a codon specifying a "stabilizing" amino acid (in this experiment, the Met residue) before the first codon of βgal at the Ub–βgal junction (Table II). Insertion of a stabilizing residue (Met) before either another stabilizing residue (Thr) or a variety of destabilizing residues (Gln, Lys, and Arg) at the Ub–βgal junction invariably results in a *long-lived* deubiquitinated βgal (Table II). Furthermore, in contrast to Ub–Pro–βgal, which is not only short-lived but also resistant to deubiquitination (Figure 4, lanes j–p, and Table I), Ub–Met–Pro–βgal is efficiently deubiquitinated *in vivo* to yield a long-lived Met–Pro–βgal (Table II).

These results[31] show that both the identity of an amino acid residue and its amino-terminal location (presumably the presence of a free α-amino group) are essential for its effect on βgal's half-life. Furthermore, these results (Table II) support the expectation that cleavage of Ub from the fusion protein occurs immediately after the last Gly residue of Ub (Figure 3A). In addition to showing by direct sequencing that the amino-terminal residue in the putative (deubiquitinated) Met–βgal is indeed the

Table II. Amino-Terminal Location of an Amino Acid Is Essential for Its Effect on βgal Half-Life[a]

Fusion protein	$t_{1/2}$ of deubiquitinated fusion protein
Ub–Thr–βgal	>20 hr
Ub–Met–Thr–βgal	>20 hr
Ub–Gln–βgal	≈10 min
Ub–Met–Gln–βgal	>20 hr
Ub–Lys–βgal	≈3 min
Ub–Met–Lys–βgal	>20 hr
Ub–Arg–βgal	≈2 min
Ub–Met–Arg–βgal	>20 hr
Ub–Pro–βgal	≈7 min
Ub–Met–Pro–βgal	>20 hr

[a] The insertion mutants of βgal were obtained, and the *in vivo* half-lives of the corresponding βgal proteins were determined as described in ref. 31.

expected Met residue[31] (see Section 2), we have recently sequenced amino-terminal regions of the putative (deubiquitinated) Ile–βgal and Tyr–βgal proteins. The results (I. Wünning, W. Lane, A. Bachmair, D. Gonda, and A. Varshavsky, unpublished data) directly confirmed the presence of the expected Ile and Tyr residues at the amino termini of the corresponding purified βgals, which are short-lived *in vivo* (Table I; see also Section 5).

We note that while destabilizing amino acids (Table I) can be either hydrophobic, uncharged hydrophilic, or charged, they share the property of having larger radii of gyration[42] than any of the stabilizing amino acids except Met. Furthermore, all the stabilizing amino acids are uncharged (Table I).

2.4. A Long-Lived Cleavage Product of βgal Forms during Decay of Short-Lived βgal Proteins

The electrophoretic patterns of short-lived (but not of long-lived) βgal proteins invariably contain a specific ~90 kDa cleavage product of βgal (Figure 2, lanes n–u), which, unlike the parental βgal species, accumulates during the postlabeling (chase) period (Figure 4, lanes m–p). The ~90 kDa βgal fragment constitutes a relatively small proportion of the initial amount of the pulse-labeled βgal. Nonetheless, its existence implies that an *in vivo* endoproteolytic cleavage can rescue a protein fragment from the metabolic fate of its short-lived parental protein. It remains to be seen whether the resulting possibility of multiple half-lives within a single protein species is exploited in the design of naturally short-lived proteins.

Interestingly, the above metabolically stable *in vivo* cleavage product of βgal in yeast (Figure 2, lanes n–u) is not observed with the same βgal substrates in an *in vitro* Ub-dependent proteolytic system derived from mammalian reticulocytes (see Section 7).

2.5. Ubiquitin–βgal Is Short-Lived When Not Deubiquitinated

Ub–Pro–βgal, the only Ub–βgal fusion that is not efficiently deubiquitinated *in vivo* (Figure 4, lanes j–p), has a half-life of approximately 7 min (Table I), which is less than 1% of the half-life of metabolically stable βgal proteins (Table I). One interpretation of this result is that a metabolically stable Ub attachment to protein amino termini is sufficient to signal degradation of acceptor proteins. This interpretation is consistent with earlier biochemical and genetic evidence that ubiquitination of short-lived proteins in a mammalian cell is essential for their degradation.[1,2,26–28] At the same time, all Ub–βgal fusion proteins other than Ub–Pro–βgal

Figure 5. Both prokaryotic and eukaryotic long-lived intracellular proteins have stabilizing residues at their amino termini, whereas secreted proteins exhibit a complementary bias. (A) Relatively long-lived ($t_{1/2} \gg 1$ hr), directly sequenced, intracellular (noncompartmentalized) proteins with unblocked amino termini from both prokaryotes (77 proteins) and eukaryotes (131 proteins) were distributed into three groups according to the nature of their amino-terminal residues as defined by the N-end rule (Table I). All 208 of the long-lived intracellular proteins examined bear exclusively stabilizing residues at their amino termini. (B–D) Analogous diagrams are presented for 243 secreted eukaryotic proteins (B), for 37 light and heavy immunoglobulin chains (C), and for 94 secreted eukaryotic toxins (D). Entries in (C) and (D) are subsets of entries in (B). For proteins in (B)–(D), the amino termini compiled correspond, whenever the assignment is possible, to the most processed form of a protein that is still located within a secreting cell. The data in (A) and (D) were manually compiled from the entire set of complete protein sequences available before 1981 (see ref. 31). The amino-terminal residues of Asn, Cys, His, and Trp were excluded from the compilation shown in (A)–(D) because *in vivo* half-lives of the corresponding βgal proteins were still unknown at the time of compilation.[31] Inclusion of these residues (Table I) into a recent, more extensive survey did not change the original conclusions (see also main text). Although the amino-terminal Pro was also excluded from the original compilation, both the earlier evidence[31] and the more recent *in vitro* data (see Section 7) indicate that Pro is a stabilizing residue for βgal. The single-letter amino acid abbreviations are given in the legend to Figure 3. (From ref. 31, with permission. Copyright 1986 by AAAS.)

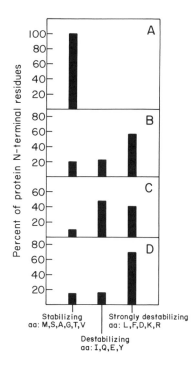

are rapidly deubiquitinated *in vivo* (Table I). Thus, it remains to be determined whether *post-translational* amino-terminal ubiquitination of proteolytic substrates occurs at later stages of the targeting pathway (see also Section 2.6).

That a slow *in vivo* deubiquitination of Ub–Pro–βgal does take place in yeast is indicated by the data of Figure 4 (lanes m–p).[31] The resulting Pro–βgal is metabolically stable (Figure 4 and data not shown), in agreement with the fact that a significant proportion of noncompartmentalized, relatively long-lived proteins in both prokaryotes and eukaryotes have Pro at their mature amino termini (see Figure 5 and discussion in Section 2.6). Addition of the purified Ub–Pro–βgal to the Ub-dependent *in vitro* system from mammalian reticulocytes also results in a slow deubiquitination reaction, yielding Pro–βgal, which is stable in the reticulocyte extract (see Section 7).

2.6. Short-Lived βgal Proteins Are Multiply Ubiquitinated *in Vivo*

Overexposures of the pulse-chase fluorograms reveal that the major band of a deubiquitinated, short-lived βgal protein coexists with a "ladder" of larger molecular mass, βgal-containing bands irregularly spaced at 4–7 kDa intervals (Figure 4, lanes c–g). No such larger species appear when the fluorograms of long-lived βgal proteins are similarly overexposed (Figure 4, lanes a and b). Immunological analysis with antibodies to both βgal and Ub demonstrated[31] that the ladder βgal species contain Ub.

At least 15 distinct Ub–βgal bands can be distinguished in the patterns of ubiquitinated βgal proteins in Figure 4. Since βgal used in this work contains 22 Lys residues (see Section 8), the above results are consistent with the possibility that most if not all of the Lys residues of βgal can be conjugated *in vivo* to Ub in the more heavily ubiquitinated βgal species. At the same time, at least *in vitro* some Ub–protein conjugates are larger than could be expected from the number of Lys residues in a target protein[43-46] and appear to include branched Ub–Ub conjugates.[43] Thus, the multiply ubiquitinated βgal species (Figure 4) may also contain such Ub–Ub conjugates. Strikingly, the results of a recent direct mapping of Ub within multiply ubiquitinated, amino-terminally targeted βgal proteins do indicate that most if not all of their multiple Ub moieties reside in branched Ub–Ub structures that are attached to βgal within its first ~20 amino-terminal residues (V. Chau, A. Bachmair, D. Gonda, and A. Varshavsky, unpublished results). Since this region of βgal contains, in addition to the amino-terminal α-amino group, two Lys residues (Figure 3B), it is still unknown which of these three potential Ub acceptor sites are actually conjugated to these Ub–Ub moieties. The functional significance of multiple, branched Ub–Ub conjugates in an amino-terminally targeted protein remains to be understood. One possibility, consistent with the above results, is that the "downstream" Ub-dependent protease[1,4,7] (see Chapter 4) specifically recognizes *branched* Ub–Ub conjugates rather than single Ub moieties in a ubiquitinated protein (for further discussion, see Sections 8 and 9).

3. THE N-END RULE

3.1. Amino-Terminal Residues in Sequenced Intracellular Proteins

The unblocked amino-terminal residues in relatively long-lived ($t_{1/2} \gg 1$ hr), noncompartmentalized proteins from both prokaryotes and eukaryotes are virtually exclusively (Figure 5A) of the stabilizing class (Met,

Ser, Ala, Thr, Val, Cys, Gly, Pro), that is, the class that confers long *in vivo* half-lives on βgal (Table I). The one *short-lived* intracellular protein for which the mature amino terminus is known is the cII protein of phage λ, the central component of a trigger that determines whether λ grows lytically or lysogenizes an infected cell.[16-18] The half-life of cII in λ-infected *E. coli* is less than 3 min.[16] Strikingly, the mature amino terminus of cII starts with the third encoded residue, Arg,[16] a destabilizing amino acid in the N-end rule (Table I). Moreover, several previously characterized mutations in cII that *increase* its *in vivo* half-life have been mapped to the second amino acid position in the nascent protein.[16,47] For instance, the *can-1* mutation changes the second encoded residue of cII from Val to Ala. The mature amino terminus of the overproduced *can-1* cII protein starts predominantly with the second residue, Ala, a stabilizing amino acid in the N-end rule, in contrast to Arg, a third residue, that is present at the mature amino terminus of the wild-type cII protein.[16]

Our recent, more extensive survey of the amino-terminal residues in the directly sequenced, noncompartmentalized, relatively long-lived intracellular proteins (data not shown) confirmed and extended the result of Figure 5A, with only two apparent exceptions encountered so far among more than 300 protein sequences scored.

3.2. Amino-Terminal Residues in Compartmentalized Proteins Are Largely of the Destabilizing Class

Figure 5 illustrates a striking difference between the choice of amino-terminal residues in long-lived, noncompartmentalized intracellular proteins (A) and in compartmentalized proteins, such as secreted proteins (B), many of which are also long-lived in their respective extracellular compartments. One implication of this finding is that a single intracellular degradation pathway operating according to the N-end rule could underlie both the diversity of *in vivo* half-lives of intracellular proteins and the selective destruction of compartmentalized proteins that are aberrantly introduced into the intracellular space.

Several specific compartmentalized proteins have been shown to be short-lived upon microinjection into the cytoplasm of *Xenopus* oocytes.[48] At the same time, at least one compartmentalized protein (RNase A) with a destabilizing amino-terminal residue (Lys) is long-lived upon microinjection into mammalian cells (see Section 8 for the discussion of these and related data). Thus, it remains to be seen whether the physiological meaning and evolutionary origin of the bias in Figure 5B have any direct relation to amino-terminal recognition in selective protein degradation.

3.3. The N-End Rule Pathway

Most nascent proteins (with the exception of natural or engineered Ub fusions) apparently lack Ub moieties. The *in vivo* amino-terminal processing of nascent, noncompartmentalized proteins generates their mature amino termini via the action of proteases whose substrate specificities have been partially characterized.[49-55] In particular, the consistent absence of 12 amino acids that are destabilizing according to the N-end rule (Table I) from the mature amino termini of relatively long-lived, noncompartmentalized proteins (see Figure 5A and Section 3.1) is largely due to the substrate specificity of the enzyme methionine aminopeptidase.[52-55] In both bacteria and eukaryotes, this enzyme[52-55] cleaves off the amino-terminal Met residue if and only if it is *not* followed either by another Met residue or by any of the 12 residues that are destabilizing according to the N-end rule. Such cleavage specificity would be obviously relevant functionally if the amino-terminal exposure of a (penultimate) destabilizing residue were to confer short half-lives on at least some of the otherwise long-lived proteins that normally retain their amino-terminal methionine. The striking "inverse" correspondence between the N-end rule and the specific substrate requirements[49-55] of the methionine aminopeptidase provides a partial functional explanation for the properties of this enzyme: a methionine-clipping aminopeptidase that is involved in processing of relatively long-lived proteins would be expected *not* to expose a residue whose presence at the amino terminus might destabilize the substrate protein.

We suggest that analogous proteases may be responsible for the generation of amino termini bearing *destabilizing* residues in specific proteins whose amino-terminal sequences contain recognition sites for such proteases. An example of a noncompartmentalized protein whose amino-terminal processing yields a destabilizing amino-terminal residue is the cII protein of phage λ (see Sections 1 and 3.1), in which the third residue, Arg, is found at the mature amino terminus.[16]

We have previously suggested[31] that the mature amino termini bearing destabilizing residues are recognized by an "N-end-reading" enzyme. One specific model of amino-terminal targeting is that a commitment to degrade a protein is made as a result of the recognition of the protein's amino-terminal residue by an enzyme whose probability of "clamping" at the target's amino terminus is determined at least in part by the N-end rule (Table I). Once the commitment is made, a Ub–protein ligase complex is assembled at the commitment site (in a way that may be mechanistically analogous to the stepwise assembly of transcription complexes at the promoter sites[56]), followed by a highly processive ubiquitination of the target protein, which in the case of βgal involves the conjugation of more than 15 Ub moieties per molecule of βgal (see Section 2.6). The ubiquitinated target protein is then degraded by a "downstream" enzyme (see Chapter 4) for which the Ub moieties of the target serve as either recognition signals or denaturation (unfolding) devices or both (see Sections 2.6, 5, and 8 for further discussion).

We note that the properties of a mammalian protein E3, whose presence is required for ubiquitination of proteolytic substrates by Ub-conjugating enzymes *in vitro*,[4,57] are consistent with E3 being a component of the N-end-recognizing enzyme.[31]

The Ub-dependent degradative pathway whose initial steps involve amino-terminal recognition of proteolytic substrates is called the *N-end rule pathway* (see also Section 5), to distinguish it from other proteolytic pathways and also from other Ub-dependent processes, some of which do not involve degradation of target proteins (see Section 9).

4. POSSIBLE ROLE OF THE N-END RULE PATHWAY IN THE TURNOVER OF LONG-LIVED PROTEINS

Long-lived intracellular proteins with destabilizing (Table I) penultimate residues generally retain their initial amino-terminal methionine.[49-55] As discussed above (Figure 5A and Sections 3.1 and 3.3), amino-terminal residues in long-lived intracellular proteins that do undergo amino-terminal processing are almost invariably of the stabilizing class (Table I). An interesting possibility that would involve the N-end rule pathway in the turnover of long-lived proteins is that the rate-limiting step in the *in vivo* degradation of long-lived proteins may be a slow aminopeptidase cleavage that exposes a destabilizing residue, followed by rapid degradation via the N-end rule pathway (Figure 6). Fine tuning of the rate of degradation may in this case be a function of the rate of aminopeptidase cleavage that exposes a destabilizing residue (Figure 6), rather than a function of the residue's destabilizing capacity according to the N-end rule.

Whether or not the above hypothesis (Figure 6) will prove to be correct for at least some long-lived proteins, *in vivo* proteolytic targeting of long-lived intracellular proteins is likely to involve several distinct pathways, including lysosome-mediated ones.[6,58-60]

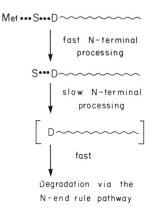

Figure 6. A model for the targeting of long-lived proteins via the N-end rule pathway. Designations: S and D, amino-terminal residues that are stabilizing and destabilizing, respectively, according to the N-end rule (see Table I and main text).

5. FUNCTION OF POST-TRANSLATIONAL CONJUGATION OF AMINO ACIDS TO AMINO TERMINI OF PROTEINS: PRIMARY AND SECONDARY DESTABILIZING AMINO ACIDS

It has been known for many years that in both bacteria and eukaryotes there exists an unusual class of enzymes, aminoacyl–transfer RNA–protein transferases, which catalyze post-translational conjugation of specific amino acids to the mature amino termini of acceptor proteins *in vitro*.[61-63] It has also been found that post-translational conjugation of amino acids to proteins *in vivo* is greatly enhanced in a stressed or regenerating tissue, for example, after physical injury to axons of neurons.[64] Furthermore, earlier *in vitro* studies[65] of a Ub-dependent proteolytic system from mammalian reticulocytes have shown that the degradation of some proteolytic substrates in this system depends on the presence of tRNA. Although the Ub-dependent *in vitro* degradation of bovine serum albumin and α-lactalbumin (which contain Asp and Glu, respectively, at their amino termini) was inhibited by the depletion of tRNAs,[65] no such effect was seen with another proteolytic substrate, lysozyme, whose mature amino-terminal residue is Lys.

The N-end rule provided a mechanistically straightforward explanation for the above observations and for the function of amino-terminal amino acid conjugation.[31] Specifically, Asp and Glu, the amino-terminal residues of proteolytic substrates whose *in vitro* degradation is dependent on the presence of tRNA (see Section 4), are both destabilizing according to the N-end rule (Table I). We have therefore suggested[31] that certain amino-terminal residues in proteins may not be directly destabilizing as such but only through their ability to be conjugated to other, *primary* destabilizing residues. Strikingly, the known reactions of post-translational conjugation of amino acids to proteins[61-64] involve largely those amino acids (Arg, Lys, Leu, Phe, and Tyr) that are destabilizing according to the N-end rule (Table I). Thus, the function of currently known aminoacyl–tRNA–protein transferases could be to conjugate a *primary* destabilizing amino acid to a *secondary* destabilizing residue at the amino terminus of a protein, thereby converting the protein into a direct substrate of the N-end rule pathway (Figure 7). The transferases, which participate in the above reactions (Figure 7), are expected to be constitutively synthesized enzymes whose substrate specificity is limited largely to the recognition of an unblocked *secondary* destabilizing residue at the amino terminus of a target protein. Properties of the currently known aminoacyl–tRNA–protein transferases from both bacteria and eukaryotes[61-64] are consistent with this view. Recent *in vitro* evidence for the above model (Figure 7) is that a proteolytic substrate such as serum albumin, whose degradation in the reticulocyte extract is tRNA dependent, does acquire an extra, post-translationally added, *primary* destabilizing amino acid residue (Arg in the case of serum albumin) at its amino terminus.[66] Furthermore, using purified βgal proteins with different amino-terminal residues as substrates for a reticulocyte-derived *in vitro* proteolytic system, we have recently shown that the set of *secondary* destabilizing amino acids includes Asp, Asn, Glu, and Gln (D. Gonda, A. Bachmair, and A. Varshavsky, unpublished results; see also Section 7). Finally, direct *in vivo* evidence for the above model (Figure 7) was obtained recently by sequencing amino-terminal regions of Asp–βgal and Glu–βgal that have been purified from yeast cells expressing the corresponding Ub–βgal fusions. It was found that the bulk of both Asp–βgal and Glu–βgal exist *in vivo* with an extra, *post-translationally* added amino-terminal residue (largely, if not exclusively, Arg) (I. Wünning, W. Lane, A. Bachmair, D. Gonda, and A. Varshavsky, unpublished results).

Thus, as we have suggested previously,[31] the 12 destabilizing residues in the N-end rule (Table I) are expected to constitute two apparently nonoverlapping sets of *primary* and *secondary* destabilizing residues. At present, it is already clear that, in eukaryotes, Arg, Lys, and His, and apparently also Tyr and Ile, are primary destabilizing amino acids, and Asp, Asn, Glu, and Gln are secondary ones (refs. 31 and 66 and the unpublished data cited above; see also Section 2.3).

A different, hypothetical function for aminoacyl–tRNA–protein transferases stems from a teleological argument that cells may need a mechanism to destabilize selectively a subset of otherwise undamaged, long-lived intracellular proteins (Figure 8). Transferases responsible for such reactions would have to conjugate destabilizing amino acids to *sta-*

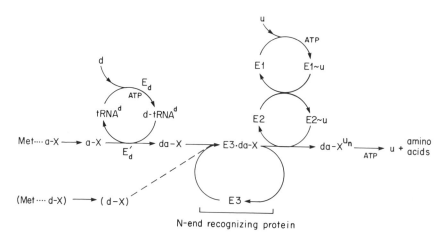

N-end recognizing protein

Figure 7. A model of the N-end rule pathway that includes the amino acid conjugation step. Designations: a, "secondary" destabilizing amino-terminal amino acid residue, which destabilizes only through its ability to be post-translationally conjugated (via the aminoacyl–tRNA cycle on the left) to the "primary" destabilizing residue d. E_d, aminoacyl–tRNA synthetase specific for d. E'_d, aminoacyl-tRNA-protein transferase, which conjugates d to a-containing amino termini. The N-end-recognizing protein is denoted as E3, since the properties of a mammalian protein E3, whose presence is required for ubiquitination of proteolytic substrates by Ub-conjugating enzymes E1 and E2 in $vitro$,[57] are consistent with it being a component of the N-end-recognizing protein.[31] E1 and E2, Ub-activating enzyme and Ub-conjugating enzymes, respectively.[1,2,4,7,32,37] u, ubiquitin. We use a designation X^{u_n} to denote a protein X that contains n ubiquitin moieties branch-conjugated to it (see Section 2.6 for discussion of the structure of multiply ubiquitinated βgal proteins). In the example shown, a newly formed protein, Met···a-X, is processed in $vivo$ to yield an a-containing mature amino terminus, which is then conjugated to d. The amino-terminal d residue in the protein da-X is recognized by the N-end-recognizing protein (E3) as a $primary$ destabilizing residue, and da-X is targeted for multiple ubiquitination and subsequent selective degradation. An alternative pathway (dashed line) starts with a newly formed protein Met···d-X whose amino-terminal processing yields d as its mature amino-terminal residue. Since d is a $primary$ destabilizing residue, the protein d-X is targeted for ubiquitination directly, bypassing the post-translational amino acid conjugation step. Not shown is the pathway for a protein whose amino-terminal processing yields a stabilizing residue (Table I) at its mature amino terminus. Such proteins would not be targets for the N-end rule pathway. Note that by not specifying the relative positions of residues a and d within newly formed proteins, this scheme implicitly incorporates the possibility of endoproteolytic processing, which may expose destabilizing residues far away from the initial amino-terminal Met residue (for a further discussion of this point, see Section 6, ref. 31, and Figure 9). See Sections 5 and 8 for a further discussion of the amino-terminal targeting.

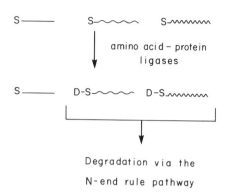

Figure 8. A hypothetical pathway for selective destabilization of proteins via amino-terminal amino acid conjugation. Shown are three long-lived proteins with stabilizing amino-terminal residues. Two of these proteins are selectively converted to short-lived ones by the amino-terminal conjugation of destabilizing residues. Designations are as in Figure 6. See Figure 7 and Section 5 for details and discussion.

bilizing residues at the amino termini of target proteins and would therefore be expected to recognize specific features in their protein substrates that are distinct from those recognized by the known transferases discussed above. Unlike the constitutively produced transferases, the hypothetical transferases of Figure 8 are expected to be inducible and otherwise tightly regulated. Erythroid maturation in animals and sporulation in yeast are potential examples of differentiation pathways in which the selective destabilization mechanism of Figure 8 may play a role.

6. AMINO TERMINAL VERSUS "INTERIOR" TARGETING IN SELECTIVE PROTEIN TURNOVER

The recognition of polypeptide chain folding patterns or of local chemical features that target a protein for selective degradation *in vivo* is unlikely to be mediated exclusively by the N-end rule pathway. For instance, the amino-terminal processing of an otherwise long-lived but chemically damaged protein would not be expected *always* to differ from the amino-terminal processing of its undamaged counterpart. With regard to the N-end rule pathway, two distinct but not mutually exclusive classes of more general models of targeting can be envisioned. In one class of models, proteolytic substrates may be targeted either at the amino terminus through the N-end rule pathway or, alternatively, via a mechanism that is at no point dependent on targeting via the N-end rule pathway. Thus, in a recently identified potential example of an N-end rule-independent targeting, certain initially noncompartmentalized long-lived proteins appear to be targeted for degradation in lysosomes by a cytoplasmic receptor recognizing a specific sequence motif in the protein.[58] At the same time, *hierarchically* designed targeting pathways could also be en-

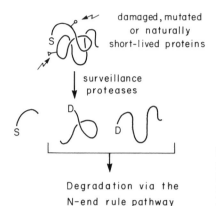

damaged, mutated
or naturally
short-lived proteins

surveillance
proteases

Degradation via the
N-end rule pathway

Figure 9. A model for selective protein degradation in which the N-end rule pathway is functionally coupled to targeting events away from the amino terminus. Designations are as in Figure 6.

visioned, whose initial, N-end rule-independent steps convert a protein into a direct substrate for the N-end rule pathway. Specifically, we suggested[31] that there exists a class of intracellular "surveillance" endoproteases (analogous in function to nucleases that recognize specific lesions in DNA) that cleave a targeted protein so as to expose a destabilizing residue at the amino terminus of one of the two products of a cut (Figure 9). Each surveillance protease is expected to recognize either a specific amino acid sequence motif or a specific class of chemical or conformational lesions in a target protein (Figure 9). This model suggests that the initial cleavage products of the degradation pathway should bear destabilizing residues at the amino termini. Preferential exposure of destabilizing residues at the amino termini of products of the initial protein cleavages may be due to either intrinsic specificites of the proteases involved or simply the fact that the majority of the amino acids belongs to the destabilizing class (Table I). Initial cleavages of a protein would be expected to destabilize aspects of its original conformation, thus increasing the probability of further internal cuts. Whether the initial cleavage products of a protein would be degraded exclusively via the N-end rule pathway or would have to be processed further by additional internal cleavages should depend on several factors, such as the accessibility of destabilizing residues at the amino termini of initial cleavage products, and the relative rates of internal cleavages. In this model[31] (Figure 9), the N-end rule pathway should be involved in degradation of most of the metabolically unstable proteins, from chemically damaged, prematurely terminated, improperly folded, and miscompartmentalized ones to those that cannot assemble into native multisubunit aggregates, and finally to otherwise normal proteins that are short-lived *in vivo*. Thus, the N-end rule pathway may be involved in the degradation of proteins whose *initial* recognition as targets for degradation is independent of the structures at their amino termini.

We note that an independently proposed hypothesis of targeting via specific internal sequences in proteolytic substrates, the PEST hypothesis,[67] is formally analogous to the first targeting step in Figure 9 but, in addition, suggests a set of specific sequence motifs in proteins that may serve as the initial signals for degradation.

7. *IN VITRO* ANALYSIS OF THE N-END RULE PATHWAY

Several Ub–βgal fusion proteins (Figure 3) have recently been overexpressed in *E. coli* and purified to homogeneity under nondenaturing conditions. The purified Ub–βgal proteins were then added to an *in vitro* proteolytic system derived from rabbit reticulocytes. In this extensively studied *in vitro* system, selective degradation of specific protein substrates is both Ub and ATP dependent.[1–4,7] One initial aim of our analysis was to determine relative stabilities of different Ub–βgal proteins in this mammalian *in vitro* system and to compare the order of stabilities obtained with the order of *in vivo* stabilities of the same proteins in yeast cells. Several conclusions can be drawn from these data (Figure 10). (1) Rapid deubiquitination of Ub–βgal fusion proteins, which has been observed previously with yeast cells *in vivo*, takes place in the above mammalian *in vitro* system as well (data not shown). (2) Deubiquitinated βgal proteins with either Gln or Arg at their amino termini, residues that are destabilizing according to the N-end rule as derived with yeast cells *in vivo*, are also unstable in the mammalian *in vitro* system (Figure 10 and data not shown). (3) An otherwise identical deubiquitinated βgal protein with Met, a stabilizing residue, at its amino terminus is completely stable in the same *in vitro* system (Figure 10 and data not shown). (4) Deubiquitination of Ub–Pro–βgal is extremely slow not only within yeast cells *in vivo*[31] (Figure 4) but in the mammalian *in vitro* system as well (data not shown). (5) Pro–βgal, the product of slow *in vitro* deubiquitination of Ub–Pro–βgal, is stable in the mammalian *in vitro* system (data not shown), in agreement with analogous *in vivo* data obtained in yeast.[31] (6) In contrast to the metabolic instability of the (nondeubiquitinated) Ub–Pro–βgal within yeast cells *in vivo* (Table I and Figure 4), the same protein is stable in the reticulocyte extract (Figure 10 and data not shown). The mechanistic significance of this difference is being examined. (7) The metabolically stable ~90 kDa cleavage product of short-lived βgals, which has been observed *in vivo* with yeast cells[31] (Figure 2), is not seen upon in-

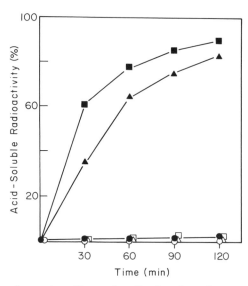

Figure 10. Selective *in vitro* degradation of deubiquitinated βgal proteins in a reticulocyte extract. Complete reaction mixtures (closed symbols) contained 50 mM Tris-HCl (pH 7.5), 5 mM $MgCl_2$, 1 mM dithiothreitol, 0.5 mM ATP, an ATP regenerating system (10 mM creatine phosphate, 0.1 mg/ml creatine phosphokinase), 70% (v/v) of ATP-depleted lysate from rabbit reticulocytes,[4] and 20 µg/ml of labeled proteolytic substrate (see below). Reactions were carried out by incubating an otherwise complete mixture in the absence of ATP and ATP-regenerating system for 10 min at 37°C to allow for the deubiquitination of a Ub–βgal fusion protein (data not shown), followed by the addition of ATP and ATP-regenerating system and further incubation at 37°C. Control reactions (open symbols) lacked ATP and ATP-regenerating system. Conversion of radioactive substrates to acid-soluble form was measured essentially as described previously.[4] Analysis of the same set of reactions by SDS–PAGE showed in particular that no detectable partial proteolysis of deubiquitinated βgal proteins occurred in the absence of ATP. The labeled protein substrates were either bovine serum albumin labeled with ^{125}I by the Chloramine T method (squares), [^{35}S]Ub–Met–βgal (circles), or [^{35}S]Ub–Gln–βgal (triangles). The latter two proteins[31] were overexpressed and metabolically labeled with [^{35}S]methionine in *E. coli* by using the pKK233-2 expression vector[79] and purified to homogeneity from an *E. coli* extract by affinity chromatography on APTG–Sepharose[80] (data not shown).

cubation of the same βgal proteins in the mammalian *in vitro* system (data not shown). (8) *In vitro* degradation of those βgal proteins that are unstable in the reticulocyte extract is both ATP dependent (Figure 10) and Ub dependent (data not shown).

Although the *in vitro* analysis of amino-terminal targeting with βgal proteins is not yet complete (see also Note added in proof), it is already clear that the N-end rule as defined with βgal in yeast *in vivo* is qualitatively similar to the partial N-end rule defined thus far with βgal in a reticulocyte-derived mammalian *in vitro* system. Strikingly, however, some of the amino-terminal residues that are stabilizing in the yeast-derived N-end rule are significantly destabilizing in the reticulocyte-derived N-end rule (D. Gonda and A. Varshavsky, unpublished results). The combination of the above *in vitro* approach and well-defined, amino-terminally targeted proteolytic substrates such as βgal (Figures 3 and 10) and dihydrofolate reductase (see below) should prove extremely helpful in further studies of the N-end rule pathway.

8. CURRENT PROBLEMS IN AMINO-TERMINAL TARGETING

The striking dependence of the *in vivo* half-life of βgal on the nature of its amino-terminal residue (Table I and Figure 7) defines one aspect of targeting in the N-end rule pathway. At the same time, as suggested before,[31] it is likely that for any given *amino-terminally* targeted protein, a variety of factors in addition to the N-end rule may combine to modulate its half-life *in vivo*. Among such factors may be the accessibility and segmental mobility of the protein's amino-terminal region, the occurrence of blocking amino-terminal modifications such as acetylation, and the presence of specific sequence motifs near the protein's amino terminus. Since amino-terminal regions of multisubunit proteins are commonly involved in the interfaces between subunits,[68] quaternary structure of proteins is yet another parameter that is expected to modulate the impact of the N-end rule pathway on protein half-lives *in vivo*.

The available evidence does indicate that the effect of an amino-terminal residue on the *in vivo* half-life of a protein can be modified by other structural features of the protein. For instance, RNase A, a secreted protein, in spite of having a destabilizing (Lys) residue at its mature amino terminus, is relatively long-lived upon microinjection into mammalian cells ($t_{1/2} \sim 60$ hr).[58,69]

In a recently constructed set of Ub fusions with mouse dihydrofolate reductase (DHFR), a monomeric ~20 kDa protein whose structure is known at atomic resolution,[70,71] the mature amino terminus of the "natural" DHFR is extended by 7 residues owing to the construction route taken (Figure 11A). After cleavage of Ub from the nascent Ub–DHFR fusion proteins *in vivo*, the deubiquitinated DHFR proteins differ exclusively in their amino-terminal residues. These constructions are analogous to the previously described set of βgal test proteins (Figure 3).[31] As expected, the DHFR proteins bearing those amino-terminal residues that are stabilizing according to the N-end rule (Table I) are long-lived in yeast (Figure 11 and data not shown). Although the presence of a residue that is destabilizing according to the N-end rule at the amino termini of an otherwise identical DHFR protein does destabilize it *in vivo*, the extent of destabilization is small (Figure 11A) in comparison to the results with βgal of analogous design[31] (Table I). To address the mechanistic significance of these findings, a 36 residue amino-terminal region of βgal was positioned upstream of the original DHFR's amino terminus (Figure 11B). The DHFR proteins bearing a destabilizing residue followed by the βgal-derived extension are approximately as short-lived *in vivo* as their unstable βgal counterparts, in striking contrast to the otherwise identical DHFR proteins that lack the βgal-specific amino-terminal extension (Fig-

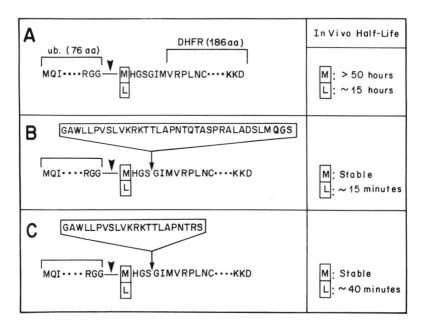

Figure 11. Structure and metabolic stability of Ub–dihydrofolate reductase (DHFR) fusions (A. Bachmair and A. Varshavsky, unpublished results). (A) The amino acid sequence of a Ub–DHFR fusion protein in which the mature amino terminus of the "natural" (mouse) DHFR is extended by 7 residues owing to a construction route taken (to be described elsewhere). Single amino acid replacements at the Ub–DHFR junction (only Met and Leu are shown) have been introduced by site-directed mutagenesis essentially as described previously for Ub–βgal (ref. 31; see also Figure 3). A panel on the right indicates the *in vivo* half-lives of the corresponding deubiquitinated DHFR proteins in the yeast *S. cerevisiae*. (B) The Ub–DHFR fusion protein in (A) was modified by inserting a 36 residue amino-terminal region of βgal (see Figure 3) 4 residues downstream from the Ub–DHFR junction. Note that due to the design of the original expression vector[72] used in our work,[31] the first 45 residues of our βgal test proteins are derived in part from an internal sequence of the *E. coli lac* repressor.[72] Thus, the 36 residue extension of DHFR that is taken from the above βgal proteins is actually derived from the sequence of the *lac* repressor (see also Section 8). (C) Same as in (B) but a shorter insertion (22 residues). The arrowhead indicates the site of the deubiquitinating *in vivo* cleavage in the nascent fusion protein (ref. 31 and data not shown). The single-letter amino acid abbreviations are given in the legend to Figure 3.

ure 11B and data not shown; cf. Figure 11A). Furthermore, the extension-bearing DHFR proteins that have *stabilizing* residues at their amino termini are long-lived *in vivo* (Figure 11B). This latter result proves that the βgal-specific extension *as such*, in the absence of a destabilizing amino-terminal residue, does not confer a short half-life on DHFR. These findings also indicate that the reason for the striking difference between half-lives of the DHFRs that either lack or contain the βgal-specific extension (and bear identical, destabilizing amino-terminal residues) is due to differences in *amino-terminal* targeting elements in these proteins and not to differences between the overall structures of DHFR and βgal.

When DHFR is fitted with a 22 residue, βgal-derived amino-terminal extension instead of the original 36 residue extension, the dependence of the *in vivo* half-life of the resulting protein on the nature of its amino-terminal residue is *intermediate* between that of the original DHFR and that of the DHFR bearing a 36 residue βgal-derived extension (Figure 11C; cf. Figure 11B). One interpretation of this and related findings (data not shown) is that sequences required for the effect of the 36 residue βgal-specific extension are not confined to a short stretch within the extension but are distributed over the length of the extension.

The above evidence indicates that the presence of an amino-terminal amino acid residue that is destabilizing according to the N-end rule[31] is a necessary but not always sufficient condition for the efficient *in vivo* destabilization of a test protein such as DHFR. These data put significant contstraints on the range of structural features additional to the amino-terminal residue that may be required for amino-terminal targeting. We discuss below those hypothetical factors that are likely to be relevant as inferred from the presently available evidence.

(1) In the deubiquitinated βgal proteins of Figure 2 and Table I, an amino-terminus-proximal internal Lys residue is located 15 residues from the amino terminus (Figure 3A). In contrast, an internal Lys residue closest to the original DHFR's amino terminus is located 25 residues away from the amino terminus, that is, 10 residues farther than in βgal. The ε-amino groups of Lys residues are the only functional groups in proteins that are known to form amide bonds, post-translationally, with the carboxyl-terminal Gly residue of Ub, yielding branched Ub conjugates.[1,2,4] Thus, it is conceivable that an otherwise "committed" complex at the amino terminus of a targeted protein (Figure 7) requires the presence of an internal Lys residue at a sufficiently small distance from the amino terminus to allow efficient *initiation* of ubiquitination of the targeted protein.

(2) An alternative but not mutually exclusive explanation of the influence of the βgal-derived extension (Figure 11), which shall be directly combined below with the "lysine proximity" hypothesis, is that efficient amino-terminal targeting requires not only a destabilizing residue at the amino terminus of a target protein but, in addition, a sufficiently high accessibility and *segmental mobility* of the target's amino-terminal region. The "segmental mobility" requirement may account for the relatively high metabolic stabilities of both the original DHFR test proteins bearing destabilizing amino-terminal residues (Figure 11A) and some natural pro-

teins with destabilizing amino-terminal residues, such as RNase A, whose amino-terminal regions are known to be highly ordered.[70,71] Moreover, we note that owing to the design of the original expression vector[72] used in our work,[31] our βgal test proteins bear a 45 residue amino-terminal extension derived in part from an *internal* sequence of the *lac* repressor encoded by the *lacI* gene.[72] Thus, the "βgal-derived" amino-terminal extension discussed above is derived not from the amino-terminal sequence of the wild-type βgal but from an unrelated sequence present at the amino termini of our βgal test proteins. It is likely that the *lac* repressor-specific extension at the amino termini of these βgals is more disordered (segmentally mobile) than the amino-terminal region of wild-type βgal. If so, this extension, while not metabolically destabilizing βgal as such, might allow the observed extreme dependence of the βgal half-life on the nature of its amino-terminal residue[31] and thereby, in hindsight, could have greatly facilitated the discovery of the N-end rule. Furthermore, if the N-end rule pathway is indeed involved in the degradation of proteins initially targeted by "surveillance" endoproteases as discussed in Section 6 (Figure 9), the hypothetical additional requirement for segmental mobility of the amino-terminal region could be physiologically meaningful because the newly generated amino-terminal regions of the cleavage products (Figure 9) are likely to be more disordered (segmentally mobile) than the natural amino termini of the parental proteins.

Thus, the N-end rule pathway (Figure 7) may require *both* a destabilizing amino-terminal residue (Table I) and a sufficiently disordered (segmentally mobile) amino-terminal region in a protein for its efficient targeting. If so, what mechanistic aspects of the pathway might account for the hypothetical "segmental mobility" requirement? As discussed below, a version of the "lysine proximity" hypothesis [see item (1) above], considered in the context of a kinetically controlled pathway, can directly account for the "segmental mobility" requirement without invoking additional *ad hoc* mechanisms. This model is shown in Figure 12. In the initial recognition step, an N-end-recognizing complex (see Section 3.3) binds d, a primary destabilizing residue that is exposed at the mature amino terminus of a processed protein target (Figure 12, step II). The crucial point of the proposed model is that this initial binding is kinetically reversible, and unless the next step is completed before decay of the initial complex, the protein will not be recognized as a target. A second essential recognition step, and possibly also the final one, is proposed to be the capture of an *internal* Lys residue in the target (Figure 12, step III). Note that although the Lys residue to be captured is shown in Figure 12 as being close to the protein's amino terminus, this constraint is not intrinsic to the model. Indeed, the "linear" distance of the critical Lys residue

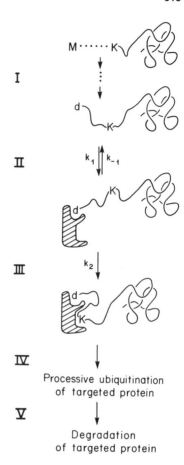

Figure 12. A hypothetical mechanism of amino-terminal targeting in the N-end rule pathway. Designations: M and K, Met and Lys, respectively; d, a primary destabilizing residue at the amino terminus of a processed target protein (see also Figure 7). For clarity, both the relative length and "segmental mobility" of an amino-terminal region have been exaggerated. This model is discussed in Section 8.

from the amino terminus of a target is expected to be less relevant than its spatial proximity to the amino terminus, as discussed below. The requirement that an internal Lys residue must be captured before the decay of the initial, reversible complex (Figure 12, steps II and III) should make recognition in the N-end rule pathway directly dependent on the segmental mobility of a corresponding region in the target protein. This is so because, even for a protein with a Lys residue that is spatially proximal to the protein's amino terminus, a lack of conformational mobility in the Lys-containing region and the (postulated) fixed mutal arrangement of the d and K binding sites in the N-end-recognizing complex (Figure 12) would in most cases, owing to steric constraints, allow the complex to bind either the residue d or the residue K(Lys) but not both. On the other hand, the more segmentally mobile a Lys-containing region that is spatially proximal to the protein's amino terminus is, the higher is the probability that during the time allotted for the Lys capture, the above Lys residue would find itself in a sufficiently close proximity to the K binding site to allow capture (Figure 12, step III). In this model, completion of the step III commits a targeted protein to processive ubiquitination, and thereby to subsequent degradation.

The first ubiquitination event in the N-end rule pathway (Figures 7 and 12) may involve either the amino-terminal α-amino group of the target, or the ε-amino group of its "captured" Lys residue, or both these groups; the only relevant experimental constraint available at present is that in a multiply ubiquitinated, amino-terminally targeted βgal protein most if not all of the multiple Ub moieties occur as branched Ub–Ub conjugates and are attached to βgal within its first $\sim$20 amino-terminal residues (see Section 2.6).

It has been recently shown that the presence of at least one of the two Lys residues within the 36 residue, βgal-specific extension of DHFR (Figures 3 and 11) is essential for the strong dependence of the half-life of the extension-bearing DHFR on the nature of its amino-terminal residue. Specifically, replacement (by site-directed mutagenesis) of the two Lys residues present in the extension (Figure 11) with two Arg residues eliminates the influence of the extension, whereas replacement of either one of the two Lys residues is without a strong effect (A. Bachmair and A. Varshavsky, unpublished results). Although these results are consistent with the above model (Figure 12), further analysis is required for its definitive testing.

(3) An additional possibility, which is not mutually exclusive with those discussed above, is that there exists an amino acid sequence motif that, when positioned in an appropriate proximity to a *destabilizing* amino-terminal residue in the same protein, would allow efficient amino-terminal targeting owing to the existence of a specific "receptor" protein for such a motif. According to this hypothesis, the targeting requirements of the N-end rule pathway are such that the receptor-binding motif and a destabilizing amino-terminal residue must *both* be present in a protein in order for it to be short-lived. The postulated amino acid sequence motif is envisioned as a highly degenerate one, as is the signal sequence motif required for protein translocation across membranes.[33–35] Thus, the complete signal for amino-terminal targeting in protein degradation may consist of *both* a destabilizing amino-terminal residue and an amino-terminus-proximal sequence motif. The latter may be recognized either by the same protein complex that recognizes the destabilizing amino-terminal residue of the target protein or by a separate, distinct protein.

In all the above models, the nature of the topologically unique amino-terminal residue in a protein (Table I) is one crucial determinant of the "complete" degradation signal, with the other, more hypothetical ingredients of a complete signal having to be present as well to allow efficient amino-terminal targeting. That the amino-terminal residue of a target protein is indeed a specific, distinct component of the complete degradation signal is indicated not only by the data with βgal that yielded the N-end rule,[31] but also by the role of amino-terminal amino acid addition in converting the *secondary* destabilizing amino termini into the *primary* ones (see Section 5) and by the recently established "inverse" correspondence between the N-end rule and the substrate requirements of methionine-specific processing aminopeptidases (see Section 3.3).

9. ON THE FUNCTIONS OF UBIQUITINATION

One established function of Ub is to mark intracellular proteins that are targeted for selective degradation *in vivo* (see Section 1). At the same time, whether damaged proteins are processed by the Ub system *exclusively* by degradation remains to be established. Alternatively, certain conformationally perturbed but chemically undamaged proteins might be restored to their native conformation through cycles of ubiquitination–deubiquitination. Indeed, the presence of even a single Ub moiety in a ubiquitinated protein is likely to perturb its conformation, either through local effects of ubiquitination or through more subtle influences such as displacement of the center of mass of the protein molecule because of the presence of Ub. Perturbations of protein conformation would be even stronger in the case of multiply ubiquitinated proteins. Thus, the attachment of a bulky modifying group such as Ub to a conformationally perturbed protein may destabilize it sufficiently to yield a new perturbed state that may more readily interconvert to the native protein conformation upon deubiquitination of the protein.

The above speculation, which is addressable experimentally, assumes that not all ubiquitinated proteins are intermediates in selective protein degradation *in vivo*. Indeed, it is clear that at least some proteins (for instance, the DNA-associated histones H2A and H2B)[1,32,75,76] are not destabilized *in vivo* by ubiquitination. What features of the Ub system could account for the existence of ubiquitinated proteins that are not degraded *in vivo*? One possibility is that the ubiquitination of such substrates may be mediated by distinct ubiquitinating enzymes that are *independent* of the N-end rule proteolytic pathway in that they recognize features of the protein structure other than the protein's amino terminus.

Furthermore, if the post-translationally formed, *branched* Ub–Ub conjugates in a ubiquitinated protein serve as specific recognition signals for the "downstream" Ub-dependent protease as proposed above (see Section 2.6), the lack of such branched Ub–Ub structures in an otherwise ubiquitinated protein could be sufficient to prevent its Ub-dependent degradation. These mutually nonexclusive possibilities are consistent with the available evidence, and in particular with the fact that only one out of several distinct Ub carrier proteins, E2s, is required for ubiquitination of proteolytic substrates in an *in vitro* Ub-dependent proteolytic system[37,57] (see Chapter 3). The other Ub carrier proteins have either no known substrates or substrates that are not destabilized by ubiquitination.[37] Recently, one Ub carrier protein, a ~20 kDa species highly conserved between yeast and mammals, that preferentially ubiquitinates *in vitro* small basic proteins such as histones[4,37] was found to be a product of the yeast *RAD6* gene[32] (see also Chapter 2). The *RAD6* gene is known to function in the DNA repair and sporulation pathways.[73,74] This discovery,[32] while possibly revealing at least one function of protein ubiquitination in chromosomes, also underscores the functional diversity of the Ub system.

Spatial sequestration of ubiquitinated proteins into specific multiprotein aggregates such as chromosomal nucleoprotein fibers,[1,75,76] inclusion bodies,[4,7] or Ub-containing intracellular filamentous aggregates of the type found in the neurons of patients with Alzheimer's disease[77,78] is yet another potential way to uncouple protein ubiquitination from protein degradation. While some of the above aggregates result from normal cellular metabolism, others, such as inclusion bodies[4,7] or neuronal lesions in Alzheimer's disease,[77,78] may be a manifestation of specific stress response pathways acting to contain damage by abnormal proteins whose selective proteolytic elimination is either not possible or does not proceed fast enough.

10. CONCLUDING REMARKS

Genetic approaches to mechanistic aspects of complex biochemical pathways have invariably proved essential for the detailed understanding of these pathways. In particular, the recent insights into the mechanism of substrate recognition in selective protein turnover[31] that have been described in this chapter were made possible largely by systematic use of the "reverse genetics" approach. These and other genetically based designs (see also Chapter 2 of this book), together with the already established biochemical approaches, hold the promise of rapid advances in the understanding of selective protein turnover.

11. SUMMARY

(1) A method was developed that makes it possible to expose different amino acid residues at the amino termini of otherwise identical test proteins *in vivo*. By using this method, it was shown that β-galactosidase (βgal) proteins that differ exclusively in their amino-terminal residues have strikingly different half-lives *in vivo*, from more than 20 hr to less than 3 min. The set of 20 individual amino acids can thus be ordered with respect to the half-lives that they confer on βgal when present at its amino terminus (the N-end rule).[31] Both the identity of an amino acid residue and its amino-terminal location are essential for its effect on βgal's half-life. Analysis of the Ub-dependent degradation of purified βgal proteins added to a reticulocyte-derived extract indicates that the N-end rule, as defined with βgal in yeast *in vivo*, is qualitatively similar to the partial N-end rule defined thus far with βgal in a mammalian *in vitro* system.

(2) Short-lived, but not long-lived, βgal proteins are multiply ubiquitinated *in vivo*, consistent with the biochemical and genetic evidence[1,2] that ubiquitinated proteins include obligatory intermediates in selective protein degradation.

(3) The currently known amino-terminal residues in relatively long-lived, *noncompartmentalized* proteins from both prokaryotes and eukaryotes belong virtually exclusively to the stabilizing class as predicted by the N-end rule. Strikingly, the amino-terminal residues in *compartmentalized* proteins are largely of the destabilizing class.

(4) The function of the previously described post-translational addition of single amino acids to protein amino termini can also be accounted for by the N-end rule. Specifically, we proposed[31] that certain amino-terminal residues in proteins may not be directly destabilizing as such but only through their ability to be conjugated to other, *primary* destabilizing residues. Recent experimental evidence strongly supports this model.

(5) Both the existence of apparent exceptions to the predictions of the N-end rule as defined with βgal and the more recent experiments with a different test protein, dihydrofolate reductase (DHFR), indicate that the complete signal for amino-terminal targeting in protein degradation may consist of *both* a destabilizing amino-terminal residue and an additional feature elsewhere in the protein. We discuss mechanisms that may underlie this additional requirement. We also consider possible hierarchical relations between the N-end rule pathway, in which substrates are tar-

geted amino terminally, and other degradative pathways that involve "interior" targeting.

Note added in proof. More recent work (D. Gonda, I. Wünning, A. Bachmair, W. Lane, and A. Varshavsky, submitted for publication) has extended the analysis of the N-end rule in mammalian reticulocytes (see Section 7) to the entire set of twenty βgal proteins that differ exclusively in their amino-terminal residue. The following conclusions have been reached: (1) With the exception of Ile, which is stabilizing in reticulocytes, the amino-terminal residues that are destabilizing in yeast are also destabilizing in reticulocytes. (2) Of the eight amino-terminal residues that are stabilizing in yeast, only Met, Val, Gly, and Pro are also stabilizing in reticulocytes. (3) In both yeast and reticulocytes, the destabilizing amino-terminal residues Glu, Gln, Asp, and Asn are the *secondary* destabilizing ones in that they are not destabilizing as such but only through their ability to be post-translationally conjugated to *primary* destabilizing amino acids such as Arg. (4) Cys is also a secondary destabilizing amino acid in reticulocytes but not in yeast where Cys is a stabilizing amino acid. (5) In reticulocytes, three distinct types of the N-end-recognizing activity are specific, respectively, for three subgroups of the primary destabilizing amino-terminal residues: basic (His, Lys, Arg), bulky hydrophobic (Leu, Phe, Trp, Tyr), and small (Ser, Ala, Thr). Thus, while the N-end rule is strongly conserved between a mammalian and a yeast cell, the two N-end rules are not identical. Ser, Ala, and Thr, which are stabilizing in yeast but destabilizing in reticulocytes, often occur at the unblocked amino termini of long-lived, noncompartmentalized proteins. At the same time, differentiation of reticulocytes is accompanied by selective destruction of their preexisting proteins. This raises a testable possibility that the exact form of the N-end rule may depend on the cell's physiological state, thereby providing a mechanism for selective destruction of preexisting proteins upon cell differentiation.

ACKNOWLEDGMENTS.   We thank Michael Lanzer and Hermann Bujard (Heidelberg, Germany) for the DHFR expression vectors, Gottfried Schatz (Basel, Switzerland) for the antibody to DHFR, Vincent Chau (University of Florida) for helpful discussions, John Tobias (MIT) for comments on the manuscript, and Barbara Doran for secretarial assistance. This work was supported by grants to A. V. from the National Institutes of Health (GM31530 and CA43309). A. B. was a Fellow of the European Molecular Biology Organization. D. G. is a Fellow of the Jane Coffin Childs Memorial Fund for Medical Research. I. W. was a Fellow of Deutsche Forschungsgemeinschaft.

REFERENCES

1. Finley, D., and Varshavsky, A., 1985, The ubiquitin system: Functions and mechanisms, *Trends Biochem. Sci.* **10:** 343–346.
2. Hershko, A., 1983, Ubiquitin: Roles in protein modification and breakdown, *Cell* **34:** 11–12.
3. Pontremoli, S., and Melloni, E., 1986, Extralysosomal protein degradation, *Annu. Rev. Biochem.* **55:** 455–482.
4. Hershko, A., and Ciechanover, A., 1986, The ubiquitin pathway for the degradation of intracellular proteins, *Prog. Nucleic Acids Res. Mol. Biol.* **33:** 19–56.
5. Rivett, J. A., 1986, Regulation of intracellular protein turnover: Covalent modification as a mechanism of marking proteins for degradation, in: *Current Topics in Cellular Regulation*, Academic Press, New York, pp. 291–337.
6. Mortimore, G., 1982, Mechanisms of cellular protein catabolism, *Nutr. Rev.* **40:** 1–12.
7. Rechsteiner, M., 1987, Ubiquitin-mediated pathways for intracellular proteolysis, *Annu. Rev. Cell. Biol.* **3:** 1–30.
8. Goldberg, A. L., and Goff, S. A., 1986, The selective degradation of abnormal proteins in bacteria, in: *Maximizing Gene Expression* (W. Reznikoff and L. Gold, eds.), Butterworths, Boston, pp. 287–314.
9. Gottesman, S., Gottesman, M., Shaw, J. E., and Pearson, M. L., 1981, Protein degradation in *E. coli*: The *lon* mutation and bacteriophage lambda N and cII protein stability, *Cell* **24:** 225–233.
10. Goff, S. A., and Goldberg, A. L., 1985, Production of abnormal proteins in *E. coli* stimulates transcription of *lon* and other heat shock genes, *Cell* **41:** 587–595.
11. Croy, R. G., and Pardee, A. B., 1983, Enhanced synthesis and stabilization of M_r 68,000 protein in transformed BALB/c-3T3 cells: Candidate for restriction point control of cell growth, *Proc. Natl. Acad. Sci. U.S.A.* **80:** 4699–4703.
12. Spindler, K., and Berk, A. J., 1984, Rapid intracellular turnover of adenovirus 5 early region 1A proteins, *J. Virol.* **52:** 706–710.
13. Hann, S. R., and Eisenmann, R. N., 1984, Proteins encoded by the human c-myc oncogene: Differential expression in neoplastic cells, *Mol. Cell. Biol.* **4:** 2486–2497.
14. Curan, T., Miller, A. D., Zokas, L., and Verma, I., 1984, Viral and cellular *fos* proteins: A comparative analysis, *Cell* **36:** 259–268.
15. Bahl, H., Echols, H., Strauss, D. B., Court, D., Crowl, R., and Georgopoulos, K., 1987, Induction of the heat shock response of *E. coli* through stabilization of σ^{32} by the phage cIII protein. *Genes Dev.* **1:** 57–64.
16. Ho, Y. S., Wulff, D., and Rosenberg, M., 1986, Protein–nucleic interactions involved in transcription activation by the phage λ regulatory protein cII, in: *Regulation of Gene Expression* (I. Booth and C. Higgins, eds.), Cambridge University Press, New York, pp. 79–103.
17. Banuett, F., Hoyt, M. A., McFarlane, L., Echols, H., and Herskowitz, I., 1986, *hflB*, a new *E. coli* locus regulating lysogeny and the level of bacteriophage λ cII protein, *J. Mol. Biol.* **187:** 213–224.
18. Hoyt, M. A., Knight, D. M., Das, A., Miller, H. I., and Echols, H., 1982, Control of phage λ development by stability and synthesis of cII protein: Role of the viral *cIII* and host *hfl A, him A* and *him D* genes, *Cell* **31:** 565–573.
19. Grossman, A. D., Erickson, J. W., and Gross, C. A., 1984, The *htpr* gene product in *E. coli* stimulates transcription of *lon* and heat shock genes, *Cell* **38:** 383–390.
20. Neidhardt, F. C., Van Bogelen, R. A., and Vaughn, V., 1984, The genetics and regulation of heat shock proteins, *Annu. Rev. Genet.* **18:** 295–329.

21. Grossman, A. D., Strauss, D. B., Walter, W. A., and Gross, C. A., 1987, σ^{32} synthesis can regulate the synthesis of heat shock proteins in *E. coli, Genes Dev.* **1:** 179–184.
22. Nasmyth, K., 1983, Molecular analysis of the cell lineage, *Nature* **320:** 670–674.
23. Sternberg, P. W., Stern, M. J., Clark, I., and Herskowitz, I., 1987, Activation of the yeast *HO* gene by release from multiple negative controls, *Cell* **48:** 567–577.
24. Bigelow, S., Hough, R., and Rechsteiner, M., 1981, The selective degradation of injected proteins occurs principally in the cytosol rather than in lysosomes, *Cell* **25:** 83–93.
25. Jones, E. W., 1984, The synthesis and function of proteases in *Saccharomyces*: Genetic approaches, *Annul. Rev. Genet.* **18:** 233–270.
26. Finley, D., Ciechanover, A., and Varshavsky, A., 1984, Thermolability of ubiquitin-activating enzyme from the mammalian cell cycle mutant ts85, *Cell* **37:** 43–55.
27. Ciechanover, A., Finley, D., and Varshavsky, A., 1984, Ubiquitin dependence of selective protein degradation demonstrated in the mammalian cell cycle mutant ts85, *Cell* **37:** 57–66.
28. Finley, D. Özkaynak, E., and Varshavsky, A., 1987, The yeast polyubiquitin gene is essential for resistance to high temperatures, starvation and other stresses, *Cell* **48:** 1035–1046.
29. Özkaynak, E., Finley, D., and Varshavsky, A., 1984, The yeast ubiquitin gene: Head-to-tail repeats encoding a polyubiquitin precursor protein, *Nature* **312:** 663–666.
30. Özkaynak, E., Finley, D., Solomon, M. J., and Varshavsky, A., 1987, The yeast ubiquitin genes: A family of natural gene fusions, *EMBO J.* **6:** 1429–1439.
31. Bachmair, A., Finley, D., and Varshavsky, A., 1986, In vivo half-life of a protein is a function of its amino-terminal residue, *Science* **234:** 179–186.
32. Jentsch, S., McGrath, J. P., and Varshavsky, A., 1987, The yeast DNA repair gene *RAD6* encodes a ubiquitin-conjugating enzyme, *Nature* **329:** 131–134.
33. Blobel, G., 1980, Intracellular protein topogenesis, *Proc. Natl. Acad. Sci. U.S.A.* **77:** 1496–1500.
34. Schatz, G., 1987, Signals guiding proteins to their correct locations in mitochondria, *Eur. J. Biochem.* **165:** 1–6.
35. Walter, P., and Lingappa, V., 1986, Mechanism of protein translocation across the endoplasmic reticulum, *Annu. Rev. Cell Biol.* **2:** 499–516.
36. Haas, A. L., and Bright, P. M., 1987, The dynamics of ubiquitin pools within cultured human lung fibroblasts, *J. Biol. Chem.* **262:** 345–351.
37. Pickart, C. M., and Rose, I. A., 1985, Functional heterogeneity of ubiquitin carrier proteins, *J. Biol. Chem.* **260:** 1573–1581.
38. Hershko, A., Heller, H., Elias, S., and Ciechanover, A., 1983, Components of the ubiquitin–protein ligase system, *J. Biol. Chem.* **258:** 8206–8214.
39. Hershko, A., Heller, H., Eytan, E., Kaklij, G., and Rose, I. A., 1984, Role of the α-amino group of protein in ubiquitin-mediated protein breakdown, *Proc. Natl. Acad. Sci. U.S.A.* **81:** 7021–7025.
40. Messing, J., and Vieira, J., 1982, A new pair of M13 vectors for selecting either DNA strand of double-digest restriction fragments, *Gene* **19:** 269–276.
41. Smith, M., 1985, *In vitro* mutagenesis, *Annu. Rev. Genet.* **19:** 423–462.
42. Levitt, M., 1978, A simplified representation of protein conformation for rapid simulation of protein folding, *J. Mol. Biol.* **104:** 59–107.
43. Hershko, A., and Heller, H., 1985, Occurrence of a polyubiquitin structure in ubiquitin–protein conjugates, *Biochem. Biophys. Res. Commun.* **128:** 1079–1086.
44. Hough, R., and Rechsteiner, M., 1986, Ubiquitin–lysozyme conjugates: Purification and succeptibility to proteolysis, *J. Biol. Chem.* **261:** 2391–2399.
45. Chin, D. T., Carlson, N., Kuehl, L., and Rechsteiner, M., 1986, The degradation of guanidinated lysozyme in reticulocyte lysate, *J. Biol. Chem.* **261:** 3883–3890.

46. Hershko, A., and Rose, I. A., 1987, Ubiquitin-aldehyde: A general inhibitor of ubiquitin-recycling processes, *Proc. Natl. Acad. Sci. U.S.A.* **84:** 1829–1833.

47. Jones, M. O., and Herskowitz, I., 1978, Mutants of bacteriophage λ which do not require the cIII gene for efficient lysogenization, *Virology* **88:** 199–206.

48. Lane, C., Shannon, S., and Craig, R., 1979, Sequestration and turnover of guinea-pig milk proteins and chicken ovalbumin in *Xenopus* oocytes, *Eur. J. Biochem.* **101:** 485–495.

49. Tsunasawa, S., Steward, J. W., and Sherman, F., 1985, Amino-terminal processing of mutant forms of yeast iso-1-cytochrome c: The specificities of methionine aminopeptidase, *J. Biol. Chem.* **260:** 5382–5391.

50. Sherman, F., Stewart, J. W., and Tsunasawa, S., 1985, Methionine or not methionine at the beginning of a protein, *BioEssays* **3:** 27–31.

51. Boissel, J. P., Kasper, T. J., Shah, S., Malone, J. I., and Bunn, H. F., 1985, Amino-terminal processing of proteins: Hemoglobin South Florida, a variant with retention of initiator methionine and N-acetylation, *Proc. Natl. Acad. Sci. U.S.A.* **82:** 8448–8452.

52. Ben-Bassat, A., Bauer, K., Chang, S. Y., Myambo, K., Boosman, A., and Chang, S., 1987, Processing of the initiation methionine from proteins: Properties of the *E. coli* methionine aminopeptidase and its gene structure, *J. Bacteriol.* **169:** 751–757.

53. Kasper, T. J., Boissel, J. P., and Bunn, H. F., 1987, Specificity of N-terminal aminopeptidase: Analysis by site-directed mutagenesis, *Abstr. ASBC Meeting, Fed. Proc.* **46:** 1980–1981.

54. Miller, C. G., Strauch, K. L., Kukral, A. M., Miller, J. L., Wingfield, P. T., Mazzei, G. J., Werlen, R. C., Graber, P., and Movva, N. R., 1987, N-terminal methionine-specific peptidase in *Salmonella typhimurium*, *Proc. Natl. Acad. Sci. U.S.A.* **84:** 2718–2722.

55. Ben-Bassat, A., and Bauer, K., 1987, Amino-terminal processing of proteins, *Nature* **326:** 315–316.

56. Maniatis, T., Goodbourn, S., and Fischer, J., 1987, Regulation of inducible and tissue-specific gene expression, *Science* **236:** 1237–1244.

57. Hershko, A., Heller, H., Eytan, E., and Reiss, Y., 1986, The protein substrate binding site of the ubiquitin–protein ligase system, *J. Biol. Chem.* **261:** 11992–11999.

58. Backer, J., and Dice, J., 1986, Covalent linkage of ribonuclease S-peptide to microinjected proteins causes their intracellular degradation to be enhanced during serum withdrawal, *Proc. Natl. Acad. Sci. U.S.A.* **80:** 996–1000.

59. Ahlberg, J., Berkenstam, A., Henell, F., and Glaumann, H., 1985, Degradation of short and long lived proteins in isolated rat liver lysosomes, *J. Biol. Chem.* **260:** 5847–5854.

60. Dean, R. T., 1984, Modes of access of macromolecules to the lysosomal interior, *Biochem. Soc. Trans.* **12:** 911–913.

61. Soffer, R. L., 1980, Biochemistry and biology of aminoacyl-tRNA-protein transferases, in: *Transfer RNA: Biological Aspects* (D. Soll, J. Abelson, and P. R. Schimmel, eds.), Cold Spring Harbor Laboratory, Cold Spring Harbor, NY, pp. 493–505.

62. Deutch, C., 1984, Aminoacyl-tRNA:Protein transferases, *Methods Enzymol.* **106:** 198–205.

63. Kaji, H., 1976, Amino-terminal arginylation of chromosomal proteins by arginyl-tRNA, *Biochemistry* **15:** 5121–5125.

64. Shyne-Athwal, S., Riccio, R. V., Chakraborty, G., and Ingolia, N. A., 1986, Protein modification by amino acid addition is increased in crushed siatic but not optic nerves, *Science* **231:** 603–605.

65. Ferber, S., and Ciechanover, A., 1986, Transfer RNA is required for conjugation of ubiquitin to selective substrates of the ubiquitin- and ATP-dependent proteolytic system, *J. Biol. Chem.* **261:** 3128–3134.

66. Ferber, S., and Ciechanover, A., 1987, Role of arginine-tRNA in protein degradation by the ubiquitin pathway, *Nature* **326:** 808–810.

67. Rogers, S., Wells, R., and Rechsteiner, M., 1987, Amino acid sequences common to rapidly degraded proteins: The PEST hypothesis, *Science* **234:** 364–368.

68. Thornton, J. M., and Sibanda, B. L., 1983, Amino and carboxy-terminal regions in globular proteins, *J. Mol. Biol.* **167:** 443–460.

69. Rote, K. V., and Rechsteiner, M., 1986, Degradation of proteins microinjected into HeLa cells: The role of substrate flexibility, *J. Biol. Chem.* **261:** 15430–15436.

70. Volz, K. W., Mathews, D. A., Alden, R. A., Freer, S. T., Hansch, C., Kaufman, B. T., and Kraut, J., 1982, Crystal structure of avian dihydrofolate reductase containing phenyltriazine and NADPH, *J. Biol. Chem.* **257:** 2528–2536.

71. Stubiquitiner, D., Ibrahimi, I., Cutler, D., Dobberstein, B., and Bujard, H., 1984, A novel *in vitro* transcription–translation system: Accurate and efficient synthesis of single proteins from cloned DNA sequences, *EMBO J.* **3:** 3143–3148.

72. Guarente, L., 1983, Yeast promoters and *lacZ* fusions designed to study expression of cloned genes in yeast, *Methods Enzymol.* **101:** 181–191.

73. Lawrence, C. W., 1982, Mutagenesis in *Saccharomyces cerevisiae, Adv. Genet.* **21:** 173–254.

74. Reynolds, P., Weber, S., and Prakash, L., 1985, *RAD6* gene of *Saccharomyces cerevisiae* encodes a protein containing a tract of 13 consecutive aspartates, *Proc. Natl. Acad. Sci. U.S.A.* **82:** 168–172.

75. Busch, H., and Goldknopf, I. L., 1981, Ubiquitin–protein conjugates, *Mol. Cell. Biochem.* **40:** 173–187.

76. Levinger, L., and Varshavsky, A., 1982, Selective arrangement of ubiquitinated and D1-containing nucleosomes within the *Drosophila* genome, *Cell* **28:** 375–385.

77. Perry, G., Friedman, R., Shaw, G., and Chau, V., 1987, Ubiquitin is detected in neurofibrillary tangles and senile plaque neurites of Alzheimer disease brains, *Proc. Natl. Acad. Sci. U.S.A.* **84:** 3033–3036.

78. Mori, H., Kondo, J., and Ihara, Y., 1987, Ubiquitin is a component of paired helical filaments in Alzheimer's disease, *Science* **235:** 1641–1644.

79. Amann, E., and Brosius, J., 1985, "ATG" vectors for regulated high-level expression of cloned genes in *E. coli, Gene* **40:** 183–190.

80. Ullman, A., 1984, One-step purification of hybrid proteins which have β-galactosidase activity, *Gene* **29:** 27–31.

Chapter 12

Selectivity of Ubiquitin-Mediated Protein Breakdown
Current Status of Our Understanding

Avram Hershko

1. INTRODUCTION

The reader of this book will appreciate the significant progress that has been achieved in this field in less than a decade. Since the discovery of the involvement of ubiquitin (Ub) in protein breakdown in 1978[1] and of its covalent ligation to proteins in 1980,[2] many of the enzymatic reactions in the formation and breakdown of Ub–protein conjugates have been elucidated. Techniques that have been developed to study the roles of Ub in intact cells include the use of specific antibodies, microinjection methods, a mammalian cell mutant, and cloned genes. Results obtained with such experimental techniques have confirmed the widespread occurrence and major physiological functions of the Ub proteolytic system (reviewed in refs. 3–5).

Following the delineation of some of the basic processes in the Ub pathway, we have concentrated in the last three years on the question of what determines the specificity and selectivity of this system. Intracellular

AVRAM HERSHKO • Unit of Biochemistry, Faculty of Medicine, Technion–Israel Institute of Technology, Haifa 31096, Israel.

protein breakdown is a highly selective process, and it appears reasonable to assume that the elaborate and energy-consuming Ub pathway possesses sufficient mechanisms for selectivity. In this brief chapter, we describe recent results from our and other laboratories, which may be the first steps toward the understanding of the mechanisms of selectivity of the Ub proteolytic pathway.

2. MECHANISMS OF SELECTIVITY: QUESTIONS AND ASSUMPTIONS

One important question concerns the sites in the Ub pathway at which the selection of protein substrates occurs. Since proteins conjugated to Ub are apparently committed for degradation,[6,7] it seems reasonable to assume that selection is exerted mainly at the Ub ligation step. However, other reactions that may influence selectivity, such as a correction mechanism by the action of isopeptidases on Ub–protein conjugates not suitable for degradation,[2] should also be considered.

Assuming that selectivity is achieved primarily at the ligation step, two further questions may be asked: (1) What specific features of protein structure are recognized by the Ub conjugation machinery? (2) What mechanisms allow the Ub–protein ligase to recognize such protein structures? With regard to features of protein structure that might be recognized, a reasonable expectation is that the Ub ligase system should be able to recognize two types of structural feature: one is predetermined (such as primary, secondary, or tertiary structures) and the other may arise by post-translational modification or damage. The first type is for the recognition of native protein structures and is required to distinguish between normal proteins of different half-lives, whereas the second would serve to recognize and remove damaged proteins. Recent work has identified some structural determinants in both categories.

3. ROLE OF THE NH₂-TERMINAL RESIDUE

In work done in collaboration with Irwin Rose, we found that a free N-terminal α-NH_2 group is required for the degradation of proteins by the Ub system.[8] The special importance of the NH_2-terminal residue was indicated by the following observations: (1) proteins that are naturally blocked in their NH_2 termini by acetylation are not substrates for the Ub system, although their nonacetylated counterparts from other species are good substrates; (2) selective chemical modification of the α-NH_2 groups of proteins prevented their conjugation to Ub; and (3) the "creation" of new α-NH_2 groups on such proteins by the introduction of polyalanine side chains restored their susceptibility to the action of the Ub system. It was suggested that the exposure of a normally buried NH_2-terminal residue may initiate Ub conjugation and protein breakdown.[8]

Other investigators have subsequently observed that the nature of the NH_2-terminal residue has a strong influence on susceptibility to Ub-dependent proteolysis. Ciechanover and co-workers (see Chapter 10) found that tRNA is required for the degradation of certain proteins by the Ub system[9] and for their ligation to Ub.[10] Attention was drawn to the NH_2-terminal residue by the observation that all tRNA-requiring substrates have an acidic residue (glutamic or aspartic acid) at the NH_2 terminus.[10] It was indeed found that the tRNA-mediated transfer of arginine to the NH_2 termini of such proteins converts them to good substrates for the Ub proteolytic system.[11] It thus seems that the replacement of an acidic NH_2-terminal residue by a basic NH_2 terminus is required for its recognition by the Ub system.

Furthermore, strong evidence for the importance of the NH_2-terminal residue was provided by Varshavsky and co-workers (ref. 12, and see Chapter 11). These investigators have systematically changed the NH_2-terminal residue of β-galactosidase by site-directed mutagenesis. Following the expression of the various derivatives in yeast cells, dramatic differences in their half-lives were observed. Species of β-galactosidase with NH_2-terminal Arg, Lys, Asp, Leu, Phe, Gln, or Tyr were degraded with half-lives of 2–10 min; those with NH_2-terminal Glu or Ile had half-lives of $\sim$30 min, while those with NH_2-terminal Met, Ser, Ala, Thr, Val, or Gly were essentially stable. That the rapidly degraded variants were indeed broken down by the Ub system was suggested by a direct correlation between rates of degradation and levels of Ub conjugates of the various derivatives expressed in yeast cells.[12]

4. OTHER POSSIBLE STRUCTURAL DETERMINANTS

4.1. PEST Regions

While the cumulative evidence from the above-mentioned investigations indicates that the NH_2-terminal residue is an important structural determinant, it appears reasonable to assume that it is not the only one. There is a wide spectrum of *in vivo* half-lives of specific proteins, and the variation is apparently much more than can be accounted for solely by 20 different amino acid residues at the NH_2 terminus. Further native structural signals are expected, such as preferred amino acid sequences or tertiary structures. Rechsteiner and colleagues[13] recently reported that rapidly degraded proteins contain regions enriched in the four amino acids, PEST (Pro, Glu, Ser, and Thr). It is not known whether these proteins are degraded by the Ub system. The authors suggested that another system may be involved, since the activation energy of the degradation of microinjected casein (a PEST-enriched protein) is lower than that of Ub-dependent proteolysis in reticulocyte lysates.[13] However, the activation energy of a multistep process merely reflects a rate-limiting reaction, and different proteins degraded by the same pathway may be rate limited at different stages of that pathway. It is interesting that PEST regions are almost invariably flanked by basic amino acid residues (Arg, Lys, or His).[13] Assuming the action of a hypothetical protease that recognizes the PEST region and cleaves at the amino side of the basic residue, the resulting fragments would contain basic residues at the NH_2 terminus and would be good substrates for the Ub system.[2]

4.2. Oxidation

An example of a determinant signaling damaged protein structure may be provided by our observation that oxidation of methionine residues of some proteins greatly increases their susceptibility to Ub ligation and degradation.[14] Oxidative damage of proteins probably occurs in cells, and the oxidation of a single histidine residue has been implicated in the inactivation and breakdown of glutamine synthetase.[15,16] It remains to be seen whether the oxidation of methionine residues to sulfoxide derivatives is a physiological signal for protein breakdown.

4.3. Protein Unfolding

It is interesting to note that protein unfolding and denaturation do not seem to have a major influence on susceptibility to Ub ligation. We found that drastic oxidation of ribonuclease with performic acid (which cleaves disulfide bonds and converts the protein to a random coil configuration) has much less effect on increasing susceptibility to Ub ligation than does the milder oxidation, which affects methionine residues only.[14] Likewise, Evans and Wilkinson[17] noted that complete reduction of disulfide bonds of serum albumin does not increase its degradation by the Ub system, despite extensive loss of structure. Katznelson and Kulka[18] reported that denaturation of serum albumin and β-lactoglobulin (by reduction and alkylation) did not increase their rate of degradation by reticulocyte lysates. Completely denatured[18] or locally unfolded[19] proteins are also not degraded faster than their native counterparts following microinjection into cultured cells. It is not clear, however, whether the degradation of microinjected proteins is carried out by the Ub system. The cumulative evidence therefore indicates that drastic unfolding of protein structure is not a major signal for Ub ligation. This is in contrast to the marked effects of protein unfolding on their susceptibility to the action of proteinases. It may be assumed that drastic unfolding and denaturation of proteins occur only rarely in cells.

5. SELECTIVITY OF E3 BINDING

5.1. Substrate Binding Site

We next turned our attention to the question of how the Ub–protein ligase system recognizes suitable protein structures. It seemed reasonable to assume that at least some initial selection is accomplished at a specific protein substrate binding site. Our first question was: Which component of the Ub–protein ligase system contains the binding site for the protein substrate? Previous work showed that the ligase system consists of three enzyme components: a Ub-activating enzyme (E1), Ub-carrier proteins (E2s), and a third enzyme (E3), the mode of action of which had not been defined (ref. 20, and see Chapter 3). The possibilities were that E2, E3, or both are involved in the binding of the protein substrate. E3 from rabbit reticulocytes was purified by a combination of affinity chromatography on Ub–Sepharose, hydrophobic chromatography, and gel filtration. A 180 kDa protein was identified as E3 by the coincidence of this protein with E3 activity throughout all stages of purification.[14] Two types of independent evidence indicated that E3 contains the protein substrate binding site of the ligase system. These were the chemical cross-linking of ^{125}I-labeled proteins to purified E3 and the functional conversion of E3-bound labeled proteins to Ub conjugates in pulse-chase experiments.

The same techniques also allowed a preliminary characterization of the specificity of the protein binding site of E3. Specificity of binding could be studied by the cross-linking of different labeled proteins to purified E3, by the competition of various unlabeled proteins with a labeled protein substrate for cross-linking to E3, or by the displacement of [^{125}I]lysozyme bound to E3 by unlabeled proteins, as measured by the functional isotope trapping (pulse-chase) assay. All methods showed that proteins with free NH_2 termini (such as cytochrome c or enolase from yeast) bound more strongly to E3 than their N^α-acetylated counterparts. Furthermore, proteins with oxidized methionine residues bound much more strongly to E3 than the unmodified proteins.[14] These results suggest that the protein binding site of E3 can recognize at least these two specific features of protein structure.

5.2. "Head" and "Body" Subsites

The above observations raise the possibility that the binding site of E3 is composed of two subsites: one for the binding of the free NH_2-terminal residue ("head" subsite) and another for the binding of other parts of the protein, which contain suitable recognition signals such as oxidized methionines ("body" subsite). The existence of a "body" subsite, which may bind certain proteins in the absence of a free α-NH_2 group, is suggested by the following observations. Although the chemical modification of all amino groups of lysozyme by reductive methylation decreases its binding to E3, a derivative that was subjected to both amino group modification and oxidation of methionine residues bound strongly to E3.[14] Similarly, Breslow and co-workers[21] reported that denatured proteins, in which all amino groups were blocked by succinylation or acetylation, inhibited the proteolysis and ligation to Ub of unblocked proteins, presumably by competition on a common ("body") binding site.

Evidence for the existence of a separate and specific "head" subsite for the NH_2-terminal residue is provided by our recent studies with methyl ester derivatives of amino acids.[22] The degradation of [^{125}I]lysozyme in reticulocyte extracts is specifically inhibited by methyl esters of the three basic amino acids His, Arg, and Lys. Methyl esters of other amino acids had no significant influence. The conjugation of lysozyme to Ub is inhibited by the same amino acid derivatives. Cross-linking and pulse-chase experiments indicate that these compounds inhibit the binding of lysozyme to E3. N^{α}-acetyl derivatives of methyl esters of basic amino acids are inactive, indicating that a free α-NH_2 group is essential for their action. We conclude that these derivatives inhibit a subsite of E3 that specifically binds NH_2-terminal basic residues. It should be noted that the NH_2-terminal residue of lysozyme is basic (Lys). It remains to be seen what other "head" subsites, possibly specific for other NH_2-terminal residues, exist and how they are involved in the selection of specific proteins for Ub conjugation.

6. CONCLUSION

It is evident from this brief description that although significant progress has been achieved recently in our understanding of some of the mechanisms of selectivity of the Ub system, much more remains to be elucidated. Discoveries of features of protein structure that are recognized by the Ub ligase system have been mostly accidental. It may therefore be expected that further structural determinants will be found to account for the great variability of protein half-lives. The elucidation of the mechanisms that account for the selective action of the Ub–protein ligase system is also just beginning. Further sites, or subsites of the E3 system that may bind other specific structures of proteins, are likely to be discovered. Selective recognition may also be found at later stages in the ligation reaction, including the successive addition of multiple Ub molecules.

ACKNOWLEDGMENTS. I thank Dr. Stuart M. Arfin for helpful comments on the manuscript and Mrs. Atara Katznelson for skillful secretarial assistance. This work was supported by Grant AM-25614 from the United States Public Health Service and by a grant from the United States–Israel Binational Science Foundation.

REFERENCES

1. Ciechanover, A., Hod, Y., and Hershko, A., 1978, A heat-stable polypeptide component of an ATP-dependent proteolytic system from reticulocytes, *Biochem. Biophys. Res. Commun.* **81:** 1100–1105.
2. Hershko, A., Ciechanover, A., Heller, H., Haas, A. L., and Rose, I. A., 1980, Proposed role of ATP in protein breakdown: Conjugation of proteins with multiple chains of the polypeptide of ATP-dependent proteolysis, *Proc. Natl. Acad. Sci. U.S.A.* **77:** 1783–1786.
3. Hershko, A., and Ciechanover, A., 1982, Mechanisms of intracellular protein breakdown, *Annu. Rev. Biochem.* **51:** 335–364.
4. Finley, D., and Varshavsky, A., 1985, The ubiquitin system: Functions and mechanisms, *TIBS* **10:** 343–347.
5. Hershko, A., and Ciechanover, A., 1986, The ubiquitin pathway for the degradation of intracellular proteins, *Prog. Nucleic Acid Res. Mol. Biol.* **33:** 19–56.
6. Hershko, A., Leshinsky, E., Ganoth, D., and Heller, H., 1984, ATP-dependent degradation of ubiquitin–protein conjugates, *Proc. Natl. Acad. Sci. U.S.A.* **81:** 1619–1623.
7. Hough, R., Pratt, G., and Rechsteiner, M., 1986, Ubiquitin–lysozyme conjugates. Identification and characterization of an ATP-dependent protease from rabbit reticulocyte lysates, *J. Biol. Chem.* **261:** 2400–2408.
8. Hershko, A., Heller, H., Eytan, E., Kaklij, G., and Rose, I. A., 1984, Role of the α-amino group of protein in ubiquitin-mediated protein breakdown, *Proc. Natl. Acad. Sci. U.S.A.* **81:** 7021–7025.
9. Ciechanover, A., Wolin, S. L., Steitz, J. A., and Lodish, H. F., 1985, Transfer RNA is an essential component of the ubiquitin- and ATP-dependent proteolytic system, *Proc. Natl. Acad. Sci. U.S.A.* **82:** 1341–1345.
10. Ferber, S., and Ciechanover, A., 1986, Transfer RNA is required for conjugation of ubiquitin to selective substrates of the ubiquitin- and ATP-dependent proteolytic system, *J. Biol. Chem.* **261:** 3128–3134.

11. Ferber, S., and Ciechanover, A., 1987, Role of arginine-tRNA in protein degradation by the ubiquitin pathway, *Nature* **326:** 808–810.
12. Bachmair, A., Finley, D., and Varshavsky, A., 1986, *In vivo* half-life of a protein is a function of its amino-terminal residue, *Science* **234:** 179–186.
13. Rogers, S., Wells, R., and Rechsteiner, M., 1986, Amino acid sequences common to rapidly degraded proteins: The PEST hypothesis, *Science* **234:** 364–368.
14. Hershko, A., Heller, H., Eytan, E., and Reiss, Y., 1986, The protein substrate binding site of the ubiquitin–protein ligase system, *J. Biol. Chem.* **261:** 11992–11999.
15. Fucci, L., Oliver, C. N., Coon, M. J., and Stadtman, E. R., 1983, Inactivation of key metabolic enzymes by mixed-function oxidation reactions: Possible implication in protein turnover and aging, *Proc. Natl. Acad. Sci. U.S.A.* **80:** 1521–1528.
16. Levine, R. L., 1983, Oxidative modification of glutamine synthetase. Inactivation is due to loss of one histidine residue, *J. Biol. Chem.* **258:** 11823–11827.
17. Evans, A. C., Jr., and Wilkinson, K. D., 1985, Ubiquitin-dependent proteolysis of native and alkylated bovine serum albumin: Effects of protein structure and ATP concentration on selectivity, *Biochemistry* **24:** 2915–2923.
18. Katznelson, R., and Kulka, R. G., 1985, Effects of denaturation and methylation on the degradation of proteins in cultured hepatoma cells and in reticulocyte cell-free systems, *Eur. J. Biochem.* **146:** 437–442.
19. Rote, K. V., and Rechsteiner, M., 1986, Degradation of proteins microinjected into HeLa cells: The role of substrate flexibility, *J. Biol. Chem.* **261:** 15430–15436.
20. Hershko, A., Heller, H., Elias, S., and Ciechanover, A., 1983, Components of ubiquitin–protein ligase system. Resolution, affinity purification and role in protein breakdown, *J. Biol. Chem.* **258:** 8206–8214.
21. Breslow, E., Daniel, R., Ohba, R., and Tate, S., 1986, Inhibition of ubiquitin-dependent proteolysis by non-ubiquitinable proteins, *J. Biol. Chem.* **261:** 6530–6535.
22. Reiss, Y., Kaim, D., and Hershko, A., 1988, *J. Biol. Chem.* (in press).

Index